CLIMATE, ENERGY AND WATER

With the global population expected to reach nine billion by 2050, our capacity to provide enough food, water and energy relies on our understanding and management of the complex interdependencies in the climate-energy-water nexus. This book presents a comprehensive and up-to-date analysis of the interdependencies between these sectors and the knock-on consequences of those interactions for other sectors, such as food production and biodiversity conservation. The interdisciplinary nature of the book across the three sectors of climate, energy and water means that it will be valuable for advanced students, researchers and policymakers across a broad range of fields, including environmental/energy/water/climate policy, environmental economics, climate science, hydrology, ecology and geography.

JAMIE PITTOCK is an associate professor in the Fenner School of Environment and Society at the Australian National University (ANU) and the director of International Programs for the ANU UNESCO Chair in Water Economics and Transboundary Water Governance. He is also the program leader of the Australia–United States Climate, Energy and Water Nexus Project for the US Studies Centre. Jamie worked for non-governmental environmental organisations in Australia and internationally from 1989 to 2007; previously, he was director of WWF's Global Freshwater Programme from 2001 to 2007. Jamie's research focusses on better governance of the interlinked issues of water management, energy and food supply, responses to climate change and conservation of biological diversity. His recent research includes assessments of the impacts of renewable energy and carbon sequestration policies on water resources and ecosystems globally, and also the water-use implications of emission reduction measures proposed in Australia. Jamie's work also considers the energy costs of water sector adaptation to climate change. He directs two large research programs on irrigation and water management in Africa and on hydropower, food and water in the Mekong region. Jamie is also chair of the Board of Water Stewardship Australia.

KAREN HUSSEY is an associate professor at the Fenner School of Environment and Society at ANU and an ANU Public Policy Fellow. Karen undertakes research and teaching related to environmental governance, water and energy policy, climate adaptation, resilience and disaster management, and the links between international trade and environmental regulation. Karen's most recent books include *Environment and Sustainability: A Policy Handbook* (with Stephen Dovers, 2013) and *Water Resources Planning and Management* (with Quentin Grafton, Cambridge University Press, 2011). Karen is a program director of the Australia–United States Climate, Energy and Water Nexus Project and a board member for the ANU UNESCO Chair in Water Economics and Transboundary Water Governance.

STEPHEN DOVERS is the director of the Fenner School of Environment and Society at ANU, Honorary Professorial Fellow at Charles Darwin University, an ANU Public Policy Fellow, and a Fellow of the Academy of the Social Sciences in Australia. Steve's research and teaching cover the policy and institutional dimensions of sustainable development, emergency and disaster management, and climate change adaptation. He is the co-author of *Environment and Sustainability: A Policy Handbook* (with Karen Hussey, 2013) and *The Handbook of Disaster Policies and Institutions* (with John Handmer, 2013).

CLIMATE, ENERGY AND WATER

Managing trade-offs, seizing opportunities

Edited by

JAMIE PITTOCK, KAREN HUSSEY AND
STEPHEN DOVERS

CAMBRIDGE
UNIVERSITY PRESS

CAMBRIDGE
UNIVERSITY PRESS

32 Avenue of the Americas, New York, NY 10013-2473, USA

Cambridge University Press is part of the University of Cambridge.

It furthers the University's mission by disseminating knowledge in the pursuit of education, learning and research at the highest international levels of excellence.

www.cambridge.org
Information on this title: www.cambridge.org/9781107029163

First published 2015

Printed in the United States of America

A catalogue record for this publication is available from the British Library

Library of Congress Cataloguing in Publication data
Climate, energy, and water : managing trade-offs, seizing opportunities / edited by Dr Jamie Pittock, Dr Karen Hussey and Professor Stephen Dovers.
pages cm
Includes bibliographical references.
ISBN 978-1-107-02916-3 (Hardback)
1. Climate change mitigation. 2. Power resources–Management. 3. Water resources–Management. I. Pittock, Jamie, editor. II. Hussey, Karen, editor. III. Dovers, Stephen, editor.
QC903.C5648 2015
363.738′746–dc23 2014042078

ISBN 978-1-107-02916-3 Hardback

Contents

Contributors

Neil Byron
Neil Byron is an Adjunct Professor at the University of Canberra in Canberra, Australia.

Nicole Carter
Nicole Carter was a 2011–12 US-Australia Fulbright Senior Scholar at the Australian National University in Canberra, Australia.

Albert van Dijk
Albert van Dijk is a Professor at the Fenner School of Environment and Society at the Australian National University in Canberra, Australia.

Georges Dyer
Georges Dyer is Senior Fellow at Second Nature in Boston, Massachusetts, in the United States.

David L. Harrison
David L. Harrison is a Senior Water Resources Advisor to the Nature Conservancy's Global Water Program focusing on sustainable hydropower, based in Boulder, Colorado, in the United States.

Joerg Hartmann
Joerg Hartmann is an independent water resources and hydropower consultant based in Amherst, Massachusetts, in the United States.

Karen Hussey
Karen Hussey is an Associate Professor at the Fenner School of Environment and Society at The Australian National University in Canberra, Australia.

Tom Iseman
Tom Iseman currently serves as Deputy Assistant Secretary for Water and Science at the Department of the Interior in Washington, DC, in the United States.

Steven Kenway
Steven Kenway is a Research Group Leader, Water-Energy-Carbon, in the School of Chemical Engineering at the University of Queensland in Brisbane, Australia.

Junguo Liu
Junguo Liu is a Professor of Hydrology and Water Resources in the School of Nature Conservation, Beijing Forestry University in Beijing, China.

Jakob Lundberg
Jakob Lundberg, PhD, is the Head of Development and Policy at We Effect in Stockholm, Sweden.

Jan Lundqvist
Jan Lundqvist is a Professor and Senior Scientific Advisor at the Stockholm International Water Institute in Stockholm, Sweden.

Nadia Madden
Nadia Madden is a Research Associate in the Climate and Energy Program of the Union of Concerned Scientists in Cambridge, Massachusetts, in the United States.

Marco Mancini
Marco Mancini is a research fellow in the Department of Agrifood Production and Environmental Sciences at the University of Florence in Florence, Italy.

Tek N. Maraseni
Tek N. Maraseni is a Vice-Chancellor's Research Fellow at the International Centre for Applied Climate Sciences, Institute of Agriculture and Environment, University of Southern Queensland in Toowoomba, Australia.

Anna Dalla Marta
Anna Dalla Marta, PhD, is a Research Fellow at the Department of Agrifood Production and Environmental Sciences at the University of Florence in Florence, Italy.

Shahbaz Mushtaq
Shahbaz Mushtaq is a Principal Scientist and Deputy Director at the International Centre for Applied Climate Sciences, Institute of Agriculture and Environment, University of Southern Queensland in Toowoomba, Australia.

Greg Oliver
Greg Oliver is the Manager for Sustainable Research at the Australian Water Recycling Centre and a Visiting Fellow at the Fenner School of Environment and Society at The Australian National University in Canberra, Australia.

Francesca Orlando
Francesca Orlando, PhD, is a Research Fellow at the Department of Agricultural and Environmental Sciences Production, Landscape, Agroenergy, Division of Plant Production at the University of Milan in Milan, Italy.

Simone Orlandini
Simone Orlandini is a full Professor of agronomy at the Department of Agrifood Production and Environmental Sciences at the University of Florence in Florence, Italy.

Jeff Opperman
Jeff Opperman is the lead scientist for The Nature Conservancy's Great Rivers Partnership and directs the Conservancy's program on sustainable hydropower from Chagrin Falls, Ohio, in the United States.

Martin J. Pasqualetti
Martin J. Pasqualetti is a Professor in the School of Geographical Sciences and Urban Planning, and a Senior Sustainability Scientist in the Julie Ann Wrigley Global Institute of Sustainability at Arizona State University, Tempe, Arizona, in the United States.

Jacqueline Peel
Jacqueline Peel is Professor of Law at The University of Melbourne in Melbourne, Australia.

Jamie Pittock
Jamie Pittock is an Associate Professor at the Fenner School of Environment and Society at The Australian National University in Canberra, Australia

Ajith Rao
Ajith Rao is a Senior Researcher at the USG Corporation in Chicago, Illinois, in the United States

Kathryn Reardon-Smith
Kathryn Reardon-Smith is a Research Fellow (Climate Risk Management) for the Digital Futures–Collaborative Research Network at the International Centre for Applied Climate, Institute of Agriculture and Environment, University of Southern Queensland in Toowoomba, Australia.

Michael Smith
Michael Smith is a Research Fellow at the Fenner School of Environment and Society at The Australian National University in Canberra, Australia.

Benjamin K. Sovacool

Benjamin K. Sovacool is Director of the Danish Center for Energy Technologies and Professor of Business and Social Sciences at Aarhus University in Aarhus, Denmark. He is also Associate Professor of Law at Vermont Law School.

Vince Tidwell

Vince Tidwell is a Distinguished Member of the Technical Staff in the Earth Systems Analysis Department at Sandia National Laboratories in Albuquerque, New Mexico, in the United States.

Philip Wallis

Philip Wallis is a Research Fellow at the Monash Sustainability Institute, Monash University, in Clayton, Australia.

Thomas J. Wilbanks

Thomas J. Wilbanks is Corporate Research Fellow, Climate Change Science Institute and Environmental Sciences Division, Oak Ridge National Laboratory in Oak Ridge, Tennessee, in the United States.

Acknowledgements

We have been privileged to work with outstanding individuals in the preparation of this book. The thirty-one chapter authors hail from different continents and bring diverse professional and disciplinary backgrounds and insights. The book is richer for their diversity, and our experience of editing it far more pleasurable for it. We wish to thank all the authors for their expertise, commitment and endurance over the last two years.

As with all edited collections, the single biggest challenge lies in marrying diversity with coherence, breadth with focus, and narrative with detail. We believe this book achieves this elusive goal, and it is the first volume to comprehensively 'unpack' the climate, energy and water nexus from the perspective of all three sectors. It also moves beyond describing problems to identifying solutions. That achievement is attributable in no small measure to our external reviewer, Dr Brian Spies. When we needed help to 'see the wood for the trees', Brian delivered such thorough reviews as to expose not only the wood, but also the adjacent rivers, mountains and valleys. Reviewing can often be a thankless task, so let it be on the record: thank you, Brian!

We would also like to acknowledge the tireless efforts of Vivienne Seedsman, our highly capable research assistant and project manager, whose efficiency and good humour kept us on track and in good spirits. To continue the analogy, if Brian helped us to see the wood, Vivienne made sure we all kept an eye on the diversity of animals within it.

Thank you also to the terrific editorial team at Cambridge University Press who steered us through the publishing process, demonstrating remarkable levels of patience at key junctures.

Finally, none of this would have been possible without the financial assistance of the Dow Sustainability Program of the United States Studies Centre, which supported some of the primary research in the book, and also enabled us to convene two authors' workshops. The opportunity for in-depth discussions about the subject matter proved invaluable to our greater understanding of the nexus and the subsequent dissemination of that understanding through this book.

Jamie Pittock,
Karen Hussey and Stephen Dovers
Canberra

1

Justifying, extending and applying "nexus" thinking in the quest for sustainable development

KAREN HUSSEY, JAMIE PITTOCK AND STEPHEN DOVERS

Introduction

There was a time, not very long ago, when scholars, policy makers and those engaged in private enterprise were consumed by the quest for sustainable development. Gaining prominence after the 1987 Brundtland Report and subsequent UN-sponsored conferences, the concept of sustainable development demanded and received enormous attention over several decades, and in many ways, the achievements of those heady days are remarkable. For example, the key tenets of sustainable development – notably, the principles of precaution, policy integration, inter- and intra-generational equity, and the polluter pays principle – were incorporated into significant international law and into the domestic policies of almost every country on earth. The sense of urgency and import generated in that period also ushered in a suite of initiatives and instruments to support the process towards sustainable development, from 'triple bottom line reporting' and 'corporate social responsibility', through to the mainstreaming of environmental impact assessments and third-party certification of environmental practices. Not surprisingly, the momentum generated from the late 1980s onwards was matched by a concomitant rise in the representation of pro-environment parties in many countries, the establishment of national and state departments of the environment, and the proliferation of sustainability-related professions and associated educational offerings.

But somewhere along the way sustainable development – the ideological concept, process and goal – has almost disappeared from the international stage. Indeed, notwithstanding its rebadging to a green economy by UNEP, 'green growth' and the post-Rio+20 emergence of Sustainable Development Goals, it is fair to say that the global effort expended on 'sustainable development' has faded, replaced by similar levels of zeal and activity targeted at climate change. Why is that the case? Perhaps because the all-encompassing scope of 'sustainable development' made it unwieldy and intractable in practice? Or was it a consequence of the characteristic boredom that humanity exhibits when a concept lurks too long or is too difficult to achieve? Or perhaps it was the sense of urgency generated about climate change and the comparative ease with which carbon emissions can be measured and, arguably, reduced, that was just too seductive to resist? Whatever the reason, by the mid-2000s climate change had replaced sustainable development as the big environmental issue of our time.

The subjugation of sustainable development by climate change is anathema to those who recognise the latter as a subset of the former (eg Dovers and Hezri 2010). Yet the shift to the single focus of climate change, away from the all-encompassing focus of sustainable development, quickly gave way to a curious and welcome consequence: the re-emergence of 'nexus' thinking. What started as the realisation that to combat climate change was to transform, predominantly, the energy sector, became the further realisation that changes to the energy sector would have consequences for the water sector. Most recently, that link has led to the further realisation that developments in the climate, energy and water sectors all have implications for agricultural production, specifically, food. And biodiversity, and health, and on it goes. The fundamental tenet of sustainable development remains: policy integration and co-ordination across sectors to attend to linkages and interdependencies.

In many ways, conceptual framing using a nexus 'approach' is a natural extension of the earlier, well-known concept of Integrated Water Resource Management (IWRM). Espousing as it does "the coordinated development and management of water, land and related resources, in order to maximize the resultant economic and social welfare in an equitable manner without compromising the sustainability of vital ecosystems" (GWP 2000), IWRM is, by its very definition, combining nexus language with the broader objectives of sustainable development. However, like sustainable development, IWRM is a somewhat nebulous concept that has proven hard to implement in practice. It is perhaps not surprising then that much of the research and literature on the nexus has been generated from the water sector (Hussey and Pittock 2012; Hussey et al. 2013), which has recognised that while there is evidence of broad acceptance of IWRM principles, success has been limited and a fresh approach is needed (Biswas 2004). The exception is the work generated in the United States on water demand in the energy sector, which has been driven by the extreme vulnerability felt by the energy sector in that country to increasingly dire water scarcity (Rogers et al. 2013; Hussey et al. 2013).

All that said, in the course of our research for this book and other projects, we have regularly questioned the usefulness of the 'nexus' approach. What is the benefit of restricting our thinking to two or three sectors when we know that important connections exist with other sectors, and that a holistic, systems approach is optimal? Why the inclusion of 'climate' in the climate-energy-water nexus, when it serves mostly as a driver of change in the other two, in the same way that changes in population and economic development act as drivers? And perhaps most pervasive is the genuine concern that 'nexus' research will follow the path of other conceptual frameworks by being intellectually interesting for a time only to be replaced by the next intellectual fad in a few years (leaving aside sustainable development and IWRM, the interesting discourses on environmental security, ecosystem services and green growth, everything under the rubric of 'earth systems science', and more recently the prevalence of scholarship dedicated to the concept of resilience, is also relevant.)

Our answer to the first question is simple: thinking about individual sectors and their relationships to each other is easier, more amenable to systematic analysis and can be applied more readily in policy and investment decision making. At various fora, scholars have described the nexus approach as a 'pedagogical tool' and a 'dialectical apparatus' but, again, we put it more plainly: you can 'get your arms around' the relationship between two or three sectors in a way that is impossible with the broader concept of sustainable development. Sustainable development demands integrated or 'whole system' research and policy approaches that may well extend beyond our capacities, including too many variables and imperatives: the 'nexus' between three or so sectors is within our intellectual and policy reach.

Our justification for including 'climate' as the third pillar in the climate-energy-water nexus in this book lies in the fact that developments in climate policy are so rapid, urgent and significant that it is a powerful policy driver in its own right, but with many 'levers' available to shift its direction; the same cannot be said for population growth and economic development which are both seemingly on steady trajectories. In truth, though, the inclusion or exclusion of one sector over another can easily be justified; indeed, to argue about which two or three sectors are paramount and should thus be favoured is to miss the point entirely.

Much scholarship has emerged in recent years illustrating the benefits to be gained from a 'nexus approach' or 'nexus language' (Rogers et al. 2013; IEA 2013; Pittock et al. 2013), which brings us to the focus of this book. Our purpose in this volume is to extend earlier work on the nexus in four very distinct ways.

Our first objective in this book is to extend the existing technical knowledge of the links between the climate, energy and water sectors. Much research has been undertaken in the water sector over the last few decades – thanks to the traction gained by IWRM and most recently that of water footprinting – but very little work has been done to focus attention on the interrelationships between climate change per se, climate change policy and the transformations it drives in both the energy and water sectors. Chapters 2 and 3 go a long way to filling that deficit. Chapters 3 to 7 provide detailed accounts of the implications for water resources from the full suite of fossil fuel and renewable energy sources. Importantly, these chapters focus not only on the negative or perverse trade-offs that exist in the energy-water nexus, but also on the many opportunities that can arise when the links between the sectors are better understood. The next two chapters explore the dynamics of the climate, energy and water nexus in rural (Chapter 8) and urban (Chapter 9) settings. In doing so, they expose not only new linkages between the two settings, but also the distinct policy and institutional arrangements that govern urban and rural contexts. In a similar vein, Chapter 10 provides valuable insight into 'the nexus' in four countries, specifically with respect to the generation of electricity. Combined, the first ten chapters of this book provide a comprehensive overview of the many links that exist between the climate, energy and water sectors.

In essence, efforts to pursue sustainable development are a matter of co-ordination and integration, and 'unpacking' the integration challenge into two or three sectors through a nexus approach exposes where the fundamental disjunctures between policy spheres lie, as well as where existing or new integration tools could be used to achieve better outcomes. Recognising this, our second objective for the book is to critically explore the range of governance mechanisms available to better account for, and integrate, the climate-energy-water nexus in policy and investment decisions. In various ways, Chapters 11 to 17 tackle this objective. Chapter 11 begins by examining the institutional dynamics within each sector and the implications of those dynamics for cross-sectoral co-ordination. In doing so, Chapter 11 sets the scene for the next three chapters, which explore the role of regulation in managing the nexus (Chapter 12), the opportunities and limitations of markets to integrate across the three sectors (Chapter 13), and the importance of knowledge, education and enabling skills amongst key stakeholders to make decisions that both reconcile trade-offs and seize opportunities in the nexus (Chapter 14). Importantly, Chapter 15 returns to the central question of governance to explore the extent to which a nexus approach to policy integration can or should be embedded in a broader systemic governance framework. Collectively, these chapters shed new and critically important light on the opportunities and limitations of existing policy and institutional arrangements to better govern the nexus.

Our third objective in this book is to identify and explore some of the knock-on consequences of dynamics in the climate-energy-water nexus for other sectors, most notably biodiversity (Chapter 17) and food (Chapter 18). We see this as a way of testing whether efforts to manage the nexus between two or three sectors leads to benefits in other sectors.

Our final objective is to attempt to distil this book's wealth of knowledge so as to identify where important 'win-win' opportunities can be seized in the short term, but also where gaps in our understanding remain and demand attention. Thus, in the concluding chapter, we endeavour to 'add it all up' to see where it gets us: the ultimate test of the nexus conceptual frame is whether it can lead us to positive possibilities. We believe that this volume moves beyond a catalogue of problems to identify practical means of minimising perverse actions from one sector on others and seize positive synergies across sectors to benefit people and the environment. Thus, in Chapter 19, we present our conclusions and insights for where the nexus has taken us, and where we ought to go with it next.

Which brings us back to our last concern, namely that our and others' scholarship on the nexus – including scholarship presented in this book – will mark another decade of important and intellectually exciting research but that, ultimately, the conceptual framework of 'nexus' thinking will not endure. We have come to the conclusion that to worry about the longevity of any one conceptual or analytical tool is to overlook the cumulative nature of research and the human experience it informs. The outstanding research contained in this volume builds on the sustainable development research that has come before it, and the use of 'nexus' framing serves only to provide a fresh perspective. In doing so we highlight where existing research can be useful, or conversely, expose the research gaps. In this way, our thinking ought to evolve over time with new ideas and understandings racing to overtake those before but always cognisant of the dangers of ignoring the accrued insights developed under previous frames of reference, topicality and analysis.

Our plea, if we may be so bold, is simply that the normative goal of sustainable development – ambitious, unwieldy and often intractable though it might be – ought not to be lost sight of in the race.

References

Biswas, A. (2004). Integrated Water Management: A Reassessment. *Water International* 29(2):248–256.

Global Water Partnership (GWP) (2000). *Integrated Water Resources Management TAC Background Papers No. 4*. Stockholm: Global Water Partnership.

Hussey, K., and J. (2012). The Energy-Water Nexus: Managing the Links between Energy and Water for a Sustainable Future. *Ecology and Society* 17(1): 31.
Hussey, K., Carter, N., and Reinhardt, W. (2013). The Transformation of the Energy Sector: Four Implications for Water Governance. *Australian Journal of Water Resources* 17(2), 170–9.
International Energy Agency (IEA) (2012). *World Energy Outlook.* Paris: International Energy Agency.
Rogers, J., Averyt, K., Clemmer, S., Davis, M., Flores-Lopez, F., Frumhoff, P., Kenney, D., Macknick, J., Madden, N., Meldrum, J., Overpeck, J., Sattler, S., Spanger-Siegfried, E. and Yates, D. (2013). *Water-Smart Power: Strengthening the U.S. Electricity System in a Warming World.* Cambridge, MA: Union of Concerned Scientists.

2

Water resources, climate change and energy

ALBERT I. J. M. VAN DIJK

Introduction

The objective of this chapter is to summarise our understanding of the physical interactions between water resources, energy use, and climate change and mitigation. The distribution and pressures on our water resources will be reviewed, as well as our current understanding of observed and projected changes due to climate change. By addressing a set of relevant questions, the aim is to provide a framework within which we can interpret which interactions are likely and which less likely; which are desirable and which not desirable. The questions addressed include the following:

- *What are the characteristics, drivers and challenges of water resource management?* Why do water resource management challenges vary between countries? Are concepts like green and blue water, embedded water, peak water and integrated water resources management useful in managing the relationship between water, climate and energy?
- *What are the observed and projected impacts of climate change on water resources?* Why does it seem that climate change makes water management harder everywhere? Does it matter whether climate change is man-made? How do melting ice caps and glaciers affect water resources? What is the relationship between drought and water resources? How do climate change, floods and water resources interact? Will climate change affect water use?
- *What are the potential impacts of water management on climate and energy security?* What adaptation measures are considered in water management? What is their influence on energy security and climate mitigation measures? For example, can changes in water management change climate? What do water management changes mean for energy use?
- *How can climate mitigation and energy security measures impact on water security?* Can switching between energy sources affect water security? What impact do the by-products of energy generation have on climate and the water cycle? What is the impact of climate mitigation measures such as landscape carbon storage?

In this chapter, 'water management' refers to all the activities designed to balance the supply and demand of water. It refers to structural as well as non-structural measures, and natural as well as man-made water systems. The goal is to achieve human and environmental benefits, including water supply, wastewater service, flood control, hydropower, recreation, navigation and environmental protection (Grigg 1996).

Water management: A challenge even without climate change

Water has traditionally been thought of and managed as a local issue, or at least one that does not extend beyond river or groundwater basin boundaries. While this is to a large extent still the case, the challenges to water management are increasingly taking on a global dimension, influenced by global processes like population growth and movement, international trade and climate change. For example, Vörösmarty et al. (2010) estimated that 80 per cent of the global population now lives in catchments with threatened surface water supplies and river health, with more than two billion people regularly having to face severe water scarcity (Hoekstra et al. 2012). The situation is no better for groundwater, which is mined at unsustainably high rates almost everywhere where livelihoods critically depend on it (eg Wada et al. 2010). Thus, water security has become a problem of global scale that only a few relatively wet and sparsely populated countries have been able to escape, and one that is contributing to political conflict between countries and regions within countries.

So far, the main stresses on water availability have not been from climate change but from increased water use, the modification of river channels and water pollution (UN 2003). In many countries increases in population, irrigation, urbanisation and household water consumption drive water use and the regulation of river systems. According to the United Nations, the increase in global water use over the last century has been two times greater than the rate of global population growth. Most of this increase has been to satisfy the increasing demand for crop products. Globally, irrigation accounts for about 70 per cent of water withdrawals and 90 per cent of consumptive use (ie water that evaporates or is too polluted for reuse; Shiklomanov and Rodda 2003). Global irrigated area has doubled since 1961 and is projected to increase another 10 per cent by 2030 (Bruinsma 2003); competition for water can be expected to increase further.

A local problem of global proportions

Water is essentially an inexpensive resource but one that requires much energy to transport. For this reason, natural surface and groundwater systems are often the logical unit of management. This is less true for municipal water systems, because the higher economic value of water in these cases can justify costly 'unnatural' water movements such as long-distance pipelines and desalinisation. Urban water use constitutes less than 20 per cent of total extractions in most countries, but managing these water supplies has its own challenges. For example, there may not be sufficient nearby and accessible river or groundwater, or the water may be of poor quality or polluted (for example, ironically, due to flooding). Many of the world's largest cities face occasional or on-going risks to water supply – even those in wet climates.

Because of the costs of water transport, considering water availability and use at global aggregate scale provides limited insight into the challenges that face water managers. National scale information still hides major transport obstacles but does illustrate the vast differences in water management challenges between countries (Fig. 2.1).

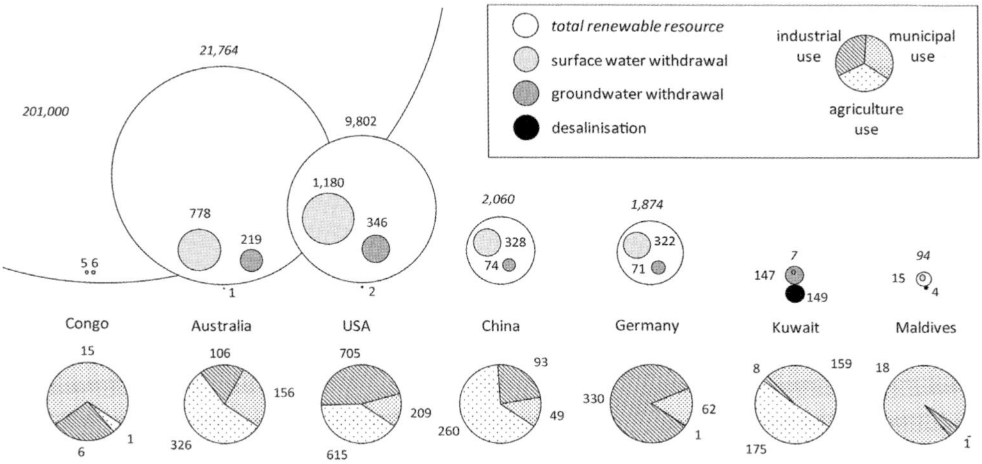

Figure 2.1 Comparison of water availability and use in a selection of countries. All numbers are expressed in m^3 per person per year. Data are from AQUASTAT (FAO n.d.).

Figure 2.1 shows the water available, withdrawn and used in seven countries, expressed in cubic metres (or thousands of litres) per person and year (FAO n.d.). The total renewable resource includes the annual average amount of water that is added naturally to surface and groundwater sources, as opposed to non-renewable water contained in glaciers or so-called fossil groundwater reservoirs. Expressed per person, the renewable resource available is the combined result of climate conditions and population density. Thus, wet and sparsely populated Congo can boost a comfortable 201,000 cubic metres (thousand litres) of renewable resources per year per person, compared to a mere 7 cubic metres in arid Kuwait. This does not include desalinisation, which theoretically creates a near infinite and renewable water source but at high-energy costs. Figure 2.1 illustrates some other points:

- Municipal water use is typically not the main part of total withdrawals. Exceptions are non-industrialised countries with a climate that is sufficiently wet to make irrigation unnecessary (eg Congo) and countries where water is so scarce that any other uses are uneconomical (eg Maldives). Usually, only municipal water use can economically justify desalinisation. It provides more than 20 per cent of total water use for several Middle Eastern countries (Kuwait, Qatar, Bahrain, United Arab Emirates, Lebanon) and small islands (Malta, Antigua and Barbuda, Maldives; FAO n.d.).
- Irrigation is often the largest use of water in countries with dry regions. Water is usually drawn from surface water sources, but groundwater can be another important water source. Groundwater is the primary water source in the Middle East and Northern Africa (where fossil groundwater reserves are used) and several countries with high yielding groundwater sources, often in limestone 'karst' aquifers (eg Slovakia, Denmark, Portugal and Jamaica).
- In most European countries, Russia and the United States, industry is the largest water user. Unlike irrigation water use, much of these withdrawals (eg for power station cooling) are not evaporated and returned to the rivers.

- At face value, several countries known to face water management challenges (eg, the United States, China, and Australia) would appear to have sufficient renewable resources available to meet demand. Indeed, at the national level, no more than ten countries withdraw water at a rate that exceeds natural renewal (nine Middle Eastern countries and Barbados). This again emphasises that the cost of water transport is a major constraint on effective water management, and one that links it directly to energy use.
- The sensitivity to climate change varies widely between countries and regions. Thus, in Congo water availability is unlikely to be a major challenge, but flooding might. In Kuwait, climate change may influence water demand but will not affect water availability. In the Maldives, climate change per se may not be the main problem but sea-level rise certainly does pose a threat to many aspects of life, including water.
- The large differences in water availability and scarcity between countries and regions can be attributed to the large amount of energy required to supply water by artificial means.

'Blue' versus 'green' water

The surface and groundwater sources from which water is drawn have been categorised conceptually as 'blue' water: water in sources from which it can in principle be extracted and transported with relative ease. This contrasts with 'green' water, which can be thought of as the water in the soil and which cannot easily be extracted other than by vegetation (eg Rockström et al. 2009). Water management tends to focus on blue water, while green water is the domain of land and crop management. This separation has a longstanding pragmatic use, but can lead to important oversights, even more so in stressed water systems. For example, green water is critical for dry land cropping, but irrigated crops, too, rely on green water; blue water is only added when rainfall is insufficient. Also, any precipitation onto the landscape will largely go towards replenishing the ('green') soil water first, before flowing over to fill ('blue') groundwater and surface water sources. The main implication of this is that an increase in green water use in the landscape can lead to a reduction in blue-water renewal. The two water types interact closely and this can lead to complex management questions. Some of these will be discussed later in this chapter.

Embedded water

A concept illustrating the link between water transportability and energy is that of 'embedded' or 'virtual' water (Allan 1997), which quantifies the large volumes of water that can be involved in the production of any particular product. The concept of a 'water footprint' captures a very similar idea. Accounting for embedded water makes it clear that transporting the product is usually more cost efficient than transporting the water required for its production. For example, for many Middle Eastern countries it makes more sense to import food and use the saved irrigation water resource to increase water security for other users. The concept also has some problems. Often no distinction is made between embedded green and blue water. This can lead to the interpretation that all embedded water that is 'freed up' can be used for other purposes, which is usually incorrect. Also, using water from geographically different sources is likely to have quite different associated outcomes, both wanted (eg food sufficiency, economic activity) and unwanted (eg environmental degradation). Accounting for these outcomes is not straightforward. Nonetheless, the concept of embedded water illustrates that water, food and energy can in fact be substituted for each other, through transport, trade and technology.

Peak water

Water resources share some characteristics with fossil fuel resources, and some resource availability concepts apply to both. For example, the concept of 'peak oil' has been extended to 'peak water' (Gleick and Palaniappan 2010). Peak oil refers to the initial exponential increase in oil production in the United States during the twentieth century (because of growing demand and extraction efficiency) and the subsequent decline after 1970 (because the remaining oil was increasingly difficult and costly to extract). The concept has been applied at local, regional, national and global scale. The concept is directly transferable to non-renewed or 'fossil' groundwater reserves in arid regions. With some modifications, the concept can be extended to renewable water resources used at an unsustainable rate. Important groundwater reservoirs that have reached 'peak water' are found in the California Central Valley, the US Great Plains (the Ogallala Aquifer), the North China Plains, and Northern India. Similarly, river water extractions can become so large that streamflow no longer reaches the river mouth for much of the time. This is the situation for the Murray River in Australia, the Colorado River in the United States, the Yellow River in China, the Jordan River in the Middle East, and the Nile River in Egypt. In a less literal interpretation of peak water, the concept may be used to describe the point at which the overall societal benefit of increased withdrawal is undone by the damage to all other water resource functions – however these are quantified.

Features of the peak oil concept that apply to water are the role of the (energy) costs of extraction and the ability to substitute one water source with another. An important difference is, again, the vastly higher cost of resource transport. Gleick and Palaniappan illustrate this with a simple sum: the oil contained in a tanker ship may be worth around US$250 million, which leaves enough margin to justify long-distance transport and encourage global trade. The same tanker but full of drinking water would be valued at US$500 thousand at most. Overall, peak water probably remains a local or regional phenomenon. In some cases, changes in climate can bring forward or delay peak water.

Integrated water resources management

Integrated water resources management (IWRM) is a widely used and broadly interpretable term. It may be used to imply that integrate management measures should be integrated, for example:

- across different water system components or management aspects (eg flood protection and water security, green and blue water, surface and groundwater, water quantity and quality);

- with other societal requirements and goals (eg land management, the protection of natural ecosystems and biodiversity, protection of human or legal rights, maintaining or creating wealth);
- across different time scales (eg the consideration of both climate variability and change, and of rapid and slow changes in the economy and society);
- within the public psyche (eg the need for community consultation, empowerment and consensus-building);
- to address competing demands and equity (eg between rich and poor, urban and rural population, businesses and individuals, people or countries upstream and downstream, or those living inside and outside the water system).

Clearly, there is no neat separation between these aspects of IWRM, and most of them raise value judgements. Within this complexity, this book emphasises only one of many integration challenges: that of integrating water management with climate mitigation and energy security measures.

Climate change and water management

It was mentioned that water management is generally a local, regional or national challenge, but one that has become harder around the world and that interacts with a wide range of management areas. Matching supply and demand is usually possible, given sufficient energy inputs, and this creates a clear physical connection between water and energy. It is with this background that climate change can make an already difficult challenge even harder. In 2008, the Intergovernmental Panel on Climate Change (IPCC) released a report on the relationship between climate change and water (Bates et al. 2008). A summary of its conclusions makes gloomy reading:

1. Global warming observed over several decades has been linked to changes in the water cycle, including changes in precipitation patterns, intensity and extremes; reduced snow cover and widespread ice melting, and changes in soil moisture and runoff. Precipitation and runoff generally have increased in temperate parts of the United States and Europe at high latitudes, and decreased in southern Europe, southernmost South America and southwest Australia.
2. Climate models consistently predict that there will be more precipitation and runoff in high latitudes and some wet tropical areas, and less in several already dry regions (eg the Mediterranean, western United States, southern Africa and northeast Brazil).
3. Due to a predicted increase in precipitation variability and extremes, both floods and drought risks are predicted to increase in much of the world.
4. Water sources stored in glaciers and snow are predicted to decline, meltwater rivers will reach their high water peak earlier in the year, and summers will be warmer and drier.
5. Higher water temperatures and greater extremes in floods and drought are predicted to aggravate many forms of water pollution (eg from sediment, nutrients, pathogens, pesticides and salt).

6. Globally, climate change will create more challenges than opportunities. Even though water quantity may increase in some regions, these often have sufficient water already, and the benefits may be offset by problems like decreased water reliability, changed flow patterns and degraded water quality.
7. Changes in water quantity and quality are expected to affect food production and security. Poor rural farmers in the dry tropics and the African and Asian deltas are particularly vulnerable.
8. Climate change will increase water management challenges, including the operation of water infrastructure, such as hydropower, flood defences and drainage and irrigation systems.
9. Current water management arrangements may not be robust under climate change: a safe prediction, as in many regions this is already true even now.
10. Climate change challenges the traditional assumption that past hydrological conditions provide a good guide for future conditions.
11. Adaptation to future water scarcity will likely make it necessary to increase water supply security as well as reduce water demand.
12. Climate change mitigation measures can ultimately help reduce the impact on water resources, but it can also have negative impacts. An example is the increased water use by carbon forestry and bioenergy crops. Conversely, some water management changes can negatively affect climate change mitigation: for example, by requiring more energy from fossil fuels.
13. Water management can affect other policy areas, including energy, health, food security and nature conservation. Less developed countries are more vulnerable, and adaptation strategies specifically may need to consider impacts on development, environment and health policies in these societies.
14. Important knowledge gaps remain and observation networks are in decline. Climate change impacts on water quality, aquatic ecosystems and groundwater, including their socioeconomic dimensions are particularly poorly understood. The current capacity to evaluate alternative adaptation and mitigation measures across all water-dependent sectors is also inadequate.

Since it was published in 2008, a number of new analyses and studies have come out. These often confirm but sometimes challenge the IPCC findings. Some of the more debated issues are briefly discussed below.

Why will climate change not make water management easier anywhere?

Reading the IPCC's findings, the news seems to be bad almost everywhere. In general, wetter areas appear set to get wetter and suffer more flooding and water quality problems, whereas drier areas will get drier and suffer more droughts. While there are likely to be some winners too, there are reasons to expect more challenging water-management conditions. Three important reasons are explained below.

First, warming of the atmosphere means that there is more energy and this energy can produce greater extremes, both in temperature and in the water cycle – the so-called intensification of the hydrological cycle (Huntington 2006). Intensification explains why the same region can be predicted to suffer from more droughts as well as from more floods (though obviously not at the same time). Water management is particularly concerned with managing extreme conditions, and increased variability inevitably makes that harder. Overall, global average evaporation and precipitation rates are predicted to increase, although how much of that increase would occur over land and how much over the oceans is not entirely clear.

Second, the increased energy of the Earth's atmosphere is predicted to 'speed up' the conveyor-belt-like air circulation between low- and mid-latitudes called Hadley cells. Hadley cells are responsible for the latitudinal climate bands: tropical moist air rises around the equator, loses its moisture and descends again over the outer tropics, where it contributes to arid conditions. As energy in the climate system increases, the Hadley cells will reach further towards the poles, bringing drier conditions to the subtropical regions around thirty degrees latitude (eg the southern USA, Caribbean, Mediterranean and southern Africa and Australia). The Hadley cell are estimated to have widened two to five degrees of latitude since 1979 towards the poles (ie about 200 to 500 kilometres), and climate models broadly agree that this will continue (Johanson and Fu 2009). The same Hadley cell strengthening can create wetter conditions in tropical monsoonal regions (eg the Sahel, Eastern Africa, India and northern Australia) and indeed, there are signs that this is occurring. Some of these regions would do well with the additional water resources. Unfortunately, most of the rainfall increases will occur in the wet season when water is often already in abundance: the ability to store water for the dry season is often the main constraint on water availability.

Third, current patterns of population and water system development around the world reflect a history of adaptation to local water availability. New opportunities for water resources development are likely to emerge, but require time, investment and innovation and population movements. For these reasons, the beneficial impacts of climate change are likely to trail the damaging impacts. Regions where climate change is likely to create new opportunities include those where economic activity is currently limited by low temperatures (eg northern Russia and Canada and at high altitudes) or marginal rainfall (eg the drier parts of tropical monsoonal regions).

Does it matter what causes the observed changes in climate?

Much debate focusses on the question of whether humans or natural processes are responsible for the observed recent changes in climate. The question is particularly important when deciding whether to accept global climate model projections. There is consensus in the scientific community about an important anthropogenic contribution to observed climate change, despite several important outstanding questions. However, when it comes to water management, the observation that climate is indeed changing may be more important than the underlying causes, at least over the next few decades. Traditionally, water managers use historic time series of climate and water resources to characterise the degree of climate variability that can be expected. Increasingly, that so-called stationarity assumption (Milly et al. 2008) creates problems. Recent conditions often no longer fit expectations based on history, and new records are being broken more often than they should. A more general problem with assessing trends in severe floods and droughts is that historic records by definition contain few of them. This makes statistical analyses difficult at the best of times.

There are other reasons why past conditions are not a good guide for the future. In many industrialised northern-latitude countries, long records show slow natural oscillations in rainfall and other aspects of climate. In recent decades, they show an unambiguous increase in temperatures that has brought forward early spring snowmelt events and changed river flows. Changes in measurement instruments, land use (which can affect climate or climate measurements) and river modification further complicate the interpretation of climate and water records. These slow background variations also exist in more recently developed countries such as Australia, but here the short records make it even more difficult to detect them (van Dijk et al. 2013). Therefore, regardless of whether one accepts the evidence that humanity is contributing to observed climate change, it is indisputable that (1) climate has always been variable and can change over many different time scales, and (2) at least certain aspects of climate (eg temperatures in most of the world) have changed well beyond historically recorded conditions. Essentially this means that it is wrong to assume that historic records adequately characterise future conditions. The implications for managing vulnerable water systems are increasingly recognised.

Are the world's glaciers melting away?

In 2010, the IPCC retracted a claim that the Himalayan glaciers would melt away by 2035. The IPCC identified a breach in its review process that some see as typical and other as an exception. There is no doubt that many of the world's glaciers are retreating and indeed some smaller ones, sometimes critical for local water supplies, are likely to disappear within a few decades. Equally, however, many of these glaciers have been in retreat for more than a century, making a direct connection to human climate change less likely.

The future of glaciers is important for water management. One-sixth of the world lives in regions supplied by high mountain ranges, and they are an important source of irrigation water resources in India, China and the United States, among other countries. For example, Immerzeel et al. (2010) showed that melt water from the Himalayas is a major source of water for the Indus and Brahmaputra Rivers, sustaining more than 270 million people. Changes in the volume and timing of runoff from glaciers could have major implications for water availability and flood risk downstream. (As an aside, the future fate of glaciers is also relevant to coastal water management, as most of the melt water ends up in the oceans, contributing to sea level rise.) Glaciers can be thought of as very slow conveyor belts, gaining snow at the top and melting at the bottom. The influence of climate change on glacier melt is quite complicated and only partly understood. Higher temperatures and rainfall (rather than snow) increase melt rates, and if melting is not compensated by greater snowfall, the glacier will retreat. The result is an initial increase in streamflow due to glacier melt, but at some point, the meltwater river will reach the equivalent of peak water, after which the shrinking glacier will release ever lesser flows. A recent review found that Himalayan glaciers are generally losing mass at a rate similar to glaciers elsewhere, but considers it unlikely that dramatic changes in total runoff will occur, and in fact the glaciers that feed the Indus River appear to be stable (Bolch et al. 2012).

Predicting climate change is important to anticipate future glacier retreat, but where retreat has already been observed, water management will need to be changed regardless. For example, smaller Andean glaciers supplying towns and cities such as La Paz in Bolivia are retreating relatively fast, and alternative water sources may need to be found before long.

Drought and water security

There is no doubt that there are regions that have experienced more frequent or more severe droughts in recent decades than in the past. Whether the number and severity of droughts are also increasing at a global aggregate scale is a matter of debate. The impact of drought depends not only on rainfall, but also on its interaction with evaporation. These together create the deficits in green and blue water that cause crop and ecosystem damage, falling groundwater levels, reduced river flows and shrinking water bodies. Droughts can have many and complicated impacts, and long records are not available for all of them. Instead, proxy indicators based on weather data are used to estimate drought severity. A popular example is the Palmer Drought Severity Index (PDSI), which uses climate station data of rainfall and temperature on the assumption that temperature is a good predictor of evaporation. This assumption creates an artificial trend in global drought frequency and severity: a more accurate method to estimate evaporation did not suggest any globally significant drought trend over the last sixty years (Sheffield, Wood and Roderick 2012). Even so, there are certainly regions that have become prone to more frequent or severe droughts, including China, the Mediterranean region and southern Australia. Changes in crop production and water stores are a more direct reflection of drought impacts, but they are also affected by other human and natural influences and therefore not always straightforward to analyse (van Dijk et al. 2013).

The complexity of droughts is also illustrated by their connection with heat waves. The 2003 heatwave in Europe turned attention from a persistent lack of rainfall as the main or only cause of drought impacts, to high temperature as a contributor. It caused heat stress in people, animals and crops, bush fires, and damage to ecosystems and crops, but also interacted with water resources: reduced river flows affected irrigation and power plant cooling, for example (García-Herrera et al. 2010). It is difficult to separate the direct effects of the heat from the indirect effects on increasing evaporation from soils and water bodies. In fact, soil drying itself appeared to have prolonged the atmospheric conditions causing the heat wave. Similar unprecedented heat waves have recently occurred nearly every year: in 2006 (Europe and the western United States), 2009 (Australia and Argentina), 2010 (European Russia, northern China and the eastern United States), 2011 (across the United States and in the Persian region), and 2012 (the United States and Australia). In almost all cases, they coincided with drought conditions. The coincidence of summer droughts and heat waves is more pronounced in continental regions with a humid or subhumid climate, where summer soil drying heats the atmosphere, which in turn prevents cooler and moister ocean air from flowing in. In dry areas, such conditions are common and will therefore not necessarily be experienced as particularly extreme.

Ocean circulation patterns have a particularly important role in creating drought conditions in tropical and subtropical regions. In particular, the El Niño/La Niña circulation pattern in the Pacific Ocean and sea surface temperatures in the Indian Ocean contribute to particularly dry and wet conditions in the countries surrounding these oceans. Countries like Australia and Indonesia are influenced by both oceans, which can amplify rainfall variations. Unfortunately, the behaviour and connections between circulation patterns in the different oceans is still rather poorly understood. So far, there is no consensus about their behaviour under future climate change.

Climate change, floods and water resources

It has been difficult to attribute the occurrence, frequency and severity of hydrological extremes such as floods and droughts unambiguously to climate change. Climate models do not provide particularly clear predictions of future changes beyond predicting a general increase in variability and therefore in extremes. Recent studies appear to disagree on whether there has already been a global increase in rainfall intensity over the last decade (Sun et al. 2012; Westra, Alexander and Zwiers 2012).

Increased rainfall intensity does not necessarily have a negative effect on water resources availability. The fraction of rainfall that is retained by soils (green water) and later evaporated is less for large rainfall events, and therefore, in proportion more of it will typically reach rivers. For similar reasons, a change in average rainfall (eg over a few years) will create a change in river flows that, in relative terms, will be two or more times greater. This is also true for flood-causing rainfall events, but whether this leads to a gain in water availability will depend on whether the flood flows can be used beneficially or stored in reservoirs or wetlands. Intense rainfall may also cause water-quality problems by washing sediment and nutrients from farmland and spilling industrial or waste-water pollutants into the river (Delpla et al. 2009).

Large rainfall events can also be more effective in recharging groundwater. For example, most climate models predict a decrease in average rainfall for the southern Australian Murray-Darling Basin, yet groundwater recharge might actually increase (McCallum et al. 2010). This is because the climate models also predict an increase in large storms, which recharge the groundwater more effectively. Groundwater systems that are recharged by river floods can also benefit from more intense rainfall. By contrast, where a period of steady, cool-season rainfall is needed to saturate the soil before groundwater recharge occurs, less reliable and more intense rains may largely evaporate and not help replenish groundwater. In turn, if groundwater contributes to river flows, the reduced recharge can threaten river water availability during dry periods. Water systems in Mediterranean climates often fall into this category.

For people living in the floodplains and deltas of the world, river floods have traditionally been an important source of water and maintained soil fertility. Some rivers still provide these functions, but many others have been contained and regulated. With more people living in flood-prone areas, floods are increasingly a danger to property and life rather than a source of water. Globally, the frequency of severe flood events doubled from around 1965 to 2000, but the associated economic damage increased five times (Kron and Berz 2007). It has not been possible to attribute the increase of large floods and damage to climate change because modifications of the water system (the catchment, river and floodplain) and the increased population in floodplains have been at least as important.

A common water management tension is that between water supply protection and flood mitigation. Protecting supplies requires water to be retained as much as possible, whereas flood mitigation requires flexibility in releasing water; water may need to be released when households, industries and agriculture cannot beneficially use it, or even when it causes minor flood damage (in order to prevent worse). This conflict in reservoir operating approach is not always understood, and reservoir construction is still regularly seen as a panacea for all water management challenges.

Water resources and sea-level rise

For low-lying islands and coastal plains, rising sea levels due to climate change presents a special challenge. Many small islands critically depend on lenses of fresh water that 'float' on top of saline groundwater. These lenses face pressure from a combination of threats. Increased water withdrawal and reduced precipitation reduces the size of the lenses and increases the risk of freshwater contamination with seawater (eg from cyclones, storm surges and tsunamis). Particularly vulnerable are small island nations, which often lack the resources for mitigating infrastructure works. Low-lying coastal regions elsewhere face similar problems, sometimes aggravated by high rates of groundwater withdrawal for urban or irrigation use. This encourages intrusion of seawater into aquifers and can cause subsidence and increase risk of flooding by sea or rivers. Subsidence problems are acute in the large and fast-growing Asian cities of Jakarta, Bangkok, Ho Chi Minh City and Manila, for example. Finally, where increased withdrawals or sea-level rise reduce river flows, seawater can move upstream into rivers and estuaries, causing damage to ecology, agriculture, fisheries and infrastructure.

Climate change and water use

So far, this overview has looked at the likely impacts of climate change on water availability. Where water systems are stressed, it is equally important to consider how climate change may affect water use. It might seem logical that reduced rainfall and increasing temperature would increase the use of irrigation water, if the same amount of crop were to be produced. In reality, this relationship is more complicated. Adaptation to reduced water availability and higher water costs and improvements in agricultural technology both mean that farmers can produce ever more 'crop per drop'. There is evidence that rising CO_2 levels may also be helping increase crop-water use efficiency. Conversely, increased crop-water use efficiency does not automatically mean reduced water use, as there may well be opportunities to use the saved water and increase total farm production.

Turning to cities, it might be assumed that more hot and dry periods will mean greater water use by households, businesses and the irrigation of public green spaces, and that indoor cooling requires more electricity and so more water for power generation. Higher surface water temperatures reduce the efficiency of thermal power generation and hence could further increase water use. However, here too there are complicating factors. Experiences with water pricing and water-use restrictions to date suggest that there is much scope to limit household and business water use. Also, the gradual replacement of existing technologies with new ones can have a large impact on water and energy use. For example, old garden watering systems and wet appliances can be replaced with more water efficient ones, old air conditioners with more energy efficient ones, and aging power generation plants with new installations requiring less cooling water. Overall, the ultimate impact of climate change on urban and industrial water use may be modest, provided overall water supply can remain within sustainable limits and population growth does not outpace the increased efficiency of water use.

One of the more straightforward climate change impacts on water use is perhaps for hydropower generation. The predicted impacts depend mainly on the projected precipitation changes. By the end of the century, hydropower potential is predicted to increase by 15 to 30 per cent in Scandinavia and northern Russia, but to decline by more than 20 per cent in Mediterranean countries (Lehner, Czisch and Vassolo 2005). For all European countries combined, a reduction of about 10 per cent is predicted. The capacity of existing and future hydropower reservoirs and turbines will determine whether any increases in power potential can be realised.

Water management changes: Implications for climate and energy

Several measures have been proposed or already implemented to deal with the current and future impacts of climate change on water resources. They vary between the type of climate impacts expected and are often integrated within plans that also deal with expected changes in water demand and measures designed to improve or protect the ecological health of rivers and wetlands. Traditional responses to changes in hydrological conditions include raising flood defences, building new dams and drilling new groundwater wells. However, such infrastructure works are not always feasible, for technical, economic, environmental or social reasons. A variety of innovative alternative measures have been proposed. Most of these have implications for energy use, and several of them have direct climate impacts. Some examples are discussed here.

In the high- to mid-latitude regions of North America, Europe and Russia, water managers typically face predictions of more intense storms, earlier and more severe snowmelt peaks, and drier and hotter summers. The development of climate adaptation measures tends to coincide with a trend towards greater urbanisation and associated water demand, and a desire to restore the health and natural functions of wetlands and catchments. In this context, adaptation measures that are being developed or have already started to be implemented include the following (Bates et al. 2008):

- Expanding floodplain areas and emergency flood reservoirs, increasing flood and storm surge protection measures such as dam walls.
- Encouraging the flood proofing or relocation of houses and industries from flood-prone areas. Examples include changes to the US National Flood Insurance to reduce the risk of multiple flood claims and the requirement for households with repeated flooding to elevate houses above the 100-year flood level.
- Improving flood forecasting, monitoring and warning systems.
- Changing watercourses to reduce flood peaks and increase flood water retention, as well as improve river ecological function and riverbank stability.
- Modifying storm drainage and sewer systems to cope with increased flooding.
- Urban development designs that reduce storm water runoff (eg through the incorporation of retaining ponds and swales and by redirecting roof runoff to infiltrate on lawns and gardens).
- Changing the operation of hydropower dams to adapt to changes in supply and demand.
- Reducing leakage and evaporation from water supply and distribution systems.
- Encouraging water conservation in households, industries and agriculture.
- Improving agricultural water conservation and conservation tillage to reduce storm runoff and increase moisture retention.

The primary focus of several of these measures is on dealing with increased flood risk, which in estuaries and deltas may be exacerbated by sea level rise and increased storm surge risks. Examples are the deltas of the Mississippi River (the United States) and Rhine River (Netherlands). Repeated summer droughts and heat waves in Europe, Russia and the United States have recently added to the water management challenge. For example, in the dry spring of 2011, Dutch water managers resorted to irrigating peat-based river dykes to prevent their shrinkage and failure. Fresh water resources from storage lakes were also used to flush intruding sea salt from the country's waterways and prevent damage to horticulture.

The common tension between flood control and water security in river and reservoir operation was already mentioned. This tension is particularly problematic in catchments that will experience both increased water stress and increased flood risk. It is a common situation in many subtropical regions faced with increasingly unreliable and intense future rainfall. An example is the Wivenhoe Dam upstream in Queensland, Australia, which has the dual role of preventing flooding and enhancing water supply. In 2011, emergency releases were necessary to avoid dam spilling and failure, causing flooding downstream in the city of Brisbane. Debate continues as to whether the dam's flood protection functions were compromised by the desire to retain water resources.

In the world's dry regions, water managers often need to balance a reduced natural supply of water with continued or increasing demands for water. Measures can be categorised as 'supply side' or 'demand side'. Examples of supply-side measures include the following (Bates et al. 2008):

- Developing new surface water and groundwater reserves.
- Increasing storage capacity in existing reservoirs and dams.

- Enhancing local retention of water through rainwater harvesting or storage in small impoundments or shallow aquifers.
- Water transfers from less-stressed water resources.
- Desalinisation of seawater and brackish water.
- Increased rainwater capture and use.
- Removal or prevention of vegetation that reduces the availability of water resources.

Developing new or less-stressed water sources is a traditional strategy that works if such 'green fields' resources can be found, but such resources are increasingly uncommon. The inevitable impacts on people, economies and ecosystems are now also less acceptable than they were in the 1970s and 1980s, when international donors such as the World Bank actively supported dam building and groundwater exploration. Nonetheless, there are still opportunities to develop new or under-used water sources – for example, largely untapped groundwater resources in parts of Africa and Southeast Asia.

Most alternative water supply measures increase energy use considerably. Desalination plants in particular have been a popular solution, and several coastal cities in the Middle East, Australia and small island nations now meet part or most of their water demand through desalinisation. A new development is the use of small-scale solar distillation processes to meet (modest) drinking water requirements – for example, on small islands. The removal of riparian vegetation has been advocated as a win-win approach that enhances water supply and suppresses the spread of exotic species (eg salt cedar in the southern United States and willow in Australia). However, there is reason to question whether such programs achieve either goal, and there can be unwanted side effects, such as high water temperatures and an increased risk of algal blooms. In some countries with stressed water sources, plantation forestry is seen as a threat to water security and is prevented, discouraged or requires a license to undertake. To date, other forms of deep-rooted and tall vegetation establishment appear to have escaped similar treatment (discussed further on).

Demand-side measures aim to increase the efficiency or reuse of water and are being pursued both in urban and in agricultural water systems. Examples include (Bates et al. 2008, Rosegrant et al. 2009):

- Improving the quality and dissemination of short-term to seasonal forecasts of rainfall, weather conditions or irrigation water allocations to water managers and farmers.
- Recycling irrigation water, grey water and waste water.
- Water conserving agricultural practices in both dry land and irrigated agriculture.
- Increasing irrigation water use efficiency (eg by changing irrigation practices and improving delivery methods).
- Reducing losses in water supply systems (eg through reducing channel seepage or replacing open channels with pipelines).
- Reducing total irrigation water use (eg by meeting food demands through trade instead).
- Encouraging urban water conserving practices (eg garden designs that require less water and introducing water use restrictions).
- Greater use of water markets to reallocate water and government buy-backs of water entitlements.
- Increasing economic incentives such as metering and pricing.

A common assumption is that increased water-use efficiency will reduce water demand, but again, the evidence for this is tenuous. Water efficiency measures need to be combined with changes in water sharing (eg water buy-backs) to prevent the saved water from being used to simply increase the irrigated area, for example.

Can water management influence climate?

Large changes in the hydrological cycle at landscape scale can affect climate directly, by changing the surface exchange of energy and water, or indirectly, by influencing the emission of greenhouse gases other than water vapour. Here, human modifications of the water cycle through water management are of particular interest. The construction of reservoirs has slowed global sea-level rise over the last half century, while conversely groundwater withdrawals have accelerated it. There is no evidence that water management affects the global climate system, but it can have a local and sometimes regional effect. For example, large-scale irrigation in California's Central Valley increases atmospheric vapour content and is estimated to increase rainfall further to the east by as much as 30 per cent, boosting streamflow in the Colorado River (Lo and Famiglietti 2013). Such dramatic effects require very large irrigation areas and a special set of landscape and climate conditions, however.

Indirect influences of water management on climate include the potential changes in the emission of methane and other GHGs more potent than CO_2. Methane is produced under anoxic conditions that can be created through the development of large water reservoirs (Tremblay et al. 2005) and from inundated fields (in particular, paddy rice fields). Conversely, draining lakes, floodplains, swamps and other wetlands can reduce methane production. Wastewater and sewage treatment are an important urban source of methane and nitrous oxide, both potent GHGs, and their emissions are expected to increase considerably over the next decades, especially in Asia.

The influence of water management on energy use

When considering the range of water management adaptation solutions that has been developed, it may be obvious that all solutions use energy, and some of them large quantities of it. The energy required to transport water against gravity or through long pipes is notorious. Desalinisation plants also require much energy, although energy efficiency continues to improve as new technologies are developed.

Urban water use requires energy to extract, transport, process, heat water, and treat and dispose of wastewater. Transport costs are an important part of the total energy requirements and this creates big differences between cities. A study of Australian cities found that the energy used to supply a person with water in Adelaide is three times greater than in Melbourne (1300 versus 400 megajoules per person per year, excluding household energy use; Gregory and Hall 2011). The difference is mainly due to the energy needed for long distance pumping. The same study predicts that a (plausible) 25 per cent growth of

Australia's population between 2007 and 2030 may double the amount of energy used in water supply, for desalinisation, recycling and pumping over ever-greater distances.

Apparent 'low energy' water security measures can be much less efficient than they seem. For example, the use of rainwater tanks to harvest and store rainwater from roofs by Australian households might reduce water demand by 30 cubic metres per person per year. However, each water tank requires a pump, and increases the associated energy requirements to around 2 kilowatt-hour per cubic metres – of similar magnitude as the energy required for desalination (Burn 2011). This emphasises the ineffectiveness and undesirable side effects that distributed adaptation solutions can have if inadequately designed or implemented.

Climate mitigation and energy management impacts on water management

Energy generation and water resources

Large volumes of water are used to meet our energy needs: the extraction and refining of fuels, electric power generation, and the manufacture and installation of power plants. Water requirements vary widely between energy sources: large for hydroelectricity generation and cooling of thermal electricity plants, but negligible for wind and photovoltaic solar power.

The extraction of raw materials and fuel can evaporate large volumes of water or make it unsuitable for other uses. Particularly large amounts of water, in the order of 100 litres per litre of oil, are used in the extraction of non-conventional oil sources (eg for hydraulic fracturing, shale processing, and cooling and disposing of the residual waste). Crude oil extraction currently accounts for about 10 per cent of water consumption globally, but this is predicted to reach almost 20 per cent by 2050 due to the growing importance of non-conventional oil sources (Gerbens-Leenes, Hoekstra and Meer 2008). The water use in natural gas, coal and uranium extraction is much less. Biomass energy production (eg for firewood or bioethanol) requires far larger amounts of water per unit energy but is typically grown with green water and does not always compete with other (green or blue) water uses. Exceptions do exist; for example, the irrigation of crops for bioethanol production can have direct implications for other water uses and processing of biomass into ethanol requires additional water.

Power generation also requires water. As much as 40 per cent of total water extractions in the United States is for cooling thermal power plants, although most of it is returned to rivers. About 3 per cent of the extracted water evaporates in the process; equal to 1 per cent of total water extractions in the United States. Hydropower plants also 'consume' water, through the evaporative losses from their supply reservoirs and conduits. For each kilowatt-hour of electricity generated in the United States, about eight litres of water evaporates from power plants. This hides a lot of variation: less than two litres per kilowatt-hour evaporates from thermoelectric plants in cooler regions, but more than seventy litres from hydropower reservoirs in dry regions (Torcellini, Long and Judkoff 2003). On a global scale, the consumption of water embedded in energy use is estimated at about 100 litres per person each day (Gerbens-Leenes et al. 2008). This accounts for about 6 per cent of total global water withdrawals and 8 per cent of water consumption (FAO n.d.).

In summary, fuel and energy generation can have implications for water management, particularly in industrialised countries. However, distinctions need to be made between evaporative use of green and blue water, pollution of water resources and non-consumptive use of water that may be reused.

A relatively poorly understood connection between energy production and water availability is the effect that their non-GHG emissions have on local or regional climate. Burning coal, oil and biomass emits soot, aerosols and other airborne particles and compounds into the atmosphere, where they promote the condensation of vapour, increase cloud cover and rainfall, and reduce temperature. The resulting increase in fog, smog and low-intensity rainfall over industrial and urban regions is well documented. Aerosols from burning have also been put forward as an important cause for 'global dimming': globally, a 2 to 4 per cent decrease in solar radiation has occurred between 1960 and 1990, and more in industrialised regions (Stanhill and Cohen 2001). Less radiation certainly means less energy for evaporation, but the influence on precipitation is more complicated. Global dimming has been linked to rainfall decreases in the Northern Hemisphere, while industrial and fire aerosol emissions from Southeast Asia may have contributed to the steady increase in monsoon rainfall in northern Australia (Rotstayn et al. 2009). Since the 1990s, cleaner energy production appears to have reversed global dimming trends, except in India and China where power generation continues to grow (Wild 2009). Whether these trends contribute to the apparent rainfall changes observed in these regions is a matter of debate.

Climate mitigation and water resources

From the preceding discussion, it is clear that climate mitigation measures changing the total amount or the mix of energy sources can have an indirect impact on water resources. Other climate mitigation options could potentially also have an impact but are currently of a scale or maturity yet to be assessed. They include carbon dioxide capture and storage techniques, geothermal energy and solid waste management. At present, the most likely climate mitigation measures to affect water resources are those that involve large-scale land cover change. Examples are the establishment of forests to store carbon or clearing of forests to plant biofuel crops. Debate about the impact of forestry on water resources goes back many decades, when it was first observed that catchment streamflow declined when trees were planted and increased again when trees were logged. The often-neglected other side of increased water use is the greater return of water to the atmosphere, which at least in theory may enhance precipitation locally or downwind – albeit not necessarily where it is needed. The influence of land cover change on local and global climate is even more complicated. Among others, the associated changes in land-surface properties can change the amount of solar energy that is absorbed, the roughness of the surface to wind, and the timing of evaporation. Our understanding of the net effect of such changes is far from complete. What is clear is that large-scale changes are required to have any significant influence on regional climate or basin-scale water availability, even though local climate and catchments can be affected (van Dijk and Keenan 2007). This line of reasoning can be extended to any vegetation change that is of sufficient importance and area to modify the exchange of energy and water between land and atmosphere.

The argument that plantation forests affect downstream water availability raises important questions about water ownership and rights. Most governments recognise the rights of landowners to use the rain that falls onto their property (green water). If at all, the separation between land and water rights only addresses blue water sources, such as surface water, groundwater or overland runoff captured in farm dams. The regulation of plantation forestry as a water user essentially removes this distinction. Taken to its logical conclusion, it could have far-reaching implications for other forms of vegetation change (eg carbon or biodiversity plantings) or indeed any land management change that promotes the retention of water on site, which includes many soil conservation measures (eg zero tillage, contouring, terracing, strip plantings). It also has implications for landscape, river and wetland restoration, which typically involve greater retention and use of water. For example, the creation of 'chain-of-pond' drainage systems is encouraged in Australia as a means to improve landscape health. Under the right conditions it may also increase surface and groundwater availability in dry periods, but not without reducing the overall volume of water released downstream. Similarly, where rainfall is in limited supply, the relationship between water use and biomass production is almost inescapable, so any measure that increases biomass may affect water availability downstream as well. This includes biodiversity plantings, shelterbelts, and other plantings that may have great ameliorative value yet can be expected to reduce water volumes delivered downstream (eg into reservoirs and from thereon to irrigation and urban users).

The reverse logic also applies: while some see the removal of introduced species from riparian areas as a win-win, one can use similar arguments to support the removal of native riparian vegetation. From there it is a small step to consider large-scale deforestation and urbanisation: a tried and tested method to increase groundwater recharge and river-flow volumes (though usually not without deteriorating water quality). At the core of this argument are questions of water ownership, the relative moral right or economic value of using water onsite or downstream and inevitable trade-offs between competing land and water management goals. It is a prime example of the integrated water resources management challenge.

Conclusion

This chapter reviewed our knowledge about the physical interactions between water management, energy use and climate mitigation measures. Some conclusions are as follows:

1. There are large differences between countries and regions in water availability, water use and water scarcity.
2. Water, food and energy can be substituted for each other through technology, transport and trade. The concept of 'peak oil' can be applied (with modifications) to water management.

3. Many of the world's developed water systems already face challenges due to increasing demands from a growing population and economy and the desire to improve environmental values.
4. Climate change will usually make water management more difficult due to increased rainfall extremes and drying in already water-scarce areas, compounded by heat waves, sea level rise, and increased vulnerability to flooding, among others. Poor countries will struggle more to adapt to changing conditions.
5. Climate variability and change occur over several time scales, regardless of the relative contribution of human and natural drivers to recently observed changes. The limitations of using historic climate records to represent future conditions are increasingly recognised.
6. Water management can directly influence climate if it leads to large changes in evaporation, but so far the effects have been limited to local and regional scales.
7. All water management changes have implications for energy use. Long-distance pipe transport and desalinisation are particularly energy intensive.
8. Energy management can have implications for water management, particularly in industrialised societies where power generation can be a large user of water. Changes towards non-conventional fossil fuel sources, bioenergy and improved power generation technologies can affect total water use. Distinctions should be made between evaporative, polluting and non-consumptive uses.
9. Electricity generation from coal-fired power stations appears to have had a measurable influence on global climate and water resources through the production of aerosols.
10. Climate mitigation measures can affect water management through changes in energy production and changes in land use. Landscape options to mitigate carbon can affect water resources but full consideration of land use impacts raises important questions about water rights and equity, and the traditional distinction made between 'green' and 'blue' water.

References

Allan, J. A. (1997). *Virtual Water: A Long Term Solution for Water Short Middle Eastern Economies?* Occasional Paper 3. London: School of Oriental and African Studies, University of London.

Bates, B. C., Kundzewicz, Z. W., Wu, S., and Palutikof, J. P. (2008). Climate Change and Water. Technical Paper of the Intergovernmental Panel on Climate Change. Geneva: IPCC Secretariat.

Bolch, T., Kulkarni, A., Kääb, A., Huggel, C., Paul, F., Cogley, J. G., Frey, H., Kargel, J. S., Fujita, K., Scheel, M., Bajracharya, S., and Stoffel, M. (2012). The State and Fate of Himalayan Glaciers. *Science* 336: 310–14.

Bruinsma, J. (2003). *World Agriculture: Towards 2015/2030. An FAO Perspective.* London: Earthscan.

Burn, S. (2011). Future Urban Water Supplies. In *Water: Science and Solutions for Australia*, ed. I. Prosser, Collingwood, Victoria (Australia): CSIRO Publishing, pp. 89–104.

Delpla, I., Jung, A. V., Baures, E., Clement, M., and Thomas, O. (2009). Impacts of Climate Change on Surface Water Quality in Relation to Drinking Water Production. *Environment International* 35: 1225–33.

García-Herrera, R., Díaz, J., Trigo, R.M., Luterbacher, J., and Fischer, E.M. (2010). A Review of the European Summer Heat Wave of 2003. *Critical Reviews in Environmental Science and Technology* 40: 267–306.

Gerbens-Leenes, P., Hoekstra, A. and Meer, T. H. (2008). *Water Footprint of Bio-Energy and Other Primary Energy Carriers.* Delft (Netherlands): UNESCO-IHE Institute for Water Education.

Gleick, P. H. and Palaniappan, M. (2010). Peak Water Limits to Freshwater Withdrawal and Use. *Proceedings of the National Academy of Sciences* 107: 11155–62.

Gregory, A. and Hall, M. (2011). Urban Water Sustainability. In *Water: Science and Solutions for Australia*, ed. I. Prosser, Collingwood, Victoria (Australia): CSIRO Publishing, pp. 75–88.

Grigg, N. S. (1996). *Water Resources Management: Principles, Cases, and Regulations.* New York: McGrawHill.

Hoekstra, A. Y., Mekonnen, M. M., Chapagain, A. K., Mathews, R. E. and Richter, B. D. (2012). Global Monthly Water Scarcity: Blue Water Footprints versus Blue Water Availability. *PLoS ONE* 7: e32688.

Huntington, T. (2006). Evidence for Intensification of the Global Water Cycle: Review and Synthesis. *Journal of Hydrology* 319: 83–95.

Immerzeel, W. W., van Beek, L. P. and Bierkens, M. F. (2010). Climate Change Will Affect the Asian Water Towers. *Science* 328: 1382–5.

Johanson, C. M. and Fu, Q. (2009). Hadley Cell Widening: Model Simulations versus Observations. *Journal of Climate*, 22, 2713–2725.

Kron, W. and Berz, G. (2007). Flood Disasters and Climate Change: Trends and Options–A (Re-)Insurer's View. In *Global Change: Enough Water for All?*, ed. by J. L. Lozán, H. Graßl, P. Hupfer, L. Menzel, and C. D. Schönwiese, Hamburg: University of Hamburg, pp. 268–273.

Lehner, B., Czisch, G., and Vassolo, S. (2005). The Impact of Global Change on the Hydropower Potential of Europe: A Model-Based Analysis. *Energy Policy* 33: 839–55.

Lo, M.-H. and Famiglietti, J. S. (2013). Irrigation in California's Central Valley Strengthens the Southwestern US Water Cycle. *Geophysical Research Letters* 40: 301–6.

McCallum, J. L., Crosbie, R. S., Walker, G. R., and Dawes, W. R. (2010). Impacts of Climate Change on Groundwater in Australia: A Sensitivity Analysis of Recharge. *Hydrogeology Journal* 18: 1625–38.

Milly, P., Betancourt, J., Falkenmark, M., Hirsch, R., Kundzewicz, Z., Lettenmaier, D., and Stouffer, R. (2008). Climate Change. Stationarity Is Dead: Whither Water Management? *Science* 319: 573.

Rockström, J., Falkenmark, M., Karlberg, L., Hoff, H., Rost, S., and Gerten, D. (2009). Future Water Availability for Global Food Production: The Potential of Green Water for Increasing Resilience to Global Change. *Water Resources Research* 45(7): http://onlinelibrary.wiley.com/doi/10.1029/2007WR006767/full.

Rosegrant, M. W., Ringler, C., and Zhu, T. (2009). Water for Agriculture: Maintaining Food Security Under Growing Scarcity. *Annual Review of Environment and Resources* 34: 205–22.

Rotstayn, L. D., Keywood, M. D., Forgan, B. W., Gabric, A. J., Galbally, I. E., Gras, J. L., Luhar, A. K., McTainsh, G. H., Mitchell, R. M., and Young, S. A. (2009). Possible

Impacts of Anthropogenic and Natural Aerosols on Australian Climate: A Review. *International Journal of Climatology* 29: 461–79.
Sheffield, J., Wood, E. F., and Roderick, M. L. (2012). Little Change in Global Drought Over the Past 60 Years. *Nature* 491: 435–38.
Shiklomanov, I. A. and Rodda, J. C. (2003). *World Water Resources at the Beginning of the Twenty-First Century*. Cambridge University Press.
Stanhill, G. and Cohen, S. (2001). Global Dimming: A Review of the Evidence for a Widespread and Significant Reduction in Global Radiation with Discussion of its Probable Causes and Possible Agricultural Consequences. *Agricultural and Forest Meteorology* 107: 255–78.
Sun, F., Roderick, M. L., and Farquhar, G. D. (2012). Changes in the Variability of Global Land Precipitation. *Geophysical Research Letters* 39: L19402.
Torcellini, P. A., Long, N., and Judkoff, R. (2003). *Consumptive Water Use for US Power Production*. Golden, CO: National Renewable Energy Laboratory.
Tremblay, A., Varfalvy, L., Roehm, C., and Garneau, M. (2005). In *Greenhouse Gas Emissions-Fluxes and Processes: Hydroelectric Reservoirs and Natural Environments*. Berlin: Springer.
United Nations (UN) (2003). World Water Development Report: Water for Life, Water for People. Paris: UNESCO.
United Nations Food and Agricultural Organisation (FAO) (n.d.). AQUASTAT Database, www.fao.org/nr/water/aquastat/main/index.stm.
van Dijk, A. I. J. M. and Keenan, R. J. (2007). Planted Forests and Water in Perspective. *Forest Ecology and Management* 251: 1–9.
van Dijk, A. I. J. M., Beck, H. E., Crosbie, R. S., de Jeu, R. A. M., Liu, Y. Y., Podger, G. M., Timbal, B. and Viney, N. R. (2013). The Millennium Drought in Southeast Australia (2001–2009): Natural and Human Causes and Implications for Water Resources, Ecosystems, Economy and Society. *Water Resources Research* 10.1002/wrcr.20123.
Vörösmarty, C., McIntyre, P., Gessner, M., Dudgeon, D., Prusevich, A., Green, P., Glidden, S., Bunn, S., Sullivan, C., and Liermann, C. (2010). Global Threats to Human Water Security and River Biodiversity. *Nature* 467: 555–61.
Wada, Y., van Beek, L. P. H., van Kempen, C. M., Reckman, J. W. T. M., Vasak, S., and Bierkens, M. F. P. (2010). Global Depletion of Groundwater Resources. *Geophysical Research Letters* 37: L20402.
Westra, S., Alexander, L. V., and Zwiers, F. W. (2012). Global Increasing Trends in Annual Maximum Daily Precipitation. *Journal of Climate* 26: 3904–18.
Wild, M. (2009). Global Dimming and Brightening: A Review. *Journal of Geophysical Research: Atmospheres* 114: D00D16.

3

Implications of climate change for energy systems in a multisectoral context

THOMAS J. WILBANKS

Introduction

For decades, energy production and use have been considered the primary culprits as observers try to assign blame for climate change. Beyond question, patterns of energy resource development and technology use, combined with growing global appetites for energy services for socioeconomic development, are important *drivers* of climate change. But energy systems are also vulnerable to *impacts* of climate change on energy activities themselves and also on other infrastructures on which energy systems depend. Energy systems are therefore *victims* of climate change, as well as culprits.

The process of understanding how climate change impacts can be a risk-management challenge for energy policy makers, decision makers and stakeholders is just beginning to emerge, because for so many years, impacts were not a focus of research – or even of serious discussion. A number of recent assessments, however, are beginning to sketch out a picture of vulnerabilities, risks and possible impacts, and many energy providers are seeing enough evidence of climate-related impacts on their operations that they are taking climate change impacts seriously.

This chapter will first summarise what is currently known about sensitivities of energy systems to climate change, including linkages with other systems and infrastructures. It will then summarise the exposures to climate change that are of the greatest concern for energy systems along with several issues raised by a less-than-perfect fit between the outputs of major climate change models and the needs of energy risk management. Next, it will summarise recent findings about the most serious implications of climate change for energy systems in both the near and the long term. Finally, it will consider possible strategies for reducing vulnerabilities and impacts of climate change on energy systems.

Several themes thread through the various parts of the chapter: the current focus of decision makers on extreme weather events; the fact that direct effects of climate change interact with other driving forces for energy-sector impacts; and capacities for major energy-sector institutions to reduce vulnerabilities and manage risks once they are recognised.

Sensitivities of energy systems to climate change

Direct effects of climate change are temperature increases, including changes in extremes as well as averages; changes in precipitation, including changes in variability and extremes and changes in their forms (eg rain vs snow); changes in extreme weather events, possibly including changes in frequency, intensity and location; and a rising sea level. Sensitivities of energy systems to these effects vary by location, differences in the climate change threat, differences in systems exposed to the effects, differences in geographic scale and differences in socioeconomic development context (ranging from institutional capacities to available financial resources).

Energy supply and use vulnerabilities

Vulnerability is defined by a combination of exposure to a threat, sensitivity to that threat and coping capacity in responding to the threat (Clark et al. 2000). Vulnerabilities of energy supply and use to climate change have received very little research attention until recently. For example, the massive 1800-page Global Energy Assessment (GEA 2012) includes only a few pages on climate change and energy systems (pp. 200–7). But a number of assessments in recent years have summarised what is known about vulnerabilities of energy systems to climate change (CCSP 2008; USGCRP 2009; World Bank 2011; ORNL 2012a and 2012b; UK 2012), based mainly on studies in the United States and other industrialised countries. In general, these studies indicate the following implications of exposure to climate change effects, moderated by sensitivities to such effects:

1) The major near-term risk for energy systems is from episodic disruptions due to extreme weather events, especially in particularly vulnerable regions. For both energy supply and energy use, extreme weather events such as hurricanes, tornados, floods, droughts, heat waves and wildfires are the most dramatic threats to sustainable energy services for social and economic development. Although such events are difficult to predict for any specific location and time period, based on available climate models, there is some evidence that frequencies, intensities and locations of extreme weather events over the past several decades are different from century-long averages, indicating effects of climate change (NCA 2013). For example, in many regions across the world, a greater proportion of total precipitation is falling in a smaller number of more intense rainfall events, contributing to both seasonal flooding and seasonal drought.
2) Both electricity demand and electricity supply will be affected by increases in temperatures. These increases, in both averages and extremes, will affect energy systems as heating requirements shrink and cooling requirements grow. Data for the United States (most likely also applying to other mid-latitude countries) indicate that overall effects on total energy consumption at the national level may not be substantial, as savings from heating offset increases for cooling. But these two energy services differ in their fuel mixes – heating is provided by liquid and gas fuels as well as electricity, but cooling is provided almost exclusively by electricity. As a result, climate change will

increase demands for electricity, especially in regions where the number of cooling days increases considerably compared with historical experience (CCSP 2008; USGCRP 2009; ORNL 2012a). At the same time, capacities for thermal electricity generation will drop as air and water temperatures increase (ORNL 2012a), which would make it more difficult to meet the growing demand, at least in some regions and in some seasons.

3) Seasonal and chronic water supply constraints pose threats to reliable energy supplies in many regions. A number of recent assessments have shown that climate change effects on seasonal and long-term average precipitation are very likely to become challenges for energy production in many areas, both in the United States and globally (eg UCS 2011; EPRI 2011; Dell and Pasteris 2010). Among the sensitivities are that many arid areas are projected to become more arid, areas dependent on snowmelt for spring surface water supplies are projected to see reduced flows, and many other areas are likely to be subject to seasonal droughts that are more frequent, longer and more intense, in some cases coupled with seasonal heat waves that would tend to increase water demands (ORNL 2012a). Energy system sensitivities can include chronic or seasonal reductions in hydropower potential, effects of reduced water flows on the ambient water temperature (which is related to allowable temperatures of power plant water emissions) and reduced effectiveness of thermal power plant cooling technologies (ORNL 2012a).
4) Geographic patterns of renewable energy supply will be affected by changes in climate. As temperature and precipitation patterns shift, environmental conditions for renewable energy supply will also shift. Wind speed and direction will be affected, impacting wind power. Changes in cloud cover can affect solar insolation, impacting solar energy generation. Changed conditions for agricultural production will affect potentials for biofuels cultivation, and changes in water availability may impact biofuel production. In nearly every case, as environmental conditions shift, some areas will see reductions in renewable energy supply potential while others will see improvements; but shifting patterns of resource development will interact with other driving forces shaping land-use decisions in localities and regions, which adds strategic challenges in increasing the renewable energy share of total energy supply (ORNL 2012a).

In nearly every case, climate change vulnerabilities of energy systems are embedded in relationships between climate change effects and other driving forces that are shaping resilience and sustainability (IPCC 2007), including linkages with other geographic areas, infrastructures and systems.

Linkages of energy system vulnerabilities with water and other systems

Energy systems are linked with and often dependent upon other systems and infrastructures that may be vulnerable to climate change effects. Figure 3.1 illustrates some of the connections, of which water and land are often especially important.

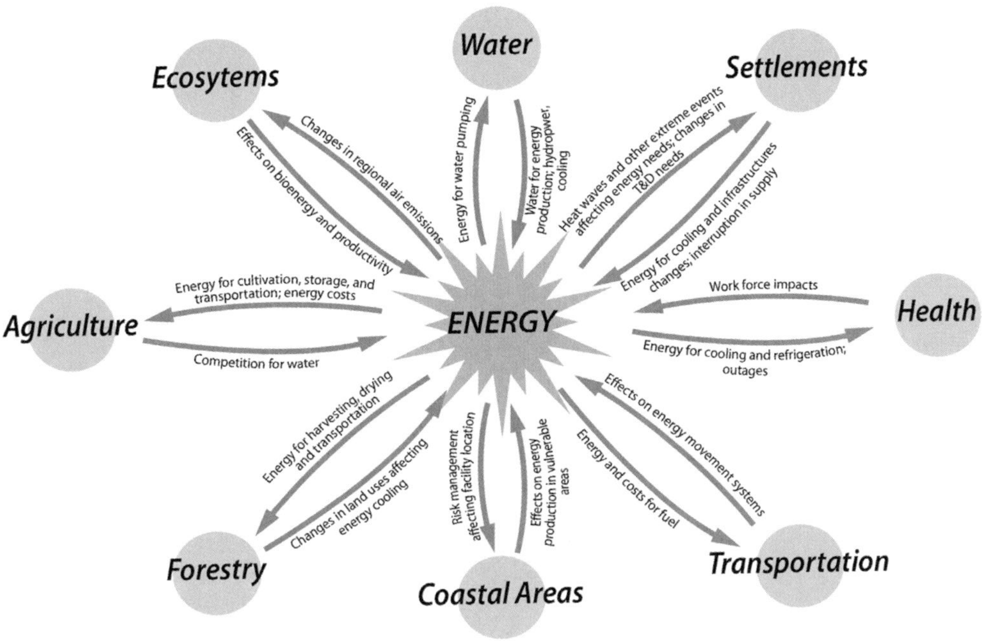

Figure 3.1

A suite of simulation and analysis tools has been developed in the United States to unravel linkages between different kinds of infrastructures and to explore their interdependencies. Intended initially to simulate the effects of terrorist acts on US infrastructures, they have mainly been applied as decision-support tools during responses to extreme weather events, such as major hurricanes (ORNL 2012b). Over a period of nearly a decade, learning from each application, the models have become remarkably accurate in depicting connections with elements of different infrastructures during major disruptive events, including cases where impacts cascade through the system of systems to cause major disasters. Figure 3.2 shows a simplified summary of interactions between energy infrastructures and some other infrastructures, based on this experience.

Examples of such interactions from the US experience include effects of electricity system interruptions due to extreme events on transportation systems (eg traffic control in cities and in air transportation), on water pumping and wastewater management where systems are powered by electricity, and on human health as space cooling for vulnerable populations is interrupted (ORNL 2012b). At the same time, interruptions of other infrastructures can affect energy systems, both within the energy sector (eg interruptions of natural gas supply because pipeline compressors are powered by electricity) and in interdependencies with other sectors (eg the reliance of Smart Grid approaches on reliable electricity services for communication, movements of fuel sources by river freight, and relationships between urban energy demands and storm impacts on vulnerable coastal area [ORNL 2012b]).

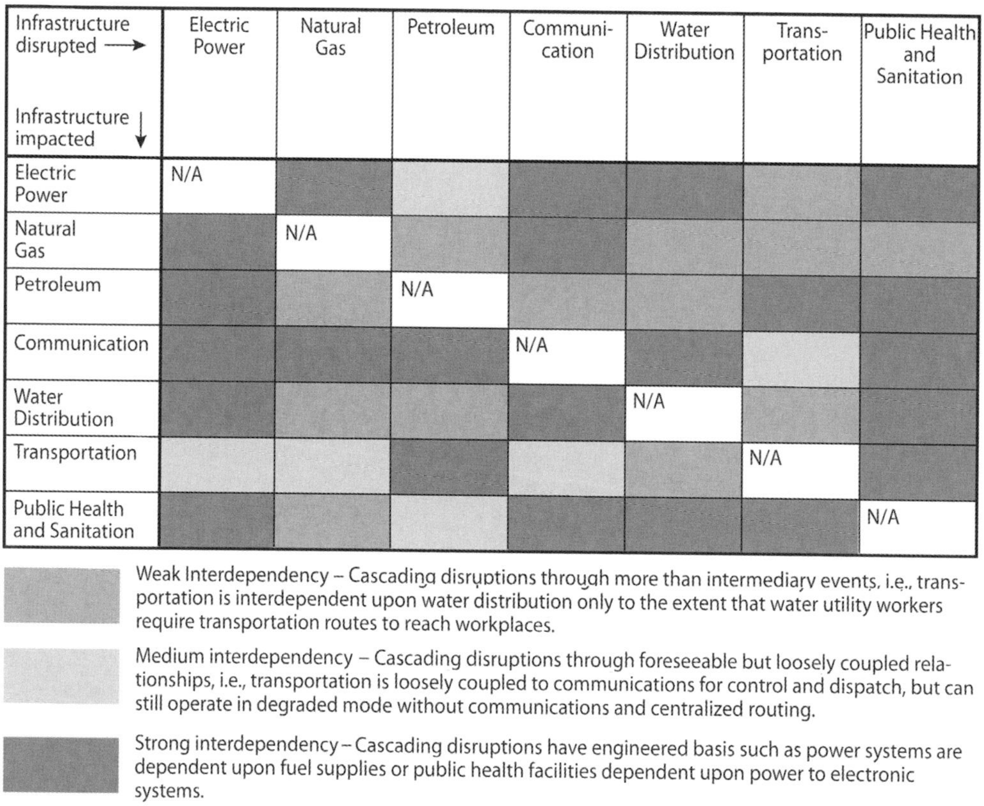

Figure 3.2 Interdependencies among sectors in the event of major disruptive weather events (ORNL 2012b).

Exposures to climate change impacts

Future exposure to climate change are projected by climate change 'scenarios', representing different possible futures under a set of assumptions. The two major families of climate change scenarios used in recent years are:

1) Scenarios developed by IPCC's Special Report on Emission Scenarios (IPCC 2000), utilised by the IPCC Fourth Assessment Report and often referred to as CMIP-3 (referring to the Climate Model Intercomparison Project – CMIP – which is the global repository for climate scenario projections); and
2) Scenarios developed more recently by the science community to provide a more current perspective on possible futures, called Representative Concentration Pathways (RCPs), now being referred to as CMIP-5 (Coupled Model Intercomparison Project). RCPs sketch out four illustrative futures defined in terms of radiative forcing, elaborated by Integrated Assessment Models (IAM) that associate illustrative forcings with greenhouse gas (GHG) emission scenarios that would lead to such alternative futures (Moss et al. 2010).

Figure 3.3 Elements of climate change science: CM/ESM refers to climate and earth system models; IAV refers to impact, adaptation and vulnerability analysis; IAM refers to integrated assessment models.

As the RCPs indicate, in principle, climate science scenarios include representations of three sets of relationships: (a) drivers of greenhouse gas emissions and accumulations, (b) changes in climate as a result of greenhouse gas emissions and (c) effects of those changes, some of which feed back towards the drivers (Figure 3.3) – although the links between effects and drivers have been less comprehensively analysed than the drivers and climate changes per se.

Existing scenarios depict a wide range of possible futures, from relatively modest climate change to very severe climate change (in both magnitude and rate of change), not generally associated with estimates of the probability of one versus another. Figure 3.4, for example, shows the futures depicted in SRES scenarios, while Figure 3.5 shows the futures depicted in RCPs. Although RCPs are becoming the standard framings of climate futures, SRES scenarios continue to be widely used because of familiarity (in some cases, easier access) and continuing development of ensemble analyses of CMIP-5 projections. As one case in point, the US National Climate Assessment scheduled for release in early 2014 uses SRES scenarios B1 and A2 to represent relatively moderate and relatively severe climate change trajectories. Downscaled versions of the RCPs are becoming available, and current attention is focused on RCP 4.5 as a relatively moderate future (although well above SRES B1) and RCP 8.5 as a relatively severe future (well above SRES A2).

Although such scenarios are useful for framing and illustrating the nature and magnitude of climate change trajectories and risks, along with what they depend upon, they are often insufficient to address major energy system concerns, especially in the next several decades. The major concerns include:

- Potential for severe climate change, given current trajectories of global GHG emissions. As extreme as RCP 8.5 appears to be – the most extreme ever modelled to date – current global greenhouse gas emissions are on a trajectory above the emission assumptions that RCP 8.5 embodies (Figure 3.6). This strongly suggests that future climate change impacts are likely to be very large indeed, often disruptive, and in many cases requiring transformational changes in the systems that support and use energy services.
- Limited current capacity to project extremes and climate-related extreme weather events. Clearly, implications of climate change for energy systems are numerous,

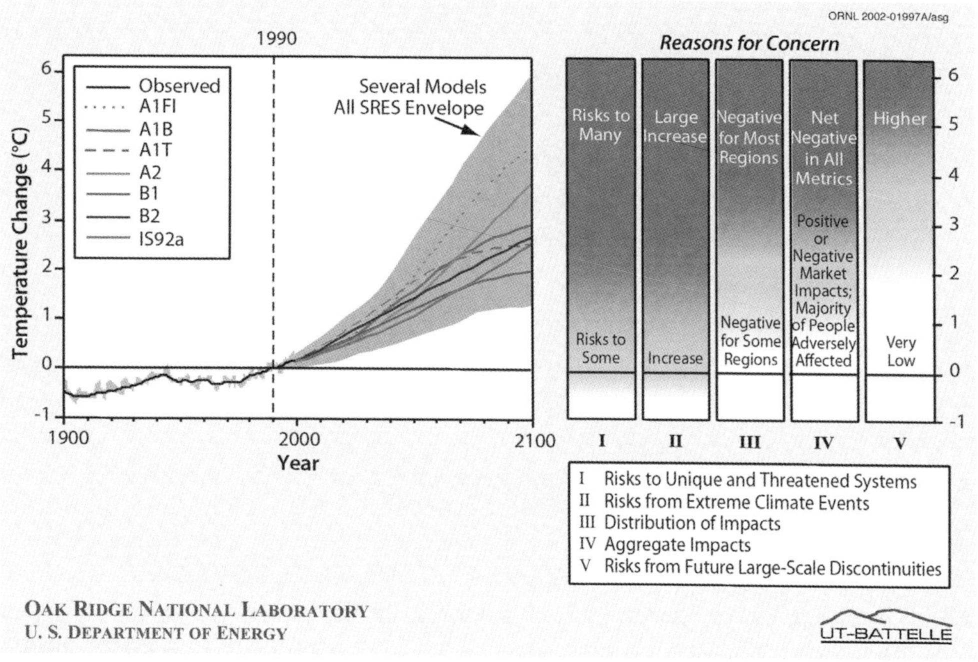

Figure 3.4 SRES scenarios (IPCC 2007).

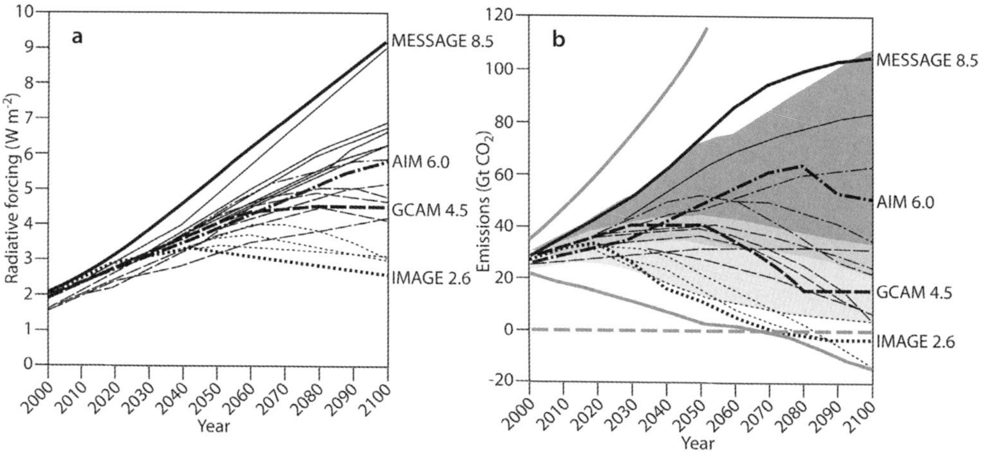

Figure 3.5 Representative concentration pathways (RCPs), both projected forcings and associated greenhouse gas emissions (Moss et al. 2010).

interconnected, and in many cases, potentially serious. Climate scenarios primarily address changes in temperature and precipitation over time, usually average changes over decadal periods. On the other hand, concerns of energy system decision makers and managers are mainly focused on climate extremes and climate-related extreme weather events

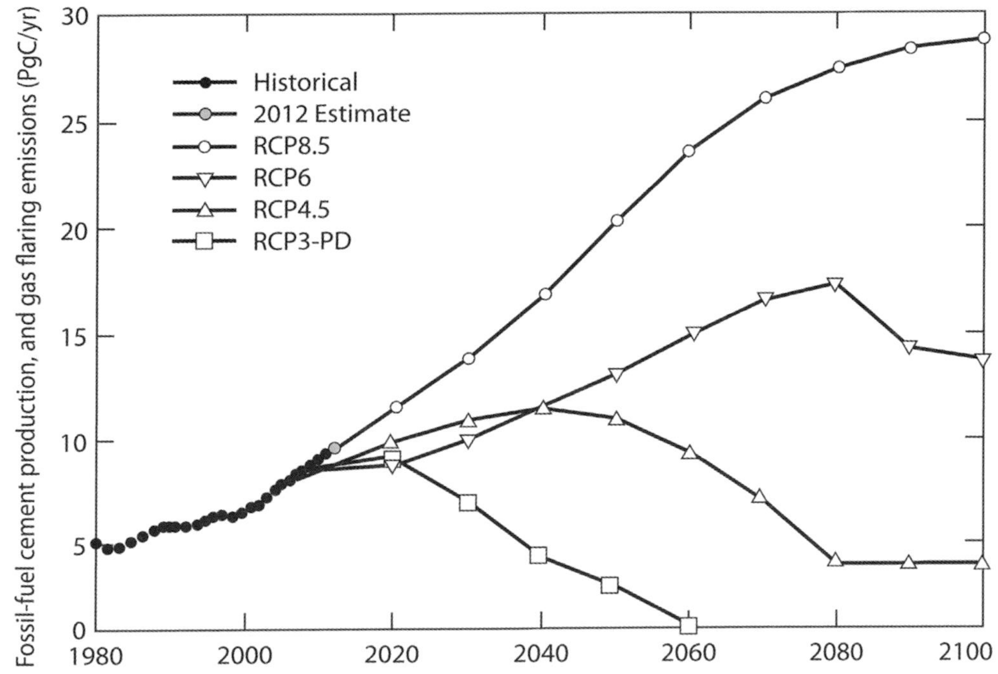

Figure 3.6 Current global emissions are above the worst scenario (Peters et al. 2013).

(ORNL 2012a), especially in the next several decades. In many cases, current climate models are not very useful as sources of information about projected climate-related phenomena such as severe storms and sea-level rise, and they are not very useful as sources of information about projections of extreme weather events that combine climate phenomena with other driving forces, such as flooding, heat waves, droughts and wildfires.

- Issues related to the sustainability of energy services for lower-income areas and populations. Most framings of climate change mitigation responses by energy systems emphasise relatively rapid decarbonisation of energy supply and use: for example, significant increases in the use of renewable energy systems (see, for instance IPCC [2012b] and Vergara [2012]). Although this issue is not always confronted directly, most alternative energy pathways would imply higher, perhaps much higher, energy costs than at present as the world transitions to a new mix of energy sources. Many observers, however, associate the sustainability of energy services for development with low energy costs (eg IPCC 2012b), which adds significantly to the challenge of assuring a smooth but relatively rapid transition to alternative energy systems worldwide.

Perspectives on impacts from climate change

Impacts of climate change on energy supply and use are potentially numerous and diverse, and if climate change is relatively severe, the impacts could be disruptive, at least seasonally and/or episodically. Figure 3.7 illustrates some of the factors that could affect energy supply systems.

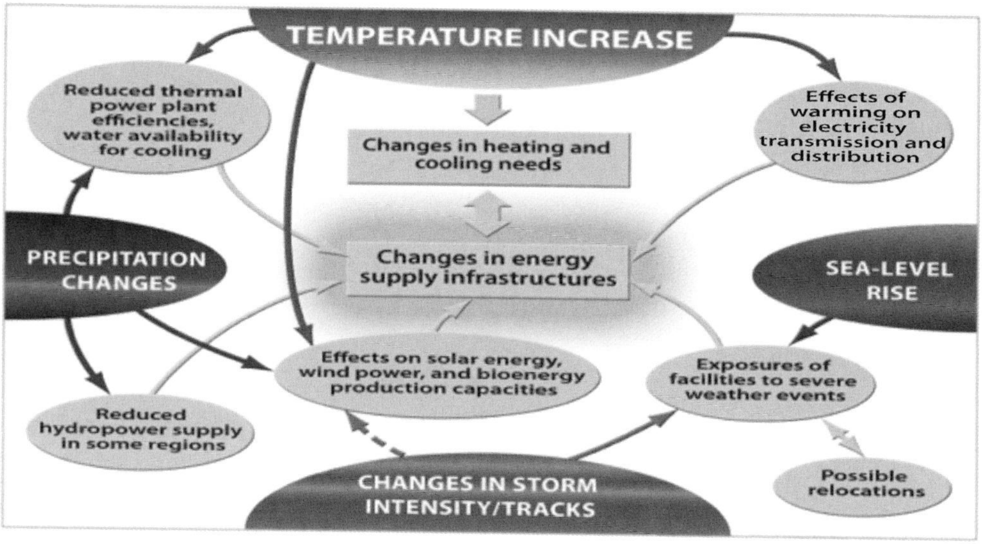

Figure 3.7 A summary of potential climate change effects on energy supply infrastructures.

Although the current body of research on implications of climate change for energy systems is still weak, there is a relatively robust consensus about the potential impacts that are of greatest concern:

- The dominance of extreme events over the next several decades. As my earlier discussion shows (see also IPCC SREX [2012], regarding this concern for most potentially impacted sectors), the major concern is for extreme events, not only for energy supply but also for energy use – where disruptions in supply not only affect such energy services as lighting and cooling but also such related infrastructures as water pumping and transportation system controls.
- Effects of temperature increases in vulnerable areas, including lower-income urban areas that lack access to air conditioning, especially heat waves during summer periods.
- Effects of changes in precipitation regimes on water availability and water temperature, both chronically and seasonally, especially droughts in areas where water is already scarce and floods during periods of extreme precipitation.
- Sea-level rise, especially in combination with storm exposures in vulnerable areas, relative to the map of energy facilities (eg facility siting) and consumer markets for regional utilities (eg longer-term population shifts away from vulnerable areas).
- Effects of climate change on patterns of energy demand and use more generally, including possible demographic and economic shifts.
- Effects of climate-change policies on energy supply and use. An issue that is seldom addressed in energy sector impact assessments, although it is a very prominent concern on the part of most energy decision makers, is the impact of climate mitigation policies on regions and energy sector institutions (CCSP 2008, chapter 4; NRC 2010: 48–50; IPCC SRREN 2012, chapter 8). Climate change policies aimed at reducing the rate of

increase in global greenhouse gas concentrations in the atmosphere have the potential to affect market shares and the financial viability of energy systems based on fossil fuels, especially coal; to raise energy prices as a result of emission controls and technology transitions; and to present challenges for regional economies linked with fossil fuel production and use. They could also affect energy technology research and development agendas and international energy technology and service markets, for example, related to innovative renewable energy alternatives. Such effects are an important aspect of interactions between adaptation and mitigation (Wilbanks and Sathaye 2007), where synergies may offer co-benefits for both.

Strategies for energy system adaptation to reduce vulnerabilities and impacts

Both globally and in most larger nations, the energy sector is large and complex, with impressive financial and management resources, capable of responding to major challenges (GCRP 2009; ORNL 2012a). And, after decades of avoiding discussions of climate change because they are connected with discussions of climate change policy responses (eg cap and trade policies, carbon emission restrictions and carbon taxes), both energy suppliers and energy users are beginning to recognise that climate change impacts may pose risk to some of their activities.

Most sizeable energy supply institutions pride themselves on their risk management, and they are well adjusted to climate variations – so well so that in some cases they have been slow to recognise that climate change presents new challenges. As climate-related changes begin to be observed, however, exposures to such effects as extreme weather events, environmental changes in polar regions and acidification in oceans where offshore production is taking place, energy institutions are adding climate change to the host of uncertainties addressed by their risk management strategies, seeking not only to reduce vulnerabilities but also to identify market opportunities (Dell and Pasteris 2010; ORNL 2012a).

For both energy supply and use, strategies for managing risks associated with climate change will vary by resource, technology, institution and threat – related to on-going processes of capital stock turnover and revitalisation. In industries where facilities and equipment have lifetimes of decades, over thirty to fifty years much of the capital stock will be changed; reducing climate change vulnerabilities can be built into the choices that must be made in any event.

Table 3.1 (from World Bank 2011) summarises examples of adaptation measures that may be considered by risk management strategies in the energy sector.

More specifically, such energy sub-sectors as the oil and gas industry have recently begun considering climate change vulnerabilities and adaptive risk management. For example, catalysed by a vulnerability assessment led by Jan Dell (Dell and Pasteris 2010; see Figure 3.8), the oil and gas industry conducted a workshop at IPIECA in London in October 2012 on climate change adaptation, at which a number of major companies made presentations about their current adaptation strategy development efforts (IPIECA 2013).

Table 3.1 Examples of adaptation measures to reduce losses/risks in energy systems (World Bank 2011).

		TECHNOLOGICAL		BEHAVIORAL		
	Energy System	"Hard" (structural)	"Soft" (technology and design)	Re(location)	Anticipation	Operation and Maintenance
Supply	Mined Resources (including oil & gas, thermal power, nuclear power)	Improve robustness of installations to withstand storms (offshore), and flooding/drought (inland)	Replace water cooling systems with air cooling, dry cooling, or recirculating systems Improve design of gas turbines (inlet guide vanes, inlet air togging, inlet air filters, compressor blade washing techniques, etc.) Expand strategic petroleum reserves Consider underground transfers and transport structures	(Re)locate in areas with lower risk of flooding/drought (re)locate to safer areas, build dikes to contain flooding, reinforce walls and roofs	Emergency planning	Manage on-site drainage and runoff Changes in coal handling due to increased moisture content Adapt regulations so that a higher discharge temperature is allowed Consider water re-use and integration technologies at refineries
	Hydropower	Build de-siting gates Increase dam height Construct small dams in the upper basins	Changes in water reserves and reservoir management	(Re)locate based on changes in flow regime		Adapt plant operations to change in river flow patterns Operational complementarities with other sources (for example, natural gas

Wind		Improve design of turbines to withstand higher wind speeds	(Re)locate based on expected changes in wind speeds (Re)locate based on anticipated sea level rise and changes in river flooding		
Solar		Improve design of panels to withstand storms	(Re)locate based on expected changes in cloud cover	Repair plans to ensure functioning of distributed solar systems after extreme events	
Biomass	Build dikes Improve drainage Expand/improve irrigation systems Improve robustness of energy plants to withstand storms and flooding	Introduce new crops with higher heat and water stress tolerance Substitute fuel sources	(Re)locate based on areas with lower risk of flooding/storms	Early warning systems (temperature and rainfall) Support for emergency harvesting of biomass	Adjust crop management and rotation Adjust planting and harvesting dates Introduce soil moisture conservation practices
Demand	Invest in high-efficiency infrastructures and equipment Invest in decentralized power generation such as rooftop PV generators or household geothermal units		Efficient use of energy through good operating practice		

Table 3.1 (cont.)

	TECHNOLOGICAL		BEHAVIORAL		
Energy System	"Hard" (structural)	"Soft" (technology and design)	Re(location)	Anticipation	Operation and Maintenance
Transmission and Distribution	Improve robustness of pipelines and other transmission and distribution infrastructure Burying or cable re-rating of the power grid		Emergency planning	Regular inspection of vulnerable infrastructure such as wooden utility poles	

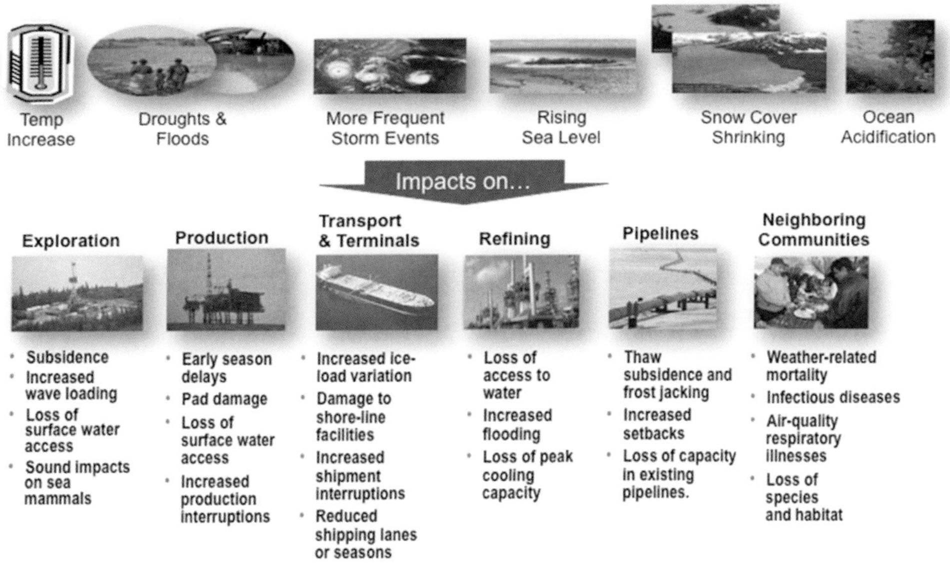

Figure 3.8 Oil and gas industry vulnerabilities to climate change impacts (Dell and Pasteris 2010).

Climate change risk management by electric utilities, at least in the United States, has been more scattered but in some cases impressive. A significant example is a recent report by Entergy, a major multistate utility in the US Gulf Coast region. The report showed the vulnerabilities of this low-lying coastal region to climate change combined with continuing economic and demographic growth and significant land subsidence (Entergy, 2012). The report estimated costs of combined effects, evaluated a range of adaptation options and recommended a nine-point plan to avoid increases in vulnerability in coming decades.

Another line of study considers ways that investments in adaptation science and technology for energy supply and use might improve adaptation capacities. Examples range from more efficient and affordable space cooling technologies to more affordable desalination, improved materials for offshore facilities exposed to ocean acidification and improved understandings of groundwater dynamics and recharge (ORNL 2012c; Wilbanks 2010). Almost certainly, a strong science base to provide viable options for adaptation will increase prospects for effective vulnerability and risk management (Figure 3.9).

Conclusion

The most prominent implications of climate change for energy supply and use systems are focused, at least in the near future, on vulnerabilities to extreme weather events, which can cause serious short-term disruptions in energy supplies and services. Longer-term implications tend to focus on water supply issues for energy production and temperature effects on electricity demand. But, as water issues illustrate, the energy sector is closely coupled

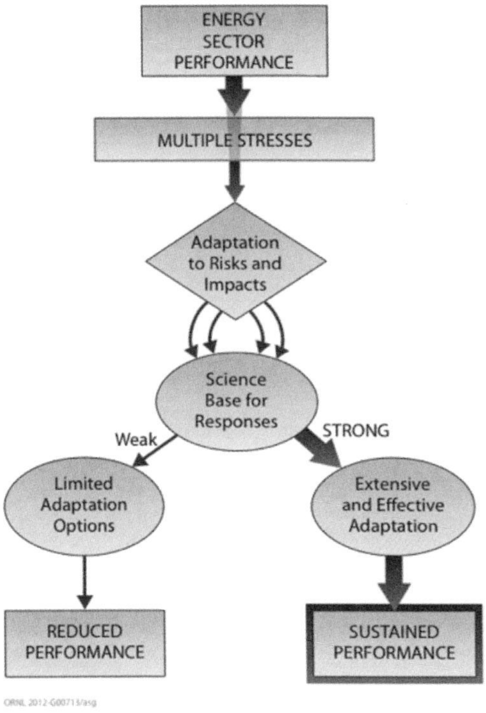

Figure 3.9 How a strong science base can enable and support effective adaptation.

with other critical infrastructures as well, including water, transportation, communication and waste management; the implications of climate change for those infrastructures are of concern.

Major energy supply institutions in the industrialised world have impressive financial and managerial resources for climate change risk management, but institutions in many developing countries are not so well equipped to cope (for instance, the capacity to access large-scale financial resources may be limited by questions about institutional performance and profitability). As a result, the greatest vulnerabilities to climate change impacts may be in the developing world, where (a) meeting rising electricity demands could be adversely affected in some regions by seasonal water scarcity, and (b) meeting needs for energy services to support economic and social development could be complicated by energy price increases to lower-income energy users.

References

Clark, W., Dickson, N., Corell, R., Cash, D., Parson, E. and Hall, M. J. (2000). *Assessing Vulnerability to Global Environmental Risks*. Discussion Paper 2000–12, Environment and Natural Resources Program, Belfer Center for Science and International Affairs, Harvard Kennedy School.

Dell, J. and Pasteris, P. (2010). *Adaptation in the Oil and Gas Industry to Projected Impacts of Climate Change*. SPE-126307. Richardson, TX: SPE International, Society of Petroleum Engineers.

Entergy (2012). *Building a Resilient Energy Gulf Coast: Executive Report*. Houston, TX.

Electric Power Research Institute (EPRI) (2011). *Water Use for Electricity Generation and Other Sectors: Recent Changes (1985–2005) and Future Projections (2005–2030)*. Palo Alto, CA: EPRI Technical Report.

Global Energy Assessment (GEA) (2012). *Global Energy Assessment*. Cambridge University Press and IIASA.

Intergovernmental Panel on Climate Change (IPCC) (2001). *Special Report on Emissions Scenarios* (SRES). Cambridge University Press.

Intergovernmental Panel on Climate Change (IPCC) (2007). *Climate Change 2007: Impacts, Adaptation, and Vulnerability. Contribution of Working Group II to the Fourth Assessment Report of the Intergovernmental Panel on Climate Change*. Cambridge University Press.

Intergovernmental Panel on Climate Change (IPCC) SREX (2012). *Special Report on Managing the Risks of Extreme Events and Disasters to Advance Climate Change Adaptation*. Geneva.

Intergovernmental Panel on Climate Change (IPCC) SSREN (2012). *Special Report on Renewable Energy Sources and Climate Change Mitigation*, Geneva.

IPIECA (2013). Addressing Adaptation in the Oil and Gas Industry. London: IPIECA, www.ipieca.org/event/20120621/addressing-adaptation-oil-and-gas-industry/.

Moss, R. et al. (2010). A New Paradigm for the Next Generation of Climate Change Scenarios. *Nature* **463**: 747–56.

National Climate Assessment (NCA) (2013). *National Climate Assessment*. US Global Change Research Program.

National Research Council (NRC) (2010). *Adapting to the Impacts of Climate Change*. Washington, DC.

Oak Ridge National Laboratory (ORNL) (2012a). *Climate Change and Energy Supply and Use*. Technical Report to the US Department of Energy in Support of the National Climate Assessment, February 29, 2012.

Oak Ridge National Laboratory (ORNL) (2012b). *Climate Change and Infrastructure, Urban Systems, and Vulnerabilities*. Technical Report to the US Department of Energy in Support of the National Climate Assessment, February 29, 2012.

Oak Ridge National Laboratory (ORNL) (2012c). *Climate Adaptation Science and the Energy Sector*. Report to the US Department of Energy, July 2012.

Peters, G. et al. (2013). The Challenge to Keep Global Warming Below 2°C. *Nature Climate Change* **3**: 4–6.

Union of Concerned Scientists (UCS) (2011). *Freshwater Use by US Power Plants: Electricity's Thirst for a Precious Resource*. Energy and Water for a Warming World Initiative (EW3).

United Kingdom (UK) (2012). *UK Climate Change Risk Assessment*. London: Department of Environment, Food, and Rural Affairs (DEFRA).

US Climate Change Science Program (CCSP) (2008). *Effects of Climate Change on Energy Production and Use in the United States*. US Climate Change Science Program, Synthesis and Assessment Product 4.5.

US Global Change Research Program (GCRP) (2009). *Global Climate Change Impacts in the United States*. US Global Change Research Program, Cambridge University Press.

Vergara, W. et al. (2012). *The Climate and Development Challenge for Latin America and the Caribbean*. Washington, DC: Inter-American Development Bank.

Wilbanks, T. (2010). Research and Development Priorities for Climate Change Mitigation and Adaptation. In *Dealing with Climate Change: Setting a Global Agenda for Mitigation and Adaptation*, ed. R. Pachauri, New Delhi: TERI, pp. 77–99.

Wilbanks, T. and Sathaye, J. (2007). Integrating Mitigation and Adaptation as Responses to Climate Change: A Synthesis, In *Mitigation and Adaptation Strategies for Global Change* (special issue), eds. T. Wilbanks, R. Klein and J. Sathaye, **12**(5): 957–962.

World Bank (2011). *Climate Impacts on Energy Systems: Key Issues for Energy Sector Adaptation*. Washington, DC: Energy Sector Management Assistance Program (ESMAP).

4

Fossil fuels and water: A complex and evolving relationship

NICOLE T. CARTER

Introduction

Where are we heading with the energy sector's water use? The energy sector's aggregate global water consumption is projected to increase; this is largely because more bioenergy and fossil fuel use is anticipated. Whether the energy sector actually uses more water in the future will depend in part on whether technologies can decouple fossil fuel use and bioenergy production from freshwater consumption and on the policies and market conditions that shape the relative attractiveness of fossil fuels and bioenergy vis-à-vis conservation, efficiency and other sources of energy. The focus herein is on the fossil fuel–water relationship.

Fossil fuels have played a significant role in global economic growth and are poised to continue their domination of the energy sector, even as increasing attention is given to their climate, water, environmental and health impacts. Water is either an essential input or one that is difficult or costly to substitute for current conventional and unconventional fossil fuel production and use. Major investments are being made or contemplated globally for how to meet future fossil fuel demand. These investments are often capital intensive, bulky and long-lived. Consequently, the energy trajectory established in the near term may influence fossil fuels' use of, dependence on and impacts on water resources for decades.

In order to understand where we are heading with the fossil fuel–water relationship, this chapter explores how the fossil fuel sector currently relies on water. It discusses how oil and natural gas from traditionally unconventional sources are raising water quantity and quality concerns, and how electric power generation is vulnerable to water availability. Finally, it presents policy options for managing the fossil fuel–water resource relationship. While this chapter presents a snapshot of the current relationship between fossil fuels and water, it is important to remember that relationships often evolve either through intentional action or through unanticipated developments.

Projections estimate fossil fuels may satisfy 75 per cent of global energy demand in 2035

In a 2012 outlook on world energy trends, the International Energy Agency (IEA) predicted 35 per cent growth in annual global energy demand from 2010 to 2035; IEA projected a global demand expansion from 12.73 million tonnes of oil equivalent (Mtoe) to 17.197 Mtoe.

Even with an anticipated decline in fossil fuels' market share due to robust deployment of renewable energy (eg seven-fold and twenty-five-fold projected increases in wind and solar photovoltaic generation, respectively), fossil fuels are expected to satisfy the majority (75 per cent) of future energy demand (IEA 2012). Under a forecast scenario that accounts for planned energy-related policies and announced commitments, IEA estimated global oil, coal and natural gas consumption to expand from 2010 to 2035 by 13 per cent, 21 per cent and 50 per cent, respectively (IEA 2012). If the IEA projections hold true, resulting carbon dioxide emissions would rise by 23 per cent, corresponding to a long-term average global temperature increase of 3.6 degrees Celsius (IEA 2012). Not only would this level of fossil fuel use be anticipated to influence the climate, it also could affect future water demand, supply and quality.

Conventional fossil fuels rely on and impact water resources

The conventional fossil fuel–water relationship is neither trivial nor anticipated to significantly diminish in the near term.

Coal. Water is essential to the coal production process. It manages dust in mining activities and is the basis of coal slurry for transport. Dewatering of some surface and underground mine sites produces wastewaters requiring management. Surface mining sites can alter runoff and infiltration patterns and may degrade water quality. Surface mine remediation often uses water for revegetation. Inadequate management techniques of mining and power plant wastewaters and mine tailings can pose risks to water supplies and aquatic ecosystems.

Oil. While the early stages of most conventional on-shore oil well development require little freshwater, oil processing requires freshwater inputs. Water flooding and some enhanced oil recovery (EOR) techniques, which are used at some conventional wells to promote production, can increase the water used per unit of oil produced if conducted using freshwater.
These techniques can make onshore oil as water-intensive as oil from oil sands (IEA 2012). In the United States, water flooding and EOR historically have represented the largest water use by the oil and natural gas sector (Matichich et al. 2012). The use of these techniques is anticipated to increase; to what extent saline, industry wastewater or freshwater will be used in the future is less clear. Spills and wastewaters from oil extraction and refining also link conventional oil development and water quality.

Natural gas. Conventional natural gas can either be produced from dedicated wells or coproduced at oil wells. These mixed-production methods make accounting for the water embedded in conventional natural gas production tricky. In general, conventional natural gas shows a wide variation in its water intensity, but remains low relative to other fossil fuels (IEA 2012).

Energy sector water use represents 15 per cent of global water withdrawals

According to IEA, the energy sector (exclusive of hydroelectricity) accounted for 15 per cent of global water withdrawals in 2010. Withdrawal represents the volume of water removed from a water source; excessive withdrawals can harm aquatic species. Generally, only a portion of withdrawn water is consumed. Water consumption represents the amount of water no longer immediately available for subsequent use. In the energy sector, water is consumed when it enters the atmosphere through evaporation (eg power plant evaporative cooling towers), is lost to geologic formations, is sufficiently degraded to require permanent disposal, or needs treatment before use in other applications or its safe return to the environment. While 89 per cent of the water withdrawn by the energy sector is returned to water bodies, 66 billion cubic metres were consumed in 2010 (IEA 2012). Evaporation associated with hydroelectric generation represents an additional, but unquantified, water consumption.

Trends portend growing fossil fuel-related water impacts

According to IEA, the energy sector is on a trajectory to increase its water withdrawals by 20 per cent and its water consumption by 85 per cent from 2010 to 2035 (IEA 2012). Bioenergy and biofuels dominate the projected withdrawal increase and represent roughly 65 per cent of the consumption increase (IEA 2012). Fossil fuel water consumption is estimated to expand by 35 per cent (IEA 2012). While these estimates provide a global reference point, they may underrepresent the impact of the energy sector's water use on the communities, watersheds, aquifers and streams where fossil fuel extraction, processing or use occurs. Notably, adoption of more freshwater-efficient fuel production and power plant cooling technologies could dampen projected growth in water consumption.

IEA's projections for water use for electric power generation reveal a continuing trend. The power sector as a whole is expected to generate 70 per cent more electricity in 2035 but only withdraw 4 per cent more water than in 2010. The relatively steady withdrawal rate captures the increased use of power plant cooling technologies that withdraw less water, and greater use of power plant generation technologies that require less water for cooling (eg combined-cycle gas turbines).

The fossil fuel sector also produces wastewaters that require management, can degrade water quality and may pose risks to water supplies. Depending on the fuel extraction process and the pollution management techniques employed, the impacts on water quality can vary widely (Allen et al. 2012). Impacts can occur quickly (eg spills) or slowly (eg contaminants leaching into waters from mine tailings). Impacts can last for months (eg stream depletion during concentrated concurrent drilling operations), years or decades (eg reductions in water stored in aquifers with slow recharge rates), or they may alter conditions permanently (eg changed hydraulics and hydrology from land disturbance or aquifer compaction).

The fossil fuel sector is vulnerable to water disruptions and constraints

During low-flow or drought conditions, which often occur during the hot summer season, water intakes and high water temperature may harm or limit thermoelectric power plant cooling, potentially forcing facilities to scale back generation during high demand. Co-occurrence of heat waves and drought can amplify impacts on both water and electric power supply and demand, producing cascading and compounding impacts for the energy sector (NCADAC 2013; US Department of Energy 2013). While alternatives are currently available to reduce the water needs of power plants, many of the available alternatives come at a higher cost and may reduce operational efficiency and electric power output.

Water supply and quality constraints also may limit the freshwater inputs available for fossil fuel extraction. Fuel transport also may be disrupted by water conditions, such as flood-induced pipeline breaks resulting from riverbed scouring, flood or storm disruption to refineries or distribution systems, and drought or flood-impaired fuel transport.

The unconventional oil and natural gas sector's use of and relationship to water is rapidly evolving

Most stakeholders agree on the overarching and aspirational objective of an affordable, secure and sustainable energy future. Water is part of the debate on how well unconventional oil and natural gas furthers this multi-faceted objective. Stakeholders often disagree on how to make trade-offs between affordability, security and sustainability.

Increasing production from traditionally unconventional fossil-energy sources is disrupting expectations about the future global energy mix and raising questions about how and where this production uses and impacts water. Unconventional oil and natural gas refer to resources in low porosity and low permeability reservoirs (eg tight oil, tight gas, organic-rich shale, coal) or resources trapped or bound to sand, tar or rock (eg bitumen from oil sands, coalbed methane). Current methods for liberating these resources generally use water, often freshwater. In contrast, conventional natural gas and oil are held in pockets beneath a rock layer and flow freely to the surface once a well is drilled. Conventional natural gas and oil generally can be extracted without water; however, water may be used if additional stimulation techniques are employed.

Unconventional oil and natural gas development generally occurs when market and infrastructure conditions and geology prompt investment, where policy and regulatory regimes are favourable, and as technological advancements reduce production costs. The combined use of hydraulic fracturing and directional drilling to liberate shale gas, tight gas and tight oil has proven to be a transformational technological development. Fracturing is a technique of injecting a mixture, typically consisting of water, sand and chemicals, into an oil- and natural gas-bearing formation. The injected mixture expands rock fissures, thereby releasing trapped oil or natural gas. Directional drilling allows wells to extend horizontally and follow a formation. The longer the drilled well, the more of the fluid mixture is needed (US Department of Energy 2009).

Unconventional gas is expected to account for nearly half of the increase in global natural gas production from 2011 to 2035; the unconventional share in global natural gas production is forecast to rise from 14 per cent to about 26 per cent (IEA 2012). Shale gas and coalbed methane contribute most of this increase, with the former making up about

14 per cent of total gas production in 2035 and the latter about 7 per cent. The bulk of the increase is projected to come from China (30 per cent), the United States (20 per cent) and Australia (12 per cent) (IEA 2012). For both global tight oil production and for oil from Canadian oil sands, projections are for increases around 200 per cent for the 2011 to 2035 period (IEA 2012). Actual production will depend on numerous factors, including production costs, market conditions and operational issues. Whether water quantity and quality concerns may constrain this growth may be influenced by whether unconventional development is able to reduce its freshwater reliance.

Changing global production and flow of fossil fuels

The rapid rise in traditionally unconventional fossil energy may shift flows of energy resources among nations and alter the geopolitical and economic relationships that derive from those flows. North America is emerging as a more significant fossil fuel producer, potentially becoming an exporter of both natural gas and coal. Tight oil production has expanded quickly in the United States since 2008, while oil sands and tight oil production has increased in Canada. Other countries, such as Argentina, France, Mexico, Poland, Russia and South Africa, also may pursue expanding their unconventional extraction. Rising demand in Asia is expected to attract more oil, coal and natural gas imports into the region from the Middle East and Australia, as well as spur extraction efforts in Asia. For example, China is attempting, albeit with some challenges, to replicate the United States' unconventional fossil fuel boom. At the same time, some countries have imposed bans or moratoria on fracturing or shale development due to environmental concerns (McKinsey Global Institute 2013).

There are a variety of viewpoints on the relative costs and benefits of unconventional oil and natural gas. Stakeholders previously concerned that a declining fossil fuel resource base might constrain future supply are relieved by these unconventional sources. Some concern, however, has merely shifted to energy prices, including the upward push from higher energy production costs and energy price volatility (McKinsey Global Institute 2013). Rising and volatile energy prices have ramifications throughout economies; for example, higher energy prices can push up fertilizer and raw materials prices.

Meanwhile, other stakeholders are concerned about the environmental and social externalities associated with the extraction, processing and use of unconventional fuels. Local water use and landscape changes are among the early and more visible impacts associated with unconventional energy. Potential groundwater depletion or contamination raises concerns about local resilience to drought, low water and high-heat conditions. Aquifers often serve as multi-year storage reservoirs that are drawn upon during dry and hot times and are often used for domestic and livestock water supplies, sometimes with no or little treatment. Thus, potential groundwater depletion or contamination could leave communities vulnerable to drought, low water and high-heat conditions. Also, for water users located farther away from surface supplies, groundwater can be difficult and costly to substitute.

The relationships between water and specific unconventional fossil fuels – shale gas, tight gas and tight oil; coalbed methane; and oil sands – are discussed below.

Shale gas, tight oil and tight gas

Basins with shale, tight gas (ie gas produced from lower permeability sandstone and carbonate reservoirs) and tight oil (ie crude oil produced from shale formations, low permeability sandstones or carbonate) are distributed worldwide. To date, the United States has led the shale gas and tight oil and gas boom. To what extent and when similar production occurs in countries like Canada, China, Poland and South Africa will depend not only on geology and market conditions but also on access to water, regulations regarding operations, and the state of the oil and natural gas services sector (eg access to drilling rigs).

While fracturing and directional drilling allow for greater extraction, these technologies typically use water, often freshwater. Water use is concentrated in the early stages of well development, usually in the first few weeks during well drilling and fracturing. Once a well is producing, little or no water inputs are required unless refracturing is necessary. How much water is used is highly variable both across and within formations. Also, interest in and use of alternative water supplies for well development is increasing (eg reusing wastewater from a nearby producing well to fracture a new well).

Water quantity issues for shale gas, tight gas and tight oil often are caused by well concentration. That is, typically many wells are located in close proximity as a formation is developed, with many wells being drilled and fractured concurrently. The cumulative and temporally concentrated water needs of multiple drilling and fracturing operations may be locally significant.

Groundwater supplies often are used for shale operations where available and when access is permitted. Surface waters also are used, but may require costly transport by truck or pipeline from the water source to the well site (Logan et al. 2012). In locations in the United States where new water access is restricted, well developers at times have obtained water by purchasing water from municipalities or paying individual landowners for their water allocations.

After fracturing and during the life of the well, wastewaters that are brought to the surface, often called produced or co-produced waters, require management. Although recent research indicates that shale gas wells may produce less wastewater per unit of recovered gas than conventional natural gas wells (Lutz et al. 2013), the geographic and temporal concentration of shale-well development nonetheless is resulting in regional wastewater-management challenges.

Produced water management is evolving rapidly with different techniques dominating in different locations. Where deep wells for permanent wastewater disposal are limited, unconventional producers are increasingly using wastewater from wells as the fracturing fluid for subsequent wells, thus relieving stress on both disposal sites and freshwater. Reducing the wastewater treatment required before reuse as fracturing fluids likely would further increase reuse. If cost-effective and safe-produced water reuse can be achieved, many of the wastewater transport and disposal costs and related environmental risks may be avoided.

Other concerns raised in association with shale gas and tight oil and gas development include the risk that fracturing and disposal of wastewaters may contaminate freshwater supplies (eg through faulty well casing, spills or well-site runoff), community disruption and air pollution from related truck traffic (eg truck transport of drilling equipment, fracturing additives, freshwater to the site and wastewater off the site), and short-term and long-term changes to the local landscape and community. Potential pathways for groundwater contamination, including methane migrating into and fracturing fluids reaching drinking water aquifers, are subjects of on-going debate that have prompted multiple studies, including a high-profile study by the US Environmental Protection Agency, which could influence operating restrictions and requirements. While these issues have resulted in some community, landowner and public pushback against developing these energy resources (especially in countries where subsurface mineral rights are not owned by the overlying landowner), other stakeholders view these concerns as overshadowed by the economic and energy security benefits of resource development.

Oil shale: An unconventional source with an uncertain future

In addition to shale gas and tight oil and gas extraction, oil shale development, particularly in the United States, also receives some attention. Oil shale is distinct from the tight oil produced from shale formations discussed above. Oil shale is a solid organic matter, known as kerogen, present in certain sedimentary rocks. The kerogen must be heated through various processes to produce synthetic oil. Most of the current technologies to convert this rock into a usable energy source require significant energy and water inputs. While research efforts to cost-effectively produce oil from these resources have been underway for decades, oil shale to date has not experienced the same technological breakthroughs as tight oil and gas and shale gas. In 2012 and 2013, two energy companies ended significant components of their long-running oil shale research efforts in the United States. Therefore, water demand for oil shale development is limited by the small scale of the anticipated near-term oil shale development (Mittal 2012).

Coalbed methane

Coalbed methane, also known as coal seam gas, is formed by the decomposition of organic material found in coal deposits that are of too poor quality or are too thin or deep (ie buried more than 1,000 metres) for commercial coal mining. Most coalbed methane formations are considerably closer to the land surface than shale gas and tight oil and gas formations.

Coalbed methane often is extracted from coal deposits using vertical wells, although horizontally drilled wells are becoming more common. Whether fracturing is employed depends on the formation's characteristics. Coalbed methane development began in the United States in earnest twenty years ago and has in recent years steadily contributed around 8 per cent of US natural gas production. Australian production and related investments ramped up starting in 2008, with plans for export, principally to Asia. Canada also produces coalbed methane. Other countries with recoverable coalbed methane resources include the United Kingdom, France and Indonesia. The extraction and comparative costs of other types of natural gas production influence the attractiveness of developing coalbed methane; production for export also is shaped by liquefied natural gas investments regionally and globally.

Water use associated with coalbed methane is complicated to evaluate (and to compare to shale or tight gas). Most coalbed methane is trapped in a coal formation by groundwater. To access the gas, the coalbed is depressurised by removing the groundwater; the amount and quality of groundwater brought to the surface as produced water varies widely depending on the geology. Some coalbed methane is produced by dewatering of freshwater aquifers that are used as regional or local water supplies (USGS 2000); other coalbed aquifers are naturally saline. The aquifers impacted may be particularly sensitive, further complicating the water impacts of coalbed methane. Some produced water from US coalbed methane is fossil water – water that has been stored in aquifers for thousands to tens of millions of years – which indicates that these aquifers are slow to replenish and their waters are essentially non-renewable resources (National Research Council 2010).

In the United States, whether a water right is needed to dewater an aquifer for coalbed methane extraction is determined largely by state law. In Australia, as of 2012, the question of whether to require a withdrawal license for the dewatering of coalbed methane aquifers was being deliberated at the state level. If a license is required and the development is located in basins where water is fully allocated, well developers may become active in water purchases from Australia's water markets; currently, these markets are dominated by water trades among agricultural water users, with limited municipal and industrial participation. The Intergovernmental Agreement on a National Water Initiative between the Commonwealth and Australian states and territories, which was first agreed to in 2004, indicated that the states and territories should include all consumptive water uses in the development of water access entitlements and related water markets (Council of Australian Governments 2004).

In some basins, the produced water from the coalbed dewatering is reinjected into the formation; in other cases other disposal options are pursued, including evaporation ponds or beneficial uses depending on the water's quality (eg land application). In Australia, efforts have been made to reuse this water (eg power plant cooling). As of 2013, the majority of the coalbed methane produced water in the largest producing US basin, the Powder River basin, is discharged into surface-water bodies. This has resulted in tensions between the state with much of the production, Wyoming, and the downstream state, Montana. In other basins in the United States (eg the San Juan Basin in Colorado and New Mexico), produced water is reinjected, usually into geologic formations that lie below the coalbed methane aquifer (Allen et al. 2012). In August 2013, the US Environmental Protection Agency proposed to discontinue its efforts to establish federal discharge standards for coalbed methane wastewaters. The agency was unable to identify a wastewater treatment technology that would be economically achievable (US Environmental Protection Agency 2013a).

Environmental concerns over coalbed methane production in Australia led to the Commonwealth's 2012 creation of a five-year effort to improve decision making and reduce water-related risks. The effort consists of an independent expert scientific committee to advise state governments about relevant coalbed methane (and large coal mining) approvals that may have significant water impacts, the development of bioregional assessments and research into potential impacts and minimisation methods, and incentive payments to the states to incorporate the resulting advice and information into decision making. In May 2013, the Commonwealth's Standing Council on Energy and Resources released its 'National Harmonized Regulatory Framework for Natural Gas from Coal Seams' (Standing Council on Energy and Resources 2013). This framework is intended to provide principles to facilitate harmonized federal, state and territory efforts to regulate coal seam development in four areas: water management and monitoring, hydraulic fracturing, well integrity and chemical use.

Oil sands

The term 'oil sands' generally refers to a mixture of sand, clay and other minerals, water and bitumen; bitumen is a dense and viscous mixture of hydrocarbons obtained naturally or as a residue from petroleum distillation. Bitumen can be processed into a fuel because it is a form of crude oil that has undergone degradation over geologic time. Resource estimates indicate that oil sands deposits are located throughout the world, but in widely varying amounts. The two largest deposits are in Canada, particularly the Province of Alberta, and in Venezuela's Orinoco Oil Belt.

Both surface mining and in situ techniques for oil sands development have water resource impacts. Surface mining operations generally withdraw water from nearby rivers (eg the north-flowing Athabasca River in Alberta, Canada). This water is heated for use in the complex separation process of liberating the oil from the sands. Process waste streams are collected in tailings ponds or lakes, which require land and may represent long-term management challenges, costs, and liabilities. In addition, water-quality issues have generated some debate, with concerns that oil sands operations discharge cadmium, copper, lead, mercury, nickel, silver and zinc into the environment. In situ processes often bring groundwater, sometimes brackish groundwater, to the surface and heat it, then reinject it for the underground steam-based separation of the oil from the sand. Notably, the freshwater intensity of in-situ oil sands production is generally 17 per cent to 25 per cent of oil sands mining. While more water-efficient than surface mining, in-situ production leaves in place (ie unrecovered) a considerable portion of the petroleum resources (Gosselin et al. 2010).

Like surface coal mining, surface oil sands mining also significantly disturbs the land and its surface and groundwater regimes; the effectiveness of mine site remediation at long-term restoration and protection of water resources is uncertain and of concern among some stakeholders.

How comparable are the water and wastewater impacts of unconventional oil and natural gas to other fuels?

As of mid-2013, there is a dearth of publically available, commonly cited and authoritative data on the freshwater consumed in fuel extraction and processing. Available comparisons can be helpful, but are often incomplete, with each source having its own shortcoming. These faults include using incomplete data (eg no differentiation between conventional natural gas and various unconventional natural gas production techniques, no differentiation in types of oil production techniques); using data from decades-old sources, thereby raising concerns that they do not reflect current practices in a rapidly evolving industry; or missing elements of the production life cycle that require water. This lack of broad, comprehensive and authoritative information has thwarted efforts to accurately evaluate and compare freshwater inputs across fuels and renewable energy alternatives for policy purposes.

While keeping in mind the shortcomings of existing datasets, a few relative relationships do appear to hold (IEA 2012; US Department of Energy 2006; Mielke et al. 2010). The amount of water needed per unit of fuel produced – referred to as the water intensity – ranges from conventional natural gas at the lowest end (less than 100 litres/toe); coal, unconventional natural gas, and uranium mining and enrichment next (roughly 100 to 1,000 litres/toe); followed by oil (1000 to 10,000 litres/toe); and ends with irrigated biofuels at the upper end (above 10,000 litres/toe).

Similarly, few data sources are available for comparing the range of produced water quantities and qualities. Produced water represents the largest by-product of fuel production. In the United States, approximately 9 billion litres of produced water are brought to surface daily from the country's 1 million active onshore oil and gas wells (Clark and Veil 2009); for perspective, US fossil fuels production consumes around 17 billion litres per day of freshwater used (Elcock 2010). For oil wells, this represents an average ratio of 7.6:1 of produced water to oil produced. By 2025, as a result of aging wells with decreasing oil production, the ratio is expected to average 12:1 for onshore crude oil (Global Water Intelligence 2011). As previously noted, research indicates that shale gas wells may produce less produced water per unit of recovered gas than conventional natural gas wells.

Challenges exist to improving the data. Much of the energy sector consists of private companies, both large and small. Maintenance of high-quality data for an evolving and dispersed industry is costly. Collating data from a wide array of sources worldwide is a problem for data consistency, accuracy and timeliness. Efforts to provide more authoritative data likely would need to combine mining the datasets of many government agencies as well as industry surveys. Due to these issues, variability in geology and variability in the availability and sensitivity of water supplies, policy decisions may be better informed by more local rather than globally aggregated data.

The electric power sector's use and relationship to water is evolving

The fossil fuel–water resource relationship is shaped by not only the water impacts of fossil fuel production, but also by the power sector's use of water and sensitivity to water conditions. Water availability problems, such as regional drought, low flow or intense competition for water, can pose a risk to electric power production, particularly hydroelectricity and thermoelectric generation.

Most thermoelectric power plants rely on water for cooling

Thermoelectric power plants, which can be operated on coal, oil, natural gas, biomass, wood, geothermal, solar thermal or nuclear fuels, produce heat to generate steam. The steam then turns a turbine connected to a generator that produces electricity. Cooling is required to condense the steam back into boiler feed water, so the process can be repeated.

The two common thermoelectric cooling methods are once-through cooling and evaporative cooling. Once-through cooling pulls large quantities of water off a water body; this water is used as a sink for the power plant's waste heat. The water temperature of the withdrawn water typically rises by 8 degrees Celsius to 12 degrees Celsius. The majority of the now-warmer water is returned to the water body (Madden et al. 2013). The intake water temperature needs to be in a range that allows the power plant's cooling system to operate efficiently; the water leaving the plant needs to be at a temperature safe for discharge to the receiving water (eg not harmful to aquatic life). Once-through cooling, while largely a non-consumptive water use, requires that large quantities of water be continuously available; this reduces the ability to put this water towards other uses and can make efficient power plant cooling vulnerable to low streamflows.

Evaporative cooling withdraws much smaller volumes of water for use in a cooling tower or pond, where evaporation dissipates the waste heat. Evaporative cooling consumes water; often, it consumes more water than once-through cooling per unit of electricity generated.

In addition to the choice of cooling technology, the amount of water withdrawn and consumed at a power plant also depends on the plant's generation technologies, efficiency and fuel. A 2011 report found that for a generic coal-fired thermoelectric power plant:

- with once-through cooling, the median withdrawal rate is 137,600 litres/MWh, and the median water consumption rate is 946 litres/MWh;
- with evaporative cooling, the median withdrawal rate is 3,804 litres/MWh, and the median water consumption rate is 3,601 litres/MWh (Macknick et al. 2011).

Many power plants operating in more water-abundant areas use once-through cooling, while newer facilities and facilities in more arid regions often use evaporative cooling. Some coastal facilities use saline water for once-through cooling. Once-through cooling using both fresh and saline waters has raised concern about the impact on the ecology (eg impingement and entrainment of aquatic species) and quality of the receiving water body (eg elevated temperature and chemicals of the discharged cooling water). Consequently, the regulatory context of a power plant often will significantly shape the choice of cooling technologies. While the location of power plants may be shaped by access to water for cooling, the siting often is driven more by access to transmission, fuel supplies and demand centres.

Development and adoption of less water-consumptive cooling technologies and treatment and use of degraded supplies for cooling may reduce future thermoelectric freshwater consumption. These options include dry cooling, hybrid dry-wet cooling, cooling with fluids other than freshwater (eg brackish groundwater, produced waters) and emerging technologies. Notably, while these alternative cooling options may reduce water consumption, these options do not reduce the greenhouse gas emissions associated with thermoelectric generation.

In addition to water for cooling purposes, many thermoelectric power plants also use water for handling solid waste, including ash, and for operating wet flue gas desulfurization scrubbers. For example, US power plants discharged 2.6 billion litres daily of these operational wastewaters (US Environmental Protection Agency 2013b).

Power from natural gas turbines is less water dependent than thermoelectric generation

Growth in natural gas production and concerns about greenhouse gas emissions are supporting the increased use of natural gas for electric power generation. Like all fossil fuels, natural gas combustion produces carbon dioxide and nitrogen oxides, though at lower rates than coal combustion (Allen et al. 2012). The burning of natural gas produces negligible solid waste and generates fewer particulates than coal combustion.

While the water-intensity of natural gas as a fuel depends on how it is extracted, electric generation using gas turbines alone or in combination with thermoelectric processes (known as combined cycle power plants because they combine the use of steam turbines and gas turbines) withdraws and consumes less water than electric power from coal. This is largely because gas turbine technologies require no water for cooling. Therefore, displacement of thermoelectric generation by gas-turbine generation may reduce the future water footprint of power generation. It is less clear that simply switching fuels from coal to natural gas at a thermoelectric facility will produce water-use benefits (Macknick 2011).

While reduced water needs may represent an advantage of natural gas turbines, the extent to which a shift to natural gas generation will produce greenhouse gas reductions over the entire life cycle of extraction, processing and use continues to be a subject of ongoing research and debate. Research indicates that electricity generated using shale gas in the United States may have life-cycle greenhouse gas emissions similar to generation using conventional gas and half of the emissions of generation using coal (Logan et al. 2012); the benefits of switching from coal to natural gas increase as fugitive methane leakages in the natural gas life cycle are reduced (Alvarez et al. 2012).

Power grid water vulnerability varies

For public policy and the electric power consumer, concerns about the power sector's water vulnerability are less about the operations of a particular facility and more about grid operations and prices. Power grid resilience depends on many factors, and excess generation and reserve capacity often are important to maintaining grid reliability during dry conditions.

Recent US drought experiences and US power reliability

Recent experiences in the United States illustrate how drought vulnerability depends on characteristics of the power grid. An assessment of the drought vulnerability of electricity in the western United States found the majority of basins showing limited risk under most scenarios and that the risk that did exist was amenable to mitigation by known strategies; excess generation and transmission capacity were particularly useful mitigation strategies (Harto et al. 2011). That is, while examples exist of low-water conditions curtailing generation from particular thermoelectric facilities in the western United States, generation shortfalls can be made up by increasing generation at other facilities or purchasing power from other sources or grids. While these actions maintain the level of service provided, they typically increase the cost of service for utilities and their customers. While identifying broad resiliency, the western US assessment revealed two regions whose electric power generation was at greater risk: the Pacific Northwest and Texas, the latter of which essentially has its own grid. The Pacific Northwest was shown to be vulnerable because of its heavy reliance on hydroelectric generation. The Texas grid was vulnerable because of heavy dependence on thermoelectric generation that relied on surface water for cooling and the region's high-drought climate hazard.

At their peak in 2012, drought conditions affected nearly 80 per cent of the contiguous United States; many locations also experienced record heat. Generation was curtailed at a number of thermoelectric facilities due to water-access problems or water-temperature issues; others pursued regulatory waivers to continue operations at higher water temperatures or made cooling system investments. Low-river conditions raised concerns about the transport of fossil fuels. However, no significant grid reliability problems developed. Lost generation at drought-impaired facilities was offset by other generation or purchasing power from other sources on the wholesale market. This contrasts to the Texas electric grid during the summer of 2011, when the state was experiencing both intense drought and heat.

In the summer of 2011, high heat in Texas resulted in increased demand for electricity, and power plants were operated for extended periods at maximum capacity. The operator of the grid that covers 75 per cent of the state and 23 million people put into effect its emergency action alert system, which eventually deemed customer conservation critical to avoid rotating outages. In the end, only one plant on the Texas grid had water-curtailed generation in the summer of 2011; others were nearing curtailment when the weather conditions improved. During and after the summer of 2011, power plant operators reduced their low-water vulnerability by building pipelines to alternative and impaired water sources, acquiring additional water rights, lowering water-intake structures and installing additional groundwater pumping capacity (Harto et al. 2011).

The Texas grid is particularly vulnerable to drought due to its poor connections to the other US grids, which reduce the ability to purchase power to offset generation curtailment. During a few days of the summer of 2011, the peak demand purchases in the real-time wholesale electric power market for the Texas grid traded at or near the market cap (i.e., US$3,000 per megawatt-hour).

Illustrations of the grid-drought relationship can be found globally, especially in relationship to hydropower-dominated systems and grids with freshwater-dependent thermoelectric facilities. Droughts in the late 1990s in Chile contributed to random and rotating electric power outages. In response, generation capacity was diversified, principally by adding more gas-fired generation to a system heavily reliant on hydropower (Karakosta et al. 2011). Australia's multi-year drought, which spanned much of the 2000s, led to curtailed hydroelectric and thermoelectric power generation and increased volatility in wholesale electricity prices (Smart and Aspinall 2009). For some Australian power plant owners, uncertainty over the security of their water supplies became an issue in their access to financing (Smart and Aspinall 2009). Australia also provides an example of grid-flood vulnerability. In 2007, a high demand resulting from cold winter weather and flood-reduced thermoelectric generation resulted in a tight electric power supply–demand balance with power spot prices reaching more than 100 times average wholesale prices (Spies and Dandy 2012).

Climate change–related alterations of historic precipitation (especially snowpack, streamflow and drought) patterns and increased heat and water temperature may exacerbate the power sector's water vulnerability in some regions; that is, how climate change may increase demand for water as well as potentially decrease supply reliability in some locations may contribute to increasingly intense competition for water resources (Foti et al. 2012). For example, higher temperatures may exacerbate peak electricity demand for air conditioning, the electricity it consumes and the water used to cool the power plant.

Energy and carbon policies alter power sector's water consumption

Researchers have illustrated that, depending on which carbon mitigation actions are taken, water consumption associated with electric power generation could either increase or decrease (Macknick 2011). Electric power sector water consumption (particularly fossil fuel-related water consumption) could fall under scenarios that manage carbon, improve energy efficiency and convert significant generation to renewable energy. Other scenarios show water withdrawals increasing under scenarios that reduce carbon emissions by expanding traditional nuclear generation with once-through cooling and applying carbon capture and storage (CCS) at coal-powered facilities (Clemmer et al. 2013). Some of the more developed carbon capture and storage technologies appropriate for power plants could nearly double water consumption for coal-fired generation (National Energy Technology Laboratory 2009; Smart and Aspinall, 2009). Innovations in CCS technologies could reduce the potential future water requirements for CCS.

Managing the energy-water relationship

Policy makers can decide whether or not to respond to the fossil fuel–water resources relationship. In many countries, fossil fuel development and use decisions are dominated by the private sector. The public sector influences these private decisions through numerous routes (eg tax incentives, loan guarantees, permits, regulations, planning and education).

In general, institutional arrangements for access to water provide few or poor economic incentives for freshwater efficiency in the energy sector; in some cases, government regulations can function as barriers to implementation or co-implementation of energy and water-efficiency programs (Alliance for Water Efficiency and the American Council for an Energy-Efficient Economy 2013). Some US states are altering state laws and regulations in an attempt to either directly or indirectly promote more freshwater-efficient oil and gas operations. For example, in 2013, the state of Texas passed a law to clarify the liability and ownership of produced waters transferred among oil and gas operations for purposes of recycling and beneficial reuse. Recent developments in energy-sector freshwater efficiency respond to water being seen as a business risk and result from overcoming an operational constraint (eg substituting freshwater with well wastewaters for fracturing in response to limited wastewater disposal opportunities; switching cooling technologies in response to regulations aimed at protecting aquatic species).

If government action to manage the fossil fuel–water resource relationship or the broader energy-water relationship is deemed appropriate, an early challenge will be to determine the aim of such a strategy (eg more watts per drop, lower energy sector water use, reducing water as a constraint on energy sector activities). Possible tools for accomplishing a water-energy strategy fall under a few broad categories, as shown in Table 4.1:

Table 4.1 Policy Responses to Energy Sector's Water Demand.

Water Demand Management	Water Supply Management	Knowledge
Minimise Energy Sector's Growth in Water Use	Improve Energy Sector Access to Water	Support Informed Decision Making
Promote *water-efficient energy* sources through standards, regulations, pricing or incentives	Allocate sustainably available water that is not otherwise allocated	Data and assessments; integrated energy-water-climate decision-support research and technical assistance; integrated resource planning; develop industry-supported standard water-energy protocols and codes
Promote *water conservation* and efficiency in the energy sector through standards, regulations, pricing or incentives	Facilitate transfer of water from non-energy sectors (eg water markets; temporary purchase of water)	Education, training, dissemination of knowledge and information
Promote *energy conservation and efficiency* to reduce demand for energy and the embedded water		Support research and development, scaling-up or early adoption of technologies to reduce energy sector freshwater use and expand water supplies

minimise water use, facilitate access to water and improve decisions and technological solutions. Perspectives on the relative attractiveness and feasibility of these policy options are influenced by long histories of regulation, management, promotion and oversight of a country's energy and water resources and infrastructure.

Fossil fuels in context: Energy choices embody complex trade-offs

Energy choices represent a complex set of trade-offs. These include environmental trade-offs: carbon and air emissions, water use, waste by-products and other effects on ecosystems and public health. They also represent energy system trade-offs: supply security, system reliability and performance, cost and associated infrastructure. For many policy makers, concerns other than water – low-cost reliable energy, energy independence and security, climate change mitigation and job creation – are more significant drivers of their positions on energy and climate policies.

Analyses quickly become complex when attempting to comprehensively evaluate energy-water-climate trade-offs. Some energy alternatives almost divorce energy and water (eg solar photovoltaics, wind turbines), but may pose other challenges, such as their land requirements or intermittent production or reduced dispatchability, which is the ability and ease with which output from an electric power generation facility can be altered. Other energy trade-offs include security and reliability of supply, transport and storage. Some fuels are easier to store and can use existing transport networks and multiple transport modes, while others may require new or expanded infrastructure investments (eg pipelines). Significantly, low-carbon energy is not necessarily low-water or low-environmental impact energy (eg new hydropower reservoirs, freshwater-cooled utility-scale concentrating solar thermal, some geothermal electric generation options), and specific carbon mitigation policies and actions may increase or decrease water consumption. Because of these complexities and the difficulty in putting comparable values on different types of impacts, analyses supporting decision making are often incomplete. It is within this complex and confusing context that policy decisions that influence future fossil fuel demand and use are being made.

Some policy options in Table 4.1 may cross the categories of both supply and demand management. For example, where water markets exist, they provide both a means for the energy sector to access water for its operations and an incentive to manage the sector's water demand by requiring expenditure to purchase either water or a water right. Each option in Table 4.1 also has consequences beyond the energy-water relationship. For example, the transfer of significant water quantities to the energy sector may have local socioeconomic impacts that may require consideration or amelioration (eg disruption of local agricultural economic communities, economies and networks).

The energy conservation and efficiency option in Table 4.1 reflects that reducing energy demand would eliminate the embedded water in the saved energy, and that water-efficient energy options (eg electricity from wind, photovoltaics or gas turbines) can reduce the energy sector's demand for water. Similarly, in some areas with energy-intensive water supplies, reducing water demand and changing water management may also be a means to reduce energy demand and its embedded water. Researchers have estimated that 4.7 per cent of the United States' annual primary energy and 6.1 per cent of the nation's electricity is used for public water delivery and services (Twomey and Webber 2011). Recognition of the interrelationship between energy and water efficiency and conservation has prompted innovative efforts to account for and reduce both energy and water consumption simultaneously (Young and Mackres 2013).

The multiple facets of the energy-water-climate relationship have prompted some to support more cost-reflective pricing for water and energy: that is, pricing that more fully accounts for the externalities of their use. For example, the National Water Commission in Australia did not support specific interventions to manage the energy-water relationship; instead it supported frameworks (eg water markets, carbon price signals) that result in individuals and firms accounting for the full water, energy and carbon impacts of their consumption and investment decisions (National Water Commission 2012). While this approach may be supported by some policy makers, and it is possible in some contexts, few locations that will experience significant near-term impacts of the fossil fuel–water resource relationship currently have both water and carbon markets or pricing in place.

Conclusion

Global fossil fuel use is anticipated to grow under existing energy policies and commitments. Along with the economic benefits derived from these fossil fuels come other impacts. These include not only consumptive water use, water quality impacts and wastewater management challenges, but also land, air quality and climate effects. These impacts and how they are managed will influence the security, affordability and sustainability of global energy consumption.

In areas with limited supplies, significant demands or environmental sensitivity, the water needs associated with fossil fuels and electric power generation may be locally significant. Rapidly expanding unconventional energy production is transforming global energy markets while also altering communities and their environments. The boom in unconventional fossil energy has created areas of concentrated water use, significant wastewater management challenges and heightened concerns about surface and groundwater contamination. The boom combines risks and opportunities.

Whether an increasing commitment of water resources to the energy sector is a reality to simply accept, or whether (and if so, how) policies and technologies could be aimed at reducing the future water needs of the energy sector is a question that confronts not only policy makers and the energy sector but also other water users and those concerned about water for species and ecosystems.

How to meet and manage the water demands of the energy sector while also meeting the water demands for domestic, industrial, agricultural and environmental purposes in the face of projected future water supplies is anticipated to be a growing public policy and resource management challenge in coming decades (Foti et al. 2012). Water, therefore, may not only be a constraint on fossil fuels but also influence how the energy sector evolves and the relative attractiveness of various climate change mitigation and adaptation measures (NCADAC 2013).

Most current institutional arrangements for water access provide few or poor incentives for freshwater efficiency. If an energy-water decoupling strategy is pursued, then a range of options become available to attenuate the water demands of the energy sector.

References

Allen, L., Cohen, M. J., Abelson, D. and Miller, B. (2012). Fossil Fuels and Water Quality. In *The World's Water*, Vol. 7, ed. Gleick, P., Washington, DC: Island Press, pp. 73–96.

Alliance for Water Efficiency and American Council for an Energy-Efficient Economy (ACEEE) (2013). *Water-Energy Nexus Research: Recommendations for Future Opportunities*. Washington, DC.

Alvarez, R. A., Pacala, S. W., Winebrake, J. J., Chameides, W. L. and Hamburg, S. P. (2012). Greater Focus Needed on Methane Leakage from Natural Gas Infrastructure. *PNAS* 109(17): 6435–40, doi:10.1073/pnas.1202407109.

Clark, C. E. and Veil, J. A. (2009). *Produced Water Volumes and Management Practices in the United States*. Washington, DC: Argonne National Laboratory.

Clemmer, S., Rogers, J., Sattler, S., Macknick, J. and Mai, T. (2013). Modeling Low-Carbon US Electricity Futures to Explore Impacts on National and Regional Water Use. *Environmental Research Letters* 8(015004).

Council of Australian Governments (2004). *Intergovernmental Agreement on a National Water Initiative Between the Commonwealth of Australia and the Governments of New South Wales, Victoria, Queensland, South Australia, the Australian Capital Territory and the Northern Territory*. Canberra, Australia.

Elcock, D. (2010). Future US Water Consumption: The Role of Energy Production. *Journal of the American Water Resources Association* 46(3): 447–80.

Foti, R., Ramirez, J. A. and Brown, T. C. (2012). *Vulnerability of US Water Supply to Shortage: A Technical Document Supporting the Forest Service 2010 RPA Assessment*. Gen. Tech. Rep. RMRS-GTR-295. Fort Collins, CO: US Department of Agriculture, Forest Service, Rocky Mountain Station.

Global Water Intelligence (2011). Water's Growing Role in Oil and Gas. *Global Water Intelligence* 12(3). www.globalwaterintel.com/archive/12/3/market-profile/waters-growing-role-oil-and-gas.html.

Gosselin, P., Hrudey, S. E., Naeth, M. A. et al. (2010). *Environmental and Health Impacts of Canada's Oil Sands Industry*, Expert Panel Report. Ottawa, Ontario: The Royal Society of Canada.

Harto, C. B., Yan, Y. W., Demissie, Y. K., et al. (2011). *Analysis of Drought Impacts on Electricity Production in the Western and Texas Interconnections of the United States*. Oak Ridge, TN: US Department of Energy.

International Energy Agency (IEA) (2012). *World Energy Outlook 2012*. Paris: OECD/IEA.

Karakosta, C., Charalampos, P. and Psarras, J. (2011). Prospects of Concentrating Solar Power to Deliver Key Energy Services in a Developing Country. *International Journal of Energy and Environment* 2(5): 771–82.

Logan, J., Heath, G., Macknick, J. et al. (2012). *Natural Gas and the Transformation of the US Energy Sector: Electricity*. Golden, CO: Joint Institute of Strategic Energy Analysis.

Lutz, B., Lewis, A. and Doyle, M. (2013). Generation, Transport, and Disposal of Wastewater Associated with Marcellus Shale Gas Development. *Water Resources Research* 49.

Macknick, J., Newmark, R., Heath, G. and Hallett, K. C. (2011). *A Review of Operational Water Consumption and Withdrawal Factors for Electricity Generating Technologies*. NREL/TP-6A20-50900. Golden, CO: National Renewable Energy Laboratory.

Madden, N., Lewis, A., and Davis, M. (2013). Thermal Effluent of the Power Plant Sector: An Analysis of Once-Through Cooling System Impacts on Surface Water Temperature. *Environmental Research Letters 8*.

Matichich, M., Mierzejewski, M., Byers, B., Pitzler, D. and Ahmed, S. (2012). *The Changing Value of Water to the US Economy: Implication from Five Industrial Sectors*. Boston: CH2M Hill.

McKinsey Global Institute (2013). *Resource Revolution: Tracking global commodity markets: Trends survey 2012*. McKinsey & Company.

Mielke, E., Diaz Anadon, L. and Narayanamurti, V. (2010). *Water Consumption of Energy Resource Extraction, Processing, and Conversion*. Cambridge, MA: Harvard Kennedy School.

Mittal, A. K. (2012). *Unconventional Oil and Gas Production: Opportunities and Challenges of Oil Shale Development*. Washington, DC: US Government Accountability Office.

National Climate Assessment and Development Advisory Committee (NCADAC) (2013) *Third Climate Assessment Report* (draft). Washington, DC: US Department of Commerce.

National Energy Technology Laboratory (2009). *Estimating Freshwater Needs to Meet Future Thermoelectric Generation Requirements*. Washington, DC: US Department of Energy.

National Research Council (NRC) (2010). *Management and Effects of Coalbed Methane Produced Water in the United States*. Washington, DC: National Academy of Sciences.

National Water Commission (NWC) (2012). *Water policy and climate change in Australia*. Canberra, Australia: National Water Commission.

Smart, A. and Aspinall, A. (2009). *Water and the electricity generation industry*. Canberra, Australia: National Water Commission.

Spies, B. R. and Dandy, G. (2012). *Sustainable Water Management: Securing Australia's Future in a Green Economy: Report of a Study by the Australian Academy of Technological Sciences and Engineering*. Melbourne: Australian Academy of Technological Sciences and Engineering.

Standing Council for Energy and Resources (2013). *National Harmonized Regulatory Framework for Natural Gas from Coal Seams*. Canberra, Australia.

Twomey, K. M. and Webber, M. E. (2011). Evaluating the Energy Intensity of the US Public Water System. Proceedings of the ASME 2011 5th International Conference on Energy Sustainability, Washington, DC, August 7.

Union of Concerned Scientists (UCS) (2009a). How Solar Energy Works. Online at www.ucsusa.org/clean_energy/our-energy-choices/renewable-energy/how-solar-energy-works.html.

Union of Concerned Scientists (UCS) (2009b). How Geothermal Energy Works. Online at www.ucsusa.org/clean_energy/our-energy-choices/renewable-energy/how-geothermal-energy-works.html/.

US Department of Energy (2006). *Energy Demands on Water Resources: Report to Congress in the Interdependency of Energy and Water*. Washington, DC: US Department of Energy.

US Department of Energy, Office of Fossil Energy and the National Energy Technology Laboratory (2009). *Modern Shale Gas Development in the United States: A Primer*, DE-FG26–04NT15455. Washington, DC: US Department of Energy.

US Department of Energy (2013). *US Energy Sector Vulnerabilities to Climate Change and Extreme Weather*. Washington, DC: US Department of Energy.

US Environmental Protection Agency (2013a). *Economic Analysts for Existing and New Projects in the Coalbed Methane Industry*, EPA-820-R-13–006. Washington, DC: US Environmental Protection Agency.

US Environmental Protection Agency (2013b), *Technical Development Document for the Proposed Effluent Limitations Guidelines and Standards for the Steam Electric Power Generating Point Source Category*, EPA-821-R-13–002. Washington, DC: US Environmental Protection Agency.

US Geological Survey (USGS) (2000). Coal-Bed Methane: Potential and Concerns. *USGS Fact Sheet FS-123-00*. Washington, DC: US Department of the Interior.

World Energy Council (2010). *Water for Energy*. London: World Energy Council.

Young, R. and Mackres, E. (2013). *Tackling the Nexus: Exemplary Programs that Save Both Energy and Water*. Washington, DC: American Council for Energy-Efficiency Economy.

5

Renewable energy and water

NADIA MADDEN AND MARTIN PASQUALETTI

Introduction

The connections between climate, energy and water have become so profound, important and expensive that we cannot avoid recognizing their impacts all around us. There are many links between energy and water, but one of the most important – albeit 'hidden' – is the thirst thermoelectric power plants have for water. Currently, this need accounts for about 15 per cent of global water withdrawals (IEA 2012), and half of domestic water withdrawals in the United States. Most of these power plants rely on fossil fuels and contribute heavily to climate change.

In turn, the spread of hydrologic variability has already become so wide that hydrologists have declared that "Stationarity is dead" in the water management sector (Milly 2008). This means we can no longer rely on the data of the past to predict the abundance or scarcity of water in the future. This has important implications for thermal power plants, as public and private attention is directed towards the greater development of low-water-use and low-carbon alternatives. Both low-water-use and low-carbon sources of energy are bountiful in the United States. As a bonus, they also emit no noxious gases nor produce a waste stream in any measure comparable to fossil fuels and nuclear power. All this, then, begs the question: How can renewable energy improve the reliability of our electricity system while not burdening our water resources?

One example of the value of renewable technologies during times of drought can be found in the state of Texas, which experienced a major drought in 2011 that heavily strained the region's energy and water resources (Hoerling et al. 2012). The impacts of the drought on conventional electricity generation have been discussed previously in this volume (see Carter, Chapter 4), but we would like to note the robustness of renewable energy generation under condition of heavy drought. While fossil-fuelled power production in Texas struggled, wind energy production reached new highs with enough capacity online to power over 17,000 homes during hot summer days (ERCOT 2012). In 2012, Texas's energy grid had over 10,000 megawatts of wind power capacity, with nearly 21,000 megawatts under review. The state is also completing high-voltage transmission lines to efficiently move wind power from the western part of the state, where wind resources are the highest, to the metropolitan areas, where electricity demand is the highest. Texas is taking the recent drought into account as it plans for its future energy needs and has found that water costs for cooling conventional power during drought make low-water energy technologies such as wind and natural gas highly competitive (ERCOT 2012).

This chapter examines future energy scenarios that explore how a shift towards renewable electricity could positively impact the electric grid's dependence on water resources. We focus on the United States, where researchers of the energy-water nexus have begun to explore this question in depth. Our review includes the life-cycle water implications of the renewable energy resources that are currently contributing meaningfully on commercial scales: wind and geothermal power[1]. Lastly, we will summarize lessons for policy makers on using renewable energy to help electricity grids adapt to changing hydrologic regimes, wherever they may occur.

The use of renewables to strengthen the electricity system: The United States

The United States is one of the world's largest electricity producers, with over one million megawatts of capacity online in 2012 (EIA 2013). It is also the world's second largest emitter of greenhouse gases, responsible for 17 per cent of estimated global carbon emissions in 2010 (UNSD 2013). The US, along with other nations, will need to substantially decarbonize its electricity production to limit climate change, including impacts on water resources. Both climate change and competing demands are affecting water availability, while increasing water temperatures can also affect the reliability of thermal electricity production that is reliant on water cooling (van Vliet et al. 2012). Whatever mix of energy technologies the country adopts will need to be not just low carbon, but able to produce reliable low cost electricity for decades. Renewables can help to meet these needs, and renewable resources are abundant within the United States (Figure 5.1). In 2010, renewable energy in the electric power sector supplied 152 gigawatts or approximately 10 per cent of total US electricity, with 6.4 per cent from hydropower, 2.4 per cent from wind energy, 0.7 per cent from biopower, 0.4 per cent from geothermal energy and 0.05 per cent from solar energy (NREL 2012). Along with a varied terrain, the United States employs a wide variety of state and federal approaches to energy and water policy. A number of researchers have assessed the potential for renewable energy technologies to provide a water smart source of low and zero carbon electricity for the United States (NREL 2012; Strzepeck et al. 2012; Clemmer et al. 2013).

Transforming the electric grid

Based on currently existing renewable energy policies and incentives in the United States, it is estimated that at least 20 per cent of US electricity generation will be from renewable resources by 2050 (NREL 2012). The US power sector is undergoing rapid transformation as coal-fired plants are being retired in favor of new power plants running on natural gas and renewable power. In 2008, almost half of US electricity came from coal. By 2012, it

[1] Hydropower, also a renewable energy resource, is covered separately in this volume.

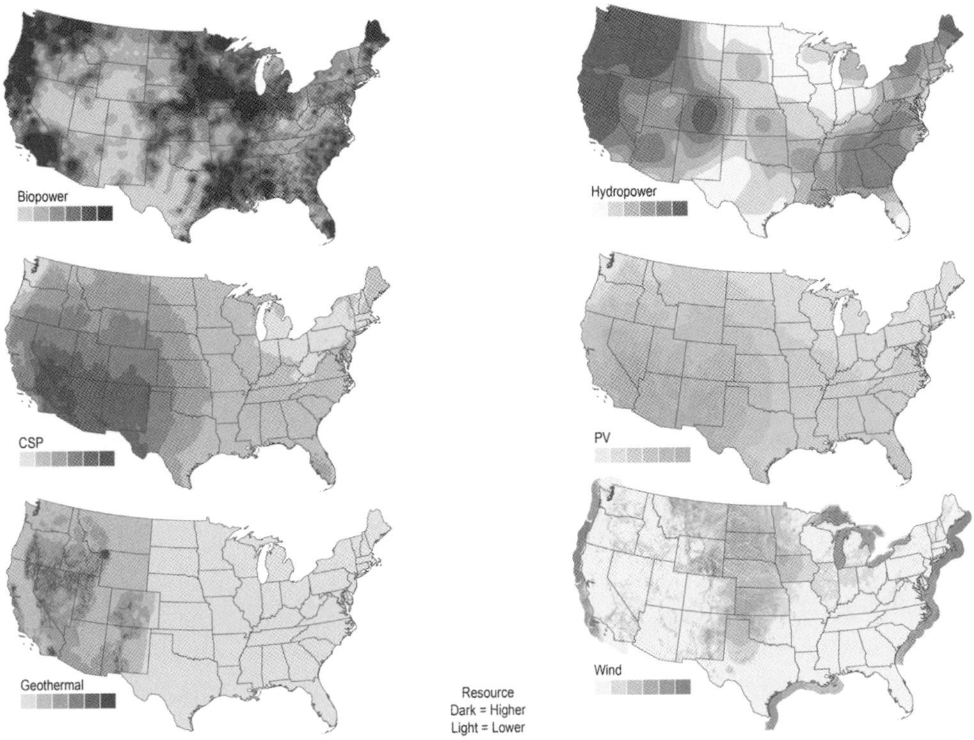

Figure 5.1 The United States has an abundant supply of renewable resources: enough to generate 212 terrawatts, which is over one hundred times all US electricity sales in 2010 (NREL 2012).

supplied only 37 per cent, while natural gas and renewable energy together supplied more than 35 per cent (Rogers et al. 2013). While low prices for natural gas have increased the proportion of gas-fired power plants, replacing coal primarily with natural gas has significant climate and water risks, as burning natural gas will continue to emit greenhouse gases, driving climate change and impacting water availability for power plant cooling.

Recent studies have found that up to 80 per cent of estimated US electricity demand in 2050 could feasibly be met with generation from renewable electricity technologies (NREL 2012; Strzepeck et al. 2012; Clemmer et al., 2013), while some have argued that it could be 100 per cent (Jacobson and Delucchi 2009). Such a shift would substantially reduce the water committed to the power sector. The reduction in water consumption ranges from 47 to 80 per cent reduction compared to 2010 (Macknick et al. 2012; NREL 2012; Strzepeck et al. 2012). The range in estimates is due to the variety of possible renewable energy portfolios and water scenarios being considered. While the above studies make varying assumptions about future electricity demand and its impact on water resources, it is clear that a shift to a renewable grid would substantially lessen the electric grid's reliance on precious water resources. Some of this change will come from retiring older power plants in favour of new technologies that are more water efficient, as can be seen from Figure 5.2. However, substantial additional reductions in water consumption

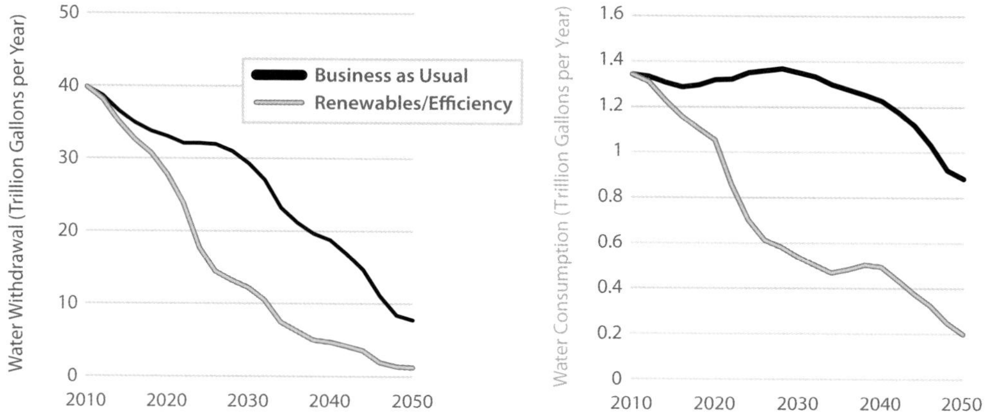

Figure 5.2 US Power Plant Water withdrawal and consumption under various scenarios from 2010–2050 (Rogers et al. 2013).

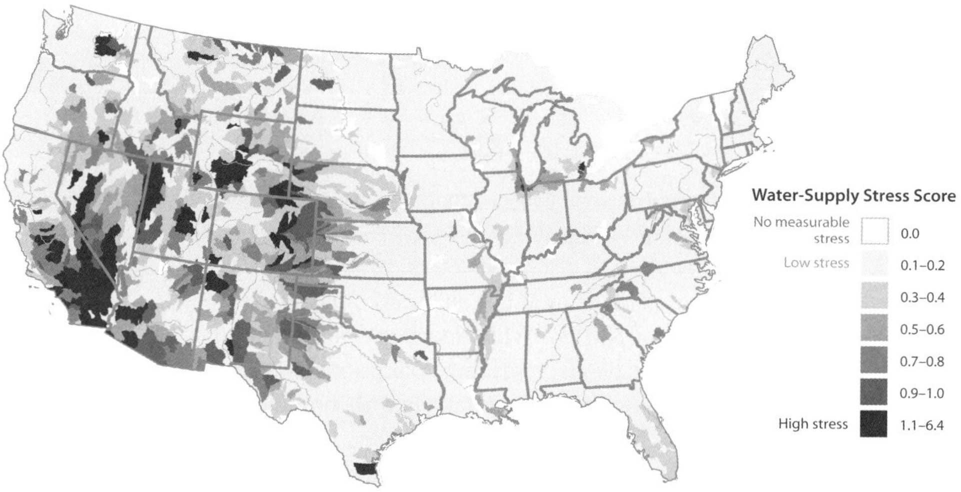

Figure 5.3 The Western United States faces considerable water supply stress (Averyt et al. 2011).

could come from large-scale implementation of wind and solar power, a change that would require changes to the electricity system to make generation more flexible and responsive, along with changes in transmission lines, grid storage and power-system operations (NREL 2012).

Impacts on current and future water stress

Reducing the need for water in electricity production is particularly important in the drier areas of the United States, where water is projected to become scarcer as the result of climate

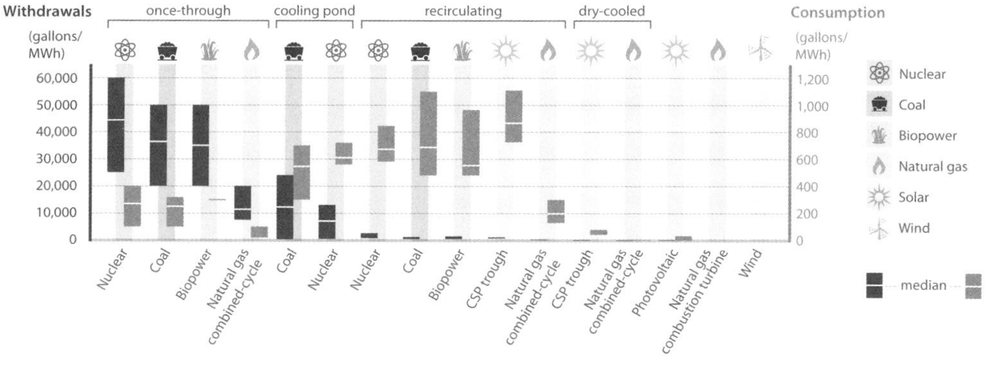

Figure 5.4 Renewable energy technologies typically withdraw and consume much less water than fossil fuels to produce electricity. The exception is concentrated solar, which can consume water at rates comparable to coal or nuclear power with recirculating cooling (Averyt et al. 2011).

change (Milly et al. 2008). In the arid Southwest, a region extending westward from the states of Colorado and New Mexico to California, almost a third of freshwater withdrawals come from groundwater, which is a key water supply during periods of drought (Yates et al. 2013a). Groundwater recharge is likely to be decreased by climate change, and the Southwest, which is heavily dependent on groundwater supplies, is quite vulnerable to any decreases in groundwater availability (Doll 2009; Gleeson et al. 2010). Transforming the electric sector could significantly reduce cumulative groundwater withdrawal in the Southwest, by the order of a million acre-feet (approximately 30 million gallons) as early as 2025, and several million acre-feet by 2050 (Yates et al. 2013b). Even in the Colorado River Basin, the driest corner of the Southwest, 330,000 acre-feet per year is evaporated by electrical generating stations (Kelley and Pasqualetti 2013). Any shift to renewables in this area would produce an important and noticeable improvement in water availability.

A shift to renewables would also substantially decrease greenhouse gas emissions, with an 80 to 83 per cent reduction by 2050 (NREL 2012; Clemmer et al. 2013). This decrease in emissions could help the United States meet its global obligations while creating an electricity sector that will be more functional during periods of drought.

Water impacts of renewable technologies

Life-cycle analysis can be used to quantify the full impact of renewable energy technologies on water resources (NAS 2010; NREL 2012; Pasqualetti 2012; Meldrum et al. 2013). Figure 5.5 presents the findings of researchers at NREL who have harmonized the life-cycle use of water over the life of a plant into gallons per megawatt-hours (Meldrum et al. 2013).[2] While the water used to operate power plants presents on-going vulnerability to constraints in local water supplies, water withdrawn for equipment manufacturing and construction can present direct and indirect impacts, depending both on water availability

[2] These are US gallons: 1 US gallon = 3.785 L

(a)

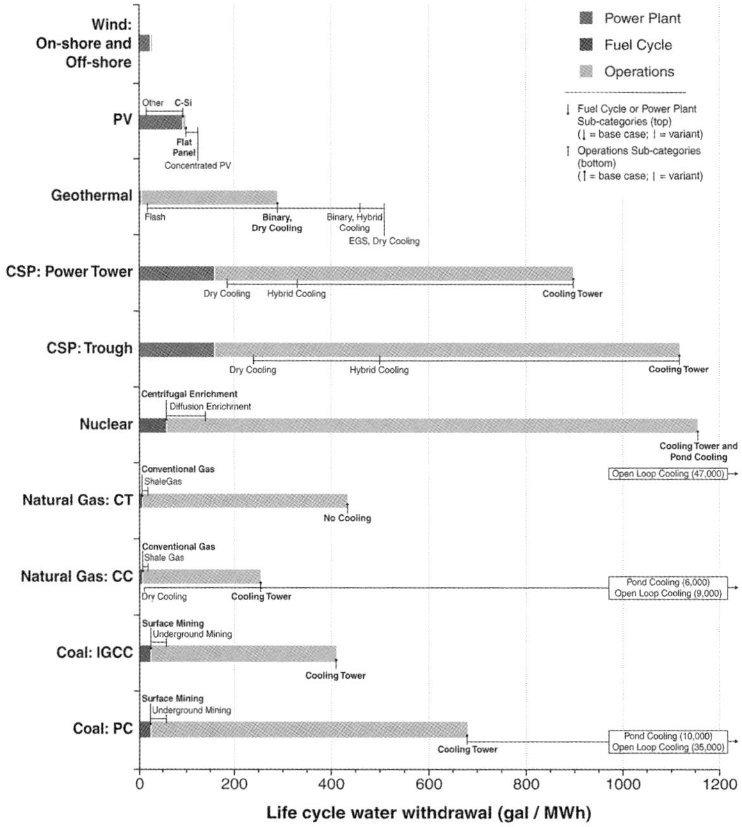

Figure 5.5 Life-cycle water withdrawals and consumption of renewable energy technologies, harmonized over the average lifetime of the plant in gal/MWh (Meldrum et al. 2013).

and on the locations where these activities occur. For wind and photovoltaic power, the largest component of life-cycle water withdrawal and consumption is for the manufacturing and construction of power facilities, where the manufacturing facilities are often in a different water basin from the power facility. For geothermal and concentrated solar power, cooling dominates the life-cycle water use, and most of the water withdrawn and consumed is in the same water basin where the production facility is located. Unlike electricity generated by fossil fuels, renewable technologies have few upstream water impacts.

Wind

Wind turbines require a small amount of water during manufacturing to make turbines, and a harmonized review of the life-cycle analysis literature found that their median

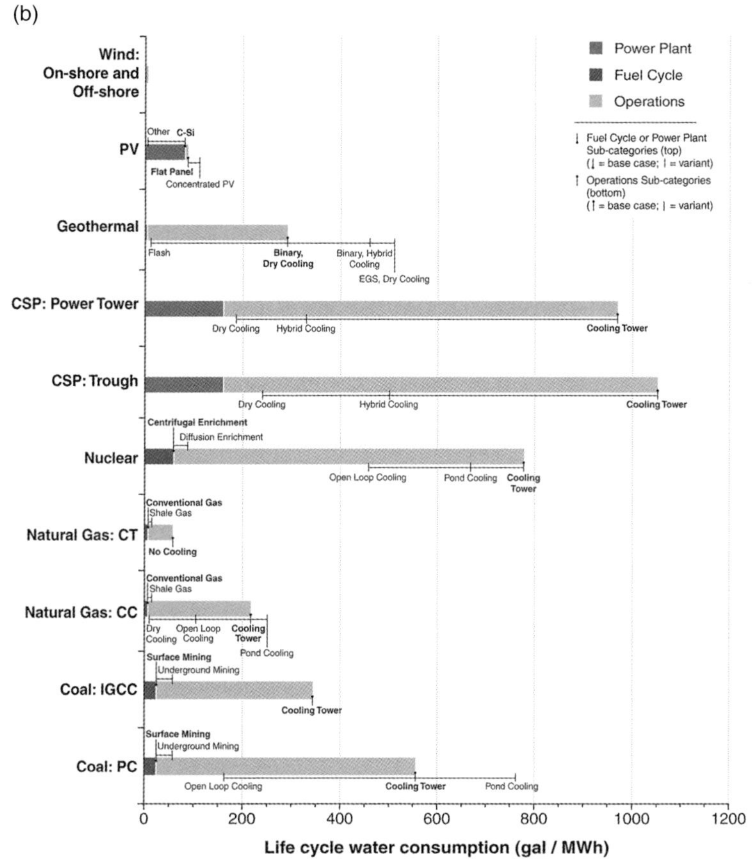

Figure 5.5 (*cont.*)

withdrawal is 13 gallons per megawatt-hours, while median consumption of water is 1 gallon per megawatt-hour (Meldrum et al. 2013). Unlike thermal power production, wind turbines require negligible amounts of water for electricity generation. Historically, wind power has been used to provide energy for near-surface groundwater extraction, and areas of the United States with substantial wind resources are often co-located with irrigated agriculture. Some operators wash their turbine blades, especially where they accumulate with crushed insects (Pasqualetti 2012).

Solar

Photovoltaic solar power (PV) uses a small amount of water to manufacture modules, with a median withdrawal of 94 gallons per megawatt-hour and a median consumption of 81 gallons per megawatt-hour for C-Si solar panels, and considerably less water for other technologies (Meldrum et al. 2013). Almost no water is consumed for PV electricity generation, usually only for cleaning and even then, not at all installations. In contrast, most concentrated solar power (CSP) technologies use a thermal cycle and thus require cooling water. Water withdrawals and consumption in these installations – whether parabolic troughs, power towers or linear fresnel – can be as much as 1,000 or more gallons per megawatt-hour. An alternate technology, Dish Stirling, uses water only for required periodic cleaning, and that must be denatured water to avoid build-up of dissolved solids.

On the commercial scale, the most common design of CSP is parabolic troughs, which are long, curved mirrors that concentrate sunlight to heat a working fluid to about 300 degrees Celsius, producing (through a heat exchanger) steam that drives an electric turbine (UCS 2009a). The turbine is cooled using technologies similar to those used at fossil-fuelled thermal plants. As a result of the comparatively low efficiencies that are achievable by this type of CSP, parabolic troughs using wet cooling technology have the highest water withdrawal and consumption rates of any renewable energy technology. However, they can also be fitted with dry or wet-dry hybrid systems (Figure 5.5). Wet cooling is the least expensive and most water intensive cooling technology. Hybrid cooling consumes less water, but makes electricity generation less efficient. One study found that hybrid cooling led to a 4.6 per cent loss in efficiency compared to wet cooling (WorleyParsons Group Inc. 2008). Dry cooling consumes the least water, but has the highest capital costs. A study of parabolic trough plants using dry cooling in the Southwestern United States found that it increased capital costs by 2 per cent to 9 per cent and reduced energy generation by 4.5 per cent to 5 per cent (DOE 2007). When ambient temperatures rise above 100 degrees Fahrenheit, the efficiency for dry cooling of parabolic troughs drops even further, reducing generation in the summers when peak electricity demand is greatest (DOE 2006; Maulbetsch and DiFilippo 2006).

Linear Fresnel reflectors are similar to parabolic troughs, using flat mirror strips to concentrate sunlight onto tubes through which working fluid is pumped to produce steam. A single study has estimated their operational water withdrawals and consumption at 1,000 gallons per megawatt-hour (Meldrum et al. 2013). Power towers use fields of mirrors to concentrate sunlight at the top of a tower. This generates heat that is used to produce steam to drive a generator at the base of the tower. Like parabolic trough plants, they consume considerable amounts of water for cooling, with the amount consumed varying by technology (Figure 5.5). A Dish Stirling consists of a stand-alone parabolic reflector that concentrates light onto a receiver positioned at the reflector's focal point. The reflector tracks the Sun along two axes. The working fluid in the receiver is heated to 250 to 700 degrees Celsius and then used by a Stirling engine to generate power. Dish Stirling technologies are highly efficient at transforming solar energy to electricity, and have a very low associated water use for periodic cleaning, estimated at 5 gallons per megawatt-hour (Meldrum et al. 2013). They continue to be challenged, however, by a variety of technical and financial problems.

Geothermal

Unlike wind and solar power, electricity generated from geothermal resources can supply continuous baseload power to electricity grids. The operational water requirements of geothermal power plants vary by more than an order of magnitude, depending on technology and local conditions. At the hottest geothermal resources, it is possible to directly pull steam through the turbine, then into a condenser where the steam is condensed into water. In 'flash' geothermal plants, very hot water is depressurized into steam, which can then be used to drive the turbine. If the water from the geothermal resources is not hot enough for direct or flash designs, a more complex process called binary generation is used. At binary geothermal plants, hot water passes through a heat exchanger where it heats isobutene in a closed loop. The isobutane boils at a lower temperature than water, so it is more efficiently converted into steam to run the turbine.

Enhanced Geothermal Systems (EGS) can utilize non-traditional geothermal resources such as dry, impermeable rock, injecting pressurized water into injection wells to create energy (UCS 2009b). EGS systems are just beginning to come on line at commercial scales in Australia, Europe and the United States. Direct plants use almost no water, while flash geothermal plants withdraw and consume a median of 11 gallons per megawatt-hour (Figure 5.5). Binary plants withdraw and consume a median of 290 gallons per megawatt-hour for dry cooling, and a median of 460 gallons per megawatt-hour for hybrid cooling. EGS systems require external water sources, and dry-cooled EGS plants withdraw and consume a median of 510 gallons per megawatt-hour (Meldrum et al. 2013).

Geothermal power production can also impact water quality, as the superheated water dissolves solids underground and brings them to the surface. The chemical impacts vary with geography and bedrock composition. A study by Argonne examined chemical data from over 50,000 samples of geofluids and found that the dissolved solids in most geofluids are predominantly sodium chloride, followed by bicarbonate, sulfate, silica, calcium and potassium. They also found that substantial percentages of geofluids did not meet US drinking water standards, and that toxic chemicals such as antimony, arsenic, lead and mercury could present a risk if geofluids are released into drinking water (Clark et al. 2011).

Indirect impacts

The intermittent nature of sunlight and wind make their use amenable to complementary or hybrid production of electricity generated by natural gas, where power plants can be brought on and off line as needed for peak production and load balancing. In the United States, low prices for natural gas have made it a currently attractive fuel for new power plants. The combination of renewable and demand-responsive natural gas electricity production requires a 'smart grid' that is flexible enough to match varying load with changing demands and communicate information about grid requirement in real time. A smart grid can also conserve water by transporting electricity generated in areas where water is abundant and move it into areas where water is not, adding reliability to electricity systems during times of drought. While wind and solar power are intermittent resources, pumped storage hydropower offers another opportunity to increase grid reliability while reducing dependence on electricity generation technologies with large water footprints.

Lessons for a water-smart energy future

As competition for water resources intensifies, it is important to consider what policies and technologies will be most effective to adapt the electricity grid to be more reliable in the face of future water shortages. The United States has provided a number of lessons for policy makers considering how to use renewable energy to make their electricity mix more reliable in the face of future changes in water availability. Some important lessons learned include the following:

- Compile reliable data into policy-relevant analyses
- Incorporate water-energy data into decision making
- Use water-smart criteria to select energy technologies

Compile reliable data into policy-relevant analyses

First, having reliable, comparable data to assess energy resources, along with their impacts on water resources and greenhouse gas emissions is critical for informed decision making. The process of quantifying the energy-water-climate nexus in the United States has improved co-ordination and communication between the government agencies that collect water and energy information and researchers who analyse it (USGAO 2012). This data has also been used within various modeling frameworks to compare future energy scenarios and their impact on water resources. A regional approach to data co-ordination, integrated modeling and planning in the arid western United States is discussed in Chapter 16 of this book.

Incorporate water-energy data into decision making

The Western United States already faces considerable pressures on scarce water resources, facilitating the process of data co-ordination and scenario analysis is being incorporated into integrated energy planning (IRP) that accounts for water impacts of new energy generation. For example, some states are adopting IRPs to assess future electric needs and plan to meet them. IRPs are typically created by electric utilities and filed with the state public utilities commission. The plans integrate reviews of both demand-side resources, such as conservation and energy efficiency, as well as the more traditional supply-side technologies of electricity generation, power plants and transmission lines.

The California Energy Commission has been participating in integrated resource planning for decades and has also integrated water resource planning into their energy plans. This process has helped them to identify renewable power production opportunities at water and wastewater facilities, which have substantial energy requirements, along with extensive watershed land holdings in areas where renewable resources such as wind, solar and hydropower are abundant. Along with land holdings, water utilities have extensive rights-of-way that could be used to transmit electricity generated by renewable resources to population centres (Klein et al. 2005).

The southwestern state of Arizona is one of the fastest-growing states in the United States. Demands for both energy and water are growing in urban areas, and the process of energy development integrates water-smart planning. Construction of new power plants greater than 100 megawatts in capacity are reviewed in a two-step certification and permitting process overseen by the Arizona Corporation Commission (ACC). Public hearings are conducted by a siting committee, which typically considers water rights, water availability for the life of the power plant and the environmental effects of water withdrawals. During the second step, the ACC reviews the overall application to ensure a balance between the state's need for energy and the plant's cost and potential environmental impacts (USGAO 2009). In 2010, the ACC ruled that a proposed 340-megawatt CSP plant in Mohave County near Kingman would have to use dry-cooling technology or treated wastewater in order to meet environmental requirements. This decision was in response to public concerns that the facility would have required 780 million gallons of groundwater annually from the Hualapai Valley Aquifer in Arizona (ACC 2010).

Use water-smart criteria to select energy technologies

While renewable technologies can have lower water requirements than fossil-fuelled power, it is important to ensure that new power plants have sufficient water to operate in the local environment as well as under any reductions in water availability that may occur under climate change. Wet cooling for concentrated solar power, with its large footprint of water consumption, is often inappropriate for deployment in arid regions where solar resources are abundant. In some specific cases, however, wet-cooled CSP could be installed if water can be reallocated from irrigated agriculture. For example, a recently commissioned 250-megawatt CSP plant near Gila Bend, Arizona, uses land and water rights purchased from farmers, evaporating 90 per cent less water than the displaced crops (Maracas 2013). However, this policy may prove maladaptive under conditions of absolute water scarcity. In desert areas where water conservation is of high priority, dry cooling may be the preferred option, as has been the case of the 392-megawatt power tower installations southwest of Las Vegas at Ivanpah Dry Lake, in the Mojave Desert (Turney and Fthenakis 2011). In March 2013, a 100-megawatt CSP plant went on line using air cooling near Madinat Zayed, Abu Dhabi. This installation will be followed by two additional 100-megawatt plants of the same size (Lenthall 2013).

Water-smart criteria are already being applied in making decisions about electricity in other parts of the world. Saudi Arabia has recently announced a goal of building 54 gigawatts of solar power that will generate a third of the country's electricity by 2032 (Beetz 2013). The country is also planning to use renewables to fuel desalination and ease freshwater scarcity (Daya 2013). Energy planners in desert regions will need to take the future impacts of climate change on water availability for cooling when making decisions about new electricity generation.

Alternately, developing economies such as India and China face growing dependence on water-intensive electricity generation fuels, such as uranium and coal. As Benjamin K. Sovacool and Ajith Rao discuss in Chapter 10 of this book, the water demands of the North China grid is expected to quadruple in size by 2040. However, China's five largest power producers are already exposed to water-supply disruptions due to the concentration of their fossil-fuelled energy portfolios in this water-scarce region (Bullard 2013). The incorporation of water-smart criteria in their energy planning, and adding low-water, low-carbon renewables into their energy mix could help to create a more reliable grid in the face of future constraints on water.

References

Arizona Corporation Commission (ACC). (2010). Hualapai Valley Solar's compliance filing and annual self-certification, Docket L-OOOOONN-09–0541–0015 1. Phoenix, AZ.

Averyt, K., J. Fisher, A. Huber-Lee, A. Lewis, J. Macknick, N. Madden, J. Rogers, and Tellinghuisen, S. (2011). *Freshwater Use by US Power Plants: Electricity's Thirst for a Precious Resource*. Cambridge, MA: Union of Concerned Scientists.

Beetz, B. (2013). White Paper Unveils Details of Saudi's 54 GW Renewable Energy Plans. *PV Magazine*, 22 February.

Bullard, N. (2013). China's Power Utilities in Hot Water. Bloomberg New Energy Finance Water Insight.

Clark C., Harto C., Sullivan J. and Wang M. (2011). Water Use in the Development and Operation of Geothermal Power Plants ANL/EVS/R-10/5. Oak Ridge, TN: Argonne National Laboratory (ANL).

Clemmer, S., et al. (2013). Modeling Low-Carbon US Electricity Futures to Explore Impacts on National and Regional Water Use. *Environmental Research Letters* 8: (015004).

Daya, A. (2013). Renewables Ease Water-Stress Problems. *Financial Times*, 16 January. www.ft.com/cms/s/0/6ed691be-5fe5-11e2-b128-00144feab49a.html#axzz2Q9m88zHn.

Döll, P. (2009). Vulnerability to the Impact of Climate Change on Renewable Groundwater Resources: A Global-Scale Assessment. *Environmental Research Letters* 4(3): 35006.

Energy Information Administration (EIA) (2013). China Report. Washington, DC: US Department of Energy, www.eia.gov/countries/cab.cfm?fips=ch/.

Electric Reliability Council of Texas, Inc. (ERCOT) (2011). 2012 Long-Term Demand and Energy Forecast.

Electric Reliability Council of Texas, Inc. (ERCOT) (2012). ERCOT Grid Operations Wind Integration Report, 11 November.

Gleeson T., Wada Y., Bierkens, M. F., van Beek, L. P. (2012). Water Balance of Global Aquifers Revealed by Groundwater Footprint. *Nature* 488(7410):197–200, doi:10.1038/nature11295.

Harto, C. B., et al. (2011). *Analysis of Drought Impacts on Electricity Production in the Western and Texas Interconnections of the United States*. Oak Ridge, TN: US Department of Energy.

Hoerling, M., Easterling, D., Perlwitz, J., Eischeid, J., Pegion, P. and Murray, D. (2011). Assessment of 2010 North American Regional Temperature [in State of the Climate in 2010]. *Bulletin of the American Meteorological Society* 92(6): S161–S163.

International Energy Agency (IEA). (2012). World Energy Outlook.

Kelley, S. and Pasqualetti, M. J. (2013). Virtual Water from a Vanishing River. *Journal – American Water Works Association*. 105:9. http://dx.doi.org/10.5942/jawwa.2013.105.0132.

Klein, G., Krebs, M., Hall, V., O'Brien, T., Blevins, B. B. (2005). California's Water-Energy Relationship. California Energy Commission. 700–2005–011-SF.

Lenthall (2013). Personal Communication.

Jacobson, M. Z. and Delucchi, Mark A. (2009). A Plan to Power 100 Percent of the Planet with Renewables. *Scientific American*, 26 October 26.

Macknick J., Newmark R., Heath G. and Hallett K. (2012). Operational Water Consumption and Withdrawal Factors for Electricity Generating Technologies: A Review of Existing Literature. *Environmental Research Letters* 7: 045802

Maracas (2013). Personal Communication.

Maulbetsch, J. S. and DiFilippo, M. N. (2006). Cost and Value of Water Use at Combined-Cycle Power Plants. California Energy Commission, PIER Energy-Related Environmental Research. CEC-500-2006-034.

Meldrum, J., Nettles-Anderson, S., Heath, G. and Macknick, J. (2013). Life Cycle Water Use for Electricity Generation: A Review and Harmonization of Literature Estimates. Environmental *Research Letters* 8: 015031.

Milly, P. C. D., Betancourt, J., Falkenmark, M., Hirsch, R. M., Kundzewicz, Z., Lettenmaier, D. and Stouffer, R. J. (2008) Stationarity is Dead: Whither Water Management? *Science* 319(5863): 573–4.

National Renewable Energy Laboratory (NREL) (2012). Renewable Electricity Futures Study. Hand, M. M., Baldwin, S., DeMeo, E., Reilly, J. M., Mai, T., Arent, D., Porro, G., Meshek, M. and Sandor, D. eds. 4 vols. NREL/TP-6A20-52409. Golden, CO: National Renewable Energy Laboratory. www.nrel.gov/analysis/re_futures/.

Pasqualetti, M. J. (2012). The Water Bargain of Solar and Wind Energy. In *The Water-Energy Nexus in the Western United States*, ed. Kenney, D. S. and Wilkinson, R. Northampton, MA: Edward Elgar Publishing, Inc., pp. 226–42.

Rogers, J., Averyt, K., Clemmer, S., Davis, M., Flores-Lopez, F., Frumhoff, P., Kenney, D., Macknick, J., Madden, N., Meldrum, J., Overpeck, J., Sattler, S., Spanger-Siegfried, E. and Yates, D. (2013). Water-Smart Power: Strengthening the US Electricity System in a Warming World. Cambridge, MA: Union of Concerned Scientists.

United Nations Statistics Division (UNSD) (2013). United Nations Statistics Division, Millennium Development Goals indicators: Carbon dioxide emissions (CO_2), Thousand Tonnes of CO_2. http://mdgs.un.org/unsd/mdg/SeriesDetail.aspx?srid=749&crid=.

United States Government Accountability Office (USGAO) (2009). *Energy-Water Nexus: Improvements to Federal Water Use Data Would Increase Understanding of Trends in Power Plant Water Use*. GAO-10-23. Washington, DC: US Government Accountability Office.

United States Government Accountability Office (USGAO) (2012). *Energy-Water Nexus: Coordinated Federal Approach Needed to Better Manage Energy and Water Tradeoffs*. GAO-12-880. Washington, DC: US Government Accountability Office.

US Department of Energy (DOE) (2007). Concentrating Solar Power Commercial Application Study: Reducing Water Consumption of Concentrating Solar Power Electricity Generation. www1.eere.energy.gov/solar/pdfs/csp_water_study.pdf.

US Department of Energy (DOE) (2006). *Energy Demands on Water Resources: Report to Congress on the Interdependency of Energy and Water*. Washington, DC: US Department of Energy.

Strzepek, K., Baker, J., Farmer, W. and Schlosser, C. A. (2012). Modeling Water Withdrawal and Consumption for Electricity Generation in the United States. MIT Joint Program on the Policy and Science of Global Change Report No. 221, June. http://globalchange.mit.edu/files/publication/MITJPSPGC_Rpt221.pdf.

Turney, D. and Fthenakis, V. (2011). Environmental Impacts from the Installation and Operation of Large-Scale Solar Power Plants. *Renewable and Sustainable Energy Reviews* 15: 3261–3270.

van Vliet, M. T. H., Yearsley, J. R., Ludwig, F., Vogele, S., Lettenmaier, D. P. and Kabat, P. (2012). Vulnerability of US and European electricity supply to climate change. *Nature Climate Change* 2, 676–81.

Western Electricity Coordinating Council (2011). Environmental Recommendations for Transmission Planning. Final Report of the Environmental Data Task Force. www.wecc.biz/committees/BOD/TEPPC/SPSG/EDTF/Shared%20Documents/Environmental_Recommendations_for_Transmission_Planning/Final_Recommendations_-Report/Environmental%20Recommendations%20for%20Transmission%20Planning%20-%20Revised%2005-27-2011.pdf.

WorleyParsons Group, Inc. (2008). FPLE – Beacon Solar Energy Project Dry Cooling Evaluation. WorleyParsons Report No. FPLS-0-LI-450-0001.

Yates, D., Averyt, K., Flores-Lopez, F., Meldrum, J., Sattler, S., Sieber, J. and Young, C. (2013a). A Water Resources Model to Explore the Implications of Energy Alternatives in the Southwestern US. *Environmental Research Letters* 8: 45004.

Yates, D., Meldrum, J. and Averyt, K. (2013b). The Influence of Future Electricity Mix Alternatives on Southwestern US Water Resources. *Environmental Research Letters* 8: 045005.

6

Hydropower within the climate, energy and water nexus

JEFFREY J. OPPERMAN, JOERG HARTMANN AND DAVID HARRISON

Introduction

Ensuring secure supplies of energy and water are among the great challenges that society confronts while seeking to maintain robust economies and raise billions of people out of poverty. The energy and water challenges are not independent, and the linkages between them are increasingly recognized. For example, the extraction and generation of energy can require enormous amounts of water while, similarly, the treatment and distribution of water can require considerable amounts of energy (Hussey and Pittock 2012; van Vliet et al. 2012). In the United States, the energy sector is the largest user (though not consumer) of water, while in California, for example, distributing water consumes nearly 20 per cent of the state's energy usage (California Energy Commission 2005).

Beyond these direct linkages – generating energy requires water, providing water requires energy – energy and water are intertwined in a variety of other ways. Hydropower is a form of energy where water itself acts as the 'fuel' for generation. Because nearly all hydropower is generated from dams on rivers, hydropower provides a very specific example of how water and energy interact. First, hydropower is a user of water that interacts with other users, such as urban water supply and irrigation. Second, the infrastructure required to generate hydropower – namely, dams and diversions – changes how rivers function. By altering river function, hydropower can affect a broad array of other resources, benefits and values produced by rivers. Third, other activities within a watershed can alter hydrologic processes, such as erosion, which can influence the performance and longevity of hydropower infrastructure (eg through reservoir sedimentation).

Climate is a third key factor, expanding the interaction between energy and water into a trio: the energy-water-climate nexus. Climate change will affect the water-energy nexus in a variety of ways. Most simply, climate change is forecast to alter both the supply and demand for water and energy (Kundzewicz et al. 2008; Ebinger and Vergara 2011), adding greater complexity and uncertainty to an already complex interaction and potentially adding increased stressors. Specifically for hydropower, climate change will likely alter the timing and amount of water that is available for generation and also increase the demand for energy, particularly low-carbon energy, while also affecting the supply and demand of other uses for water and the health of river ecosystems and the various benefits they provide.

Despite the importance of these connections, the regulatory frameworks, planning processes, and professional training and relationships for water, energy and climate are often distinct. Scientists have highlighted the need for greater integration across this nexus (Hussey and Pittock 2012) and governments, industry and civil society are responding with a broad range of regulations, standards and initiatives to promote greater integration between hydropower and other sectors. While these processes have the potential for increasing the resilience of water-energy systems to climate change, the incorporation of climate change into hydropower management and regulation is relatively nascent. For example, hydropower licensing in the United States has increasingly required energy generators to engage with a variety of other stakeholders, and regulations now require that approval of hydropower licenses strive to achieve a balance between hydropower and other users (Gillilan and Brown 1997). However, this relicensing process rarely considers climate change (Viers 2011).

In this chapter, we review how hydropower interacts with other values and resources that use water, with an emphasis on rivers (the hydropower-water nexus). In addition to considering the major sectors that are users of water, such as water supply and irrigation, we also focus on the broad range of other values and benefits produced by rivers, such as fisheries or recreation, which have generally received much less attention in discussions of the water-energy nexus (Hussey and Pittock 2012). We then discuss how climate change may affect rivers and hydropower. We conclude by reviewing frameworks intended to promote greater balance across the hydropower-water nexus and recommend how these frameworks can more fully integrate climate change into the planning, design and operation of hydropower. To achieve this greater balance, and to ensure that the overall system is resilient to climate change, we recommend a basin or regional scale for planning and management of hydropower and other water resources.

Hydropower and its role in global energy

Hydropower is power extracted from falling or moving water. In this chapter, we use hydropower synonymously with hydroelectric power, which is electricity generated by the gravitational force of water moving through turbines. The most common forms of hydropower require constructing a dam across a river. The dam maintains water at a certain elevation (the head), from which water is directed through pipes or channels to turbines that generate electricity. A dam allows water to be stored over time periods ranging from hours to years, reducing the variability of flows that otherwise can vary dramatically within a year, such as between a wet season and a dry season, and allowing a more consistent flow of water through the turbines (Figure 6.1). Large reservoirs are capable of storing water across years and can thus reduce variability between wet years and dry years. Storage reservoirs give hydropower managers the ability to release water into the turbines when energy is most needed or valuable, such as summer when the demand for energy is greater. Within a day, hydropower managers can release water into turbines to respond to rising demand, a mode of operation known as 'load following' or 'peaking' (Kumar 2011).

Hydropower dams can generally be classified into either 'storage' dams – those that impound water for use during other times of the year – and 'run-of-river' dams, in which

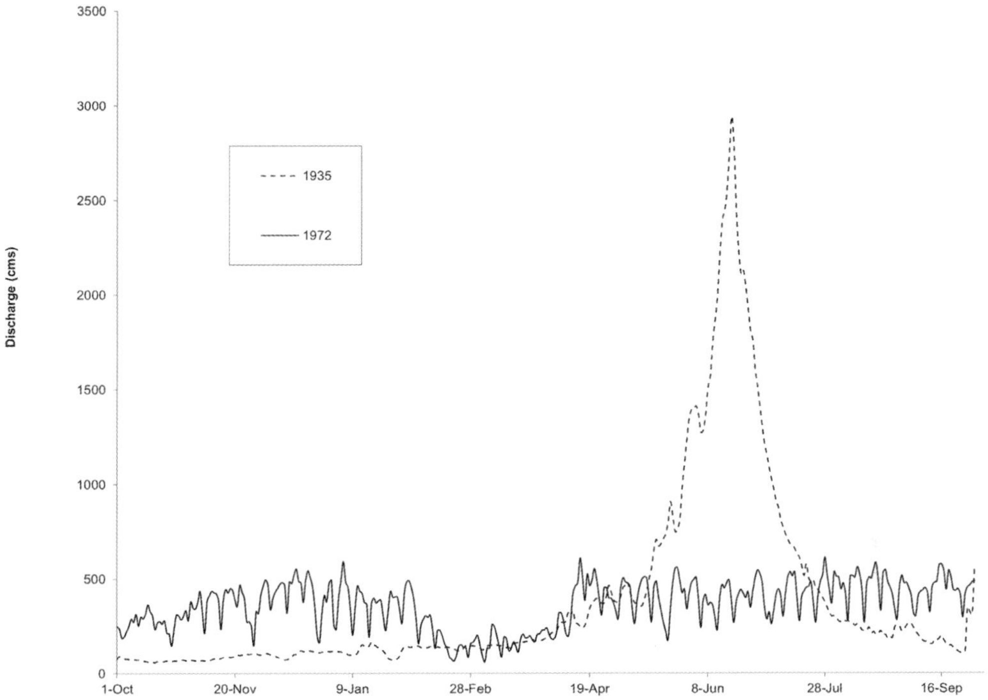

Figure 6.1 Hydrograph for the Colorado River at Lee's Ferry in a year before construction of Glen Canyon Dam (1935) and a year after construction and operation of the dam (1972). The two years had similar amounts of total discharge but the hydrograph shapes are very different because the dam is operated to store and release water for irrigation, municipal supply and hydropower. The late spring–early summer snowmelt pulse (peak in June 1935) is now stored to even out releases the rest of the year. The historic low-flow period (fall and winter) now shows higher releases with much greater variability, reflecting changing demands for electricity.

reservoir storage is held constant and outflow equals inflow (Kumar 2011). Run-of-river dams are generally considered to have a lower impact on rivers systems because they don't alter the overall flow pattern, but the actual operation associated with the term 'run-of-river' can differ by region, resulting in very different impacts. In some regions, run-of-river balances instantaneous outflow and inflows from dams. However, in other regions, run-of-river can refer to a project that stores water within a day, but in which inflow equals outflow on the basis of the daily average. This mode of operation can allow the storage of water for, say, twenty hours, with no or minimal flow release below the dam, and then a release of high flow for four hours during periods of peak demand. Although the *daily* average flow would be the same above and below the dam, the river below the dam experiences twenty hours of near-drought conditions followed by four hours of near-flood conditions each day. As described in the next section, this can have serious impacts on downstream river ecosystems.

Currently, hydropower provides approximately 16 per cent of electricity worldwide (Hamududu and Killingtveit 2012), although in some countries the proportion is much higher, including countries with high GDP (hydropower accounts for 98 per cent of Norway's electricity) and those with low GDP (Laos with 97 per cent of its electricity from hydropower). Hydropower represents approximately 7 per cent of electricity generation in both the United States and Australia (Clean Energy Council 2011).

In addition to providing a source of energy, hydropower also performs a variety of ancillary services that benefit an energy grid, including regulation and frequency response, voltage control and the ability to provide a 'black start', which refers to restarting generation following a power outage without requiring an external source of power. Among sources of electricity, hydropower is the most effective at rapidly turning on or off in response to changes in demand in the grid and thus plays an important role in 'firming up' (backing up to make reliable) the energy provided by intermittent renewables such as wind and solar (Montero and Perez 2009).

Hydropower has two important contributions in mitigating climate change: providing a direct source of low-carbon energy and facilitating a larger proportion of renewables in an energy grid than would otherwise have been possible (Pittock and Hartmann 2011). Denmark's high proportion of wind-generated electricity (20 per cent, with a stated goal of 50 per cent) is made possible in part because of grid interconnections with Norway and the ability of hydropower to quickly provide Denmark with electricity during periods of low wind. Conversely, during periods of high wind, Norway can 'store' wind power by buying electricity from Denmark and reducing flow from its hydropower reservoirs, thus increasing the potential energy stored in its reservoirs (Akhmatov et al. 2007).

These low-carbon benefits of hydropower carry two important caveats. First, some types of reservoirs – particularly shallow, tropical reservoirs – can produce high levels of methane, which can result in electricity with a carbon intensity similar to fossil fuels (Abril et al. 2005). Various studies are currently trying to quantify the overall picture of hydropower reservoirs and greenhouse gas (GHG) emissions (Barros et al. 2011), though it appears that most hydropower dams do provide electricity that is relatively low in GHG emissions (dos Santos et al. 2006). Second, although hydropower can help meet clean energy targets, the water infrastructure associated with hydropower can cause significant impacts on rivers, ecosystem services and people that depend upon them, as described in the following section.

Hydropower's ability to firm up intermittent renewables provides a specific example of both the environmental costs and benefits associated with expansion of hydropower. Although facilitating greater penetration of wind and solar certainly helps reduce GHG emissions, the necessity to rapidly ramp hydropower up and down in response to cloud cover or low-wind periods can have significant negative impacts on river ecosystems. This chapter emphasises that basin-scale planning and management of energy and water systems are key to finding solutions that can balance energy values with other water and river values.

Strategically sited pumped storage may provide a balance between the objectives of firming up intermittent renewables and maintaining healthy rivers. Pumped storage is an energy storage system in which water is pumped uphill during periods when energy is readily available and inexpensive. The higher-elevation reservoir then stores the water as potential energy that can be dispatched when needed by allowing the water to flow back downhill through turbines. Pumped storage is the most important type of energy storage in energy grids worldwide (Kumar 2011). Overall, the pumped-storage process is 70 to 85 per cent energy efficient (Pittock 2010). The environmental impacts of this pumping and rapid fluctuation can be minimised by siting the pumped-storage system between two reservoirs, or even completely off-stream, as opposed to an upper reservoir drawing from, and then discharging to, a river.

Current distribution and development of hydropower

More than 800,000 dams have been constructed worldwide, with more than 45,000 defined as 'large dams' (> 30 m tall) (Pittock and Hartmann 2011). Approximately half of all dams are located within China (World Commission on Dams 2000). More than half the world's large-river systems are fragmented by dams (Nilsson et al. 2005).

The extent to which regions have been developed for hydropower varies greatly across the world (Figure 6.2). Europe and North America have developed a relatively high proportion of their economically feasible hydropower potential, whereas South America and Asia have developed a much lower percentage; Africa has developed the least amount of its potential hydropower (Hamududu and Killingtveit 2012). Hydropower is generally expanding in those regions that have large remaining potential, including Asia (eg Mekong, Salween and Yangtze rivers), Latin America (eg Colombia and Brazil; see Figure 6.3) and Africa (eg Ethiopia and Mozambique).

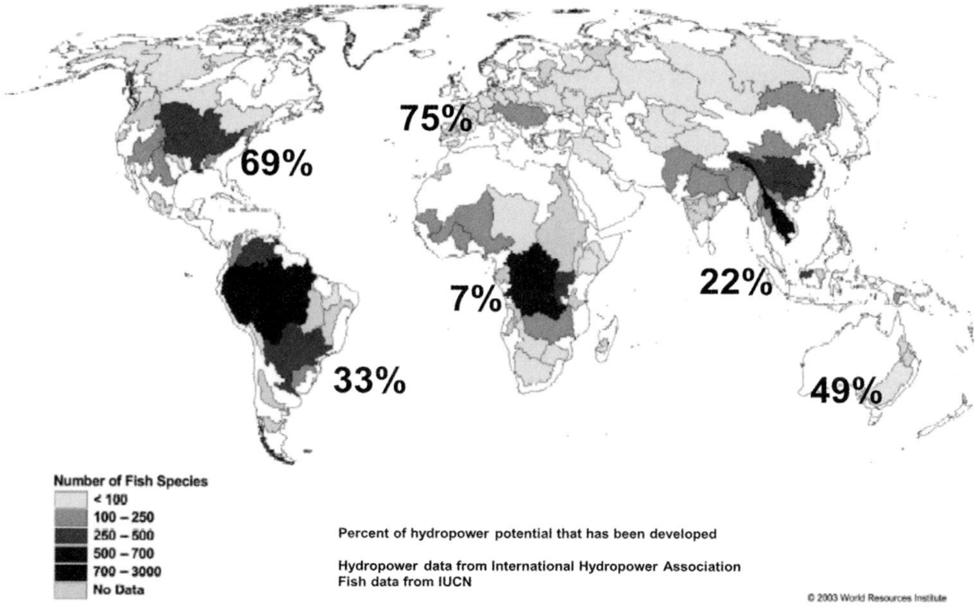

Figure 6.2 The proportion of hydropower potential that has been developed, by continent (data from the International Hydropower Association). The shading of major river basins indicates fish biodiversity (darker = more fish species; data from IUCN).

The regions likely to undergo the most hydropower development also tend to be those with high levels of aquatic biodiversity (Figure 6.2) and where communities tend to rely directly on the services provided by functioning river ecosystems (Richter et al. 2010; Turner et al. 2007), illustrating the challenge for balancing hydropower expansion with maintaining rivers' other values. Ethiopia has only developed 3 per cent of its hydropower potential, but current dam development on the Omo River will reduce considerably the annual flood pulse, which supports flood-recession agriculture for hundreds of thousands of people (EEPCO 2008).

Climate change is increasingly driving the demand for hydropower (Pittock and Hartmann 2011). For example, China cites the need for low-carbon source of energy as a driver of its plan to dam nearly all of its major rivers (Jacobs 2013). Hydropower has been eligible for Clean Development Mechanism (CDM) credits, a program launched under the Kyoto Protocol that provides a subsidy to projects that reduce global emissions (IPCC 2007). The Clean Development Mechanism recognises some hydropower as an energy source with low emissions and so CDM credits can provide an additional stream of income to eligible hydropower projects. The only specific environmental standard embedded within CDM was that, to be eligible for credits, a project must produce energy with a power density greater than four watts per square meter inundated (Pittock 2010), along with a general requirement that the project contribute to sustainable development. Hydropower projects represent a high proportion of all projects receiving CDM credits, and environmental organizations, such as International Rivers, have criticised the hydropower projects enrolled in CDM for their environmental impacts and because they believe the projects cannot meet the test for 'additionality', which means the CDM credits were essential to project viability – ie the project would not have happened on its own, so the credits provide 'additional' climate benefits rather than a payment for a benefit that would have happened anyway (International Rivers 2007).

Hydropower capacity is projected to grow at 2.5 to 3.5 per cent per year globally as part of an overall increase in renewables, which are projected to attain one-third of global energy generation by 2035. Hydropower is, and is projected to remain, the dominant source of renewable energy, although wind will increase dramatically (Hamududu and Killingtveit 2012). Policies intended to mitigate climate change will influence the rate and amount of hydropower expansion. The International Energy Agency (2011) forecasts that hydropower will increase by 75 per cent over current capacity by 2050 under a 'business-as-usual' scenario but would increase by 85 per cent in a scenario with aggressive policies to reduce GHG emissions.

Currently, major areas of expansion include South America (eg the Amazon basin), shown in Figure 6.3, and Southeast Asia (eg the Mekong basin), and these regions are the centre of major debates over the balance between hydropower energy and other water resources and values. The Mekong River supports the world's largest harvest of freshwater fish (2 million tons per year), and the most important component of the harvest is made up of long-distance migrants, vulnerable to the dozens of major dams proposed on the Mekong main stem and major tributaries (Ziv et al. 2012). The following section provides

Figure 6.3 Planned dams (black dots) in South America, based on data from government agencies from each country. Aggregate projections include 2,215 dams, including adding dams to 673 rivers that are currently undammed. Data compiled by Paulo Petry of the Nature Conservancy.

more detail on the impacts of dams on riverine resources, the fundamental challenge for the energy-water nexus of hydropower.

The energy-water nexus: Hydropower and other water resources and services from rivers

Hydropower is a major user of water and, particularly in arid environments, can also be a major consumer of water through evaporation from reservoirs (Mekonnen and Hoekstra 2012). As a user and consumer of water, hydropower interacts with other water-management sectors in a river basin, such as water supply, irrigation and flood control.

Dams can be described as being 'single purpose' or 'multipurpose'. Many hydropower dams have the single purpose of energy generation, although regulations can require dams with an initial sole management purpose of hydropower to also manage for environmental health and recreation. Single-purpose hydropower dams are most common in high elevation, mountainous river reaches. Very large dams are often multipurpose. For example, the largest dams in the United States were built by agencies with a primary purpose of either water supply (eg the US Bureau of Reclamation and the California Department of Water Resources [DWR]) or for navigation and flood control (dams built by the US Army Corps of Engineers). The largest of the dams built by these entities are generally multipurpose and include hydropower, such as Hoover (Reclamation), Oroville (DWR) and Grand Coulee (the Corps) dams.

Within multipurpose dams, the various uses can compete with each other for the allocation of storage and for flow releases at various times. For example, the ability to manage floods improves with decreasing reservoir levels, with the empty storage space available to capture and attenuate incoming flood flows, whereas full reservoirs reduce risk of drought for water-supply functions. Hydropower generation is a function of flow and head, so a full reservoir, with greater head, generates more energy. However, full reservoirs tend to 'spill' more water (water that passes over spillways and thus does not go through a turbine), so hydropower managers try to manage reservoirs to minimise spill. Within multipurpose reservoirs, climate change has the potential to increase the conflicts between the multiple purposes (Lee et al. 2009), described in the next section.

In addition to these competing uses of water within reservoirs, hydropower interacts with a number of other values from rivers that are partly or fully distinct from allocation of water within a reservoir. Below, we discuss these other values. The chapter concludes with a discussion of how multi-sectoral and basin-scale approaches are the key to managing conflicts in the water-energy nexus and will become even more important when dealing with climate change.

Environmental and social impacts of hydropower

The environmental and social impacts of hydropower projects have been described thoroughly elsewhere (Postel and Richter 2003; World Commission on Dams 2000) and thus we present only a brief review here. While impacts can be divided into environmental and social categories, these categories are highly intertwined. For example, the loss of floodplain inundation affects both fish populations and human communities dependent on floodplain fisheries. Here, we organise impacts into those affecting *upstream resources*, *connectivity* and *downstream resources.*

Impacts to upstream resources

The impacts to upstream resources have generally received the most attention in debates about dam development. Dams can create very large reservoirs that inundate upstream land, including communities, agricultural land and ecosystems. The displacement of communities by a reservoir often generates the most controversy over dam development. For example, construction of the Three Gorges Dam on the Yangtze River in China required the relocation of approximately 1.5 million people (Heggelund 2006). Those displaced by dams are often poor and lack political strength, and surveys of relocation and compensation programs have often found that they fell short of their promised benefits (Scudder 2005).

Reservoirs replace the flowing, dynamic and variable aquatic habitat of a length of river by a flatwater lake with habitat features that favour a different suite of organisms, such as reservoir-adapted fish replacing riverine fish. Often, the change from river to reservoir results in a replacement of endemic and rare species (eg the endemic fish fauna of the Colorado River in the Southwestern United States) with common and widespread 'generalist' species, such as bass or tilapia.

Impacts to connectivity

Dams and reservoirs affect the downstream transport of sediment, wood and nutrients and disrupt the up- and downstream movement of organisms, including fish and invertebrates (March et al. 2003). Dams can either be complete barriers (eg the 168 m tall Grand Coulee Dam on the Columbia River blocks all fish passage) or partial barriers: Dams with fish passage facilities can still exact considerable mortality on fish moving both up and downstream and may be impassable at some flow levels (NRC 1996). In addition to the barrier of the dam wall itself, hydropower projects can disrupt longitudinal hydrological connectivity within a river by dramatically changing hydrological conditions up- and downstream of the dam. Some hydropower dams divert most or all of a river into a canal, leaving a bypass reach with little or no downstream flow. Large storage projects can create a long lake-like reservoir with little or no current, and this reservoir can function as a barrier to the downstream movement of aquatic organisms and to the migration of terrestrial organisms across the river. Migratory fish can be particularly affected by dams that act as barriers to and from spawning habitats, and because migratory fish can be dominant within fish harvests, their loss can have major negative impacts on communities that depend on river fisheries

As a river enters a large reservoir, flow velocity can decrease towards zero and the river no longer has sufficient energy to transport much of its sediment load. Sediment sizes such as gravel and sand drop out relatively quickly, forming a delta where the river enters the reservoir. In addition to reducing the storage volume and longevity of a reservoir, this sediment-trapping function can have serious impacts on the downstream river, described below.

Impacts to downstream resources

While traditionally receiving less attention than the upstream resources affected by impoundment, dam impacts to downstream environmental resources are often far greater than the upstream impacts. Because human livelihoods and communities are often directly tied to functioning river ecosystems, these downstream environmental impacts can also have considerable social costs (Richter et al. 2010).

As described above, large reservoirs can trap nearly all sediment, except for the smallest particle sizes, and even small reservoirs can trap much of the larger sediment in transport (eg cobbles and gravels). This trapping disrupts the balance of erosion and sedimentation downstream, contributing to degradation of the river bed (incision), which can isolate the river from its floodplain, and bed 'armouring', whereby smaller sediment sizes are entrained leaving behind only large, immobile cobbles and boulders (Ligon et al. 1995).

Reservoirs capable of storing a large volume of water can significantly alter the flow regime downstream of a dam (Figure 6.1). The flow regime can be viewed as a 'master variable' that structures river ecosystems – affecting channel morphology, water quality and ecological processes – and thus disruptions to the flow regime can have serious consequences for river ecosystems. For example, fish behaviours for reproduction and migration are often triggered by changes in the flow regime, such as floods, so dam-induced changes to the magnitude, timing or frequency of flood events can therefore depress fish populations (Poff et al. 1997). Hydrological alteration, largely caused by dams, is one of the primary threats to freshwater ecosystems and their species, which are imperilled at rates that exceed those of terrestrial and marine ecosystems (Postel and Richter 2003).

As described above, the operation of hydroelectric dams to meet peak power demands results in extreme daily fluctuations in flow. The high flows can lead to 'catastrophic drift' of invertebrates, displacement of fish and channel degradation. Sudden declines in flow can 'beach' fish, and the no- or low-flow interval between periods of generation can also strand aquatic organisms within isolated pools, where they become vulnerable to high temperatures, low dissolved oxygen and predation. The effects of peaking may persist for hundreds of miles below a dam (Freeman et al. 2001).

In addition to changing the flow and sediment regimes, dams can also alter the water quality of water that is discharged from the reservoir into the river. For example, dams that discharge from the bottom of the reservoir typically release water that is colder than river water (a phenomena referred to as 'thermal pollution'), potentially contributing to changes in the fish fauna. Dams on the Murray River in Australia reduce water temperatures by 12 degrees Celsius below natural levels, preventing the spawning of native fish whose reproductive physiology requires warmer temperatures. Reduced temperatures persist for 300 kilometres downstream (Sherman et al. 2007). Reservoirs can also release water that is low in dissolved oxygen or has other water quality problems (Fearnside 2005).

Mitigating environmental impacts

The impacts described above have been known for several decades, since environmental impact assessments were first conducted for dams. Here, we briefly describe a variety of mitigation interventions that are commonly used to address the impacts of dams.

Dams can be designed to allow passage of fish or sediment. Examples of fish passage include fish ladders and elevators. Although upstream fish passage design for salmonids has received considerable study, very little is known about the ability of other fish species to successfully use fish passage structures, although research indicates that fish passage facilities may be less effective for tropical fish than they are for salmonids (Noonan et al. 2012). Fish moving downstream often pass through the turbines, which can cause injury or mortality, and engineers have recently developed 'fish friendly' turbines designed to minimise harm to fish. Engineered structures to allow passage of sediment include drawdown flushing, sluicing and bypasses. Although these techniques can be effective in some locations, they generally only provide partial compensation for losses of fish or sediment transport, and they generally have minimal or no effectiveness for very large dams, as the height of the dam wall precludes effective fish passage and the length of the reservoir impedes sediment passage methods.

To mitigate hydrological alteration, dams can release environmental flows – patterns of flow magnitude, timing and duration intended to support biological and physical processes and promote environmental and social values (Postel and Richter 2003). Environmental flow assessments are used to determine the characteristics of the flow regime, such as volume and timing of flows, which are necessary to maintain or restore certain components of the river ecosystem, and there are hundreds of methods, tools and approaches available for conducting these assessments (Tharme 2003).

Dam design can constrain or prohibit many of these mitigation methods, emphasizing the importance of dam designs that fully anticipate potential impacts and encompass mitigation into the original design (Krchnak et al. 2009). The uncertainty of climate change also increases the importance of building flexibility into dam design, as future mitigation needs may not be apparent for decades. For example, variably sized turbines may allow hydropower dams to operate more efficiently over a wider range of discharges (Balciunas and Zdankus 2007), thereby improving the downstream flow regime with less reduction of generation and revenue, and be more resilient to the uncertain hydrology with climate change.

In later sections, we discuss the limitations to mitigating impacts through these project-level interventions and how impacts can best be avoided or mitigated through basin-scale approaches.

Climate change, water resource and hydropower

Models of future climate under enhanced GHG conditions consistently predict higher average temperatures, but the models tend to have greater uncertainty with hydrology, including components such as precipitation, runoff and river discharge. There is general agreement that some currently wet regions of the world will get wetter and some dry regions will become drier (Durack et al. 2012). Scientists also generally concur that hydrographs may become 'flashier', with greater frequency of both floods and droughts (Kundzewicz et al. 2008). Recent trends suggest that precipitation patterns have already changed; for example, the frequency of intense storms (> 7.5 cm of rain in one day) has approximately doubled, compared to fifty years ago, in the Midwestern United States (Saunders et al. 2012).

Some hydrological changes are directly linked to changes in temperature and can be predicted with greater certainty. In general, because of rising temperatures, evaporation will increase, reducing water availability even with unchanged precipitation. Changes in temperature can also produce seasonal shifts in runoff due to an increase in rain relative to snow, shifting a portion of runoff from the snowmelt period to the rainfall period (Gleick and Chalecki 1999). Further, snowmelt and glacier melt will tend to begin and end earlier in the year (Barnett et al. 2005).

Climate change will have a number of direct and indirect impacts on rivers, including changes in the quantity and timing of water flows, water temperature and associated ecosystem processes. Indirectly, climate change may change sediment loads through increased bank erosion from higher magnitude floods or through changes in erosion rates from land due to changes in precipitation, land cover and fire regimes (Hamududu and Killingtveit 2012).

These hydrological changes will affect the viability of existing and future hydropower projects and complicate calculations on expected generation from new projects. The World Commission on Dams (2000) summarised a set of risks to hydropower from climate change, including:

1. *Reduced inflow*, due to changes in precipitation and evaporation in the upstream watershed.
2. *Risks from higher magnitude floods.* If projects do not accurately account for potential increases in flood magnitude, then the project risks being under-designed for resisting floods, increasing the risk of dam failure. Larger floods also increase the likelihood of emergency releases that can pose a danger to downstream communities and a greater probability of spills.
3. *More conflicts with other demands* within multipurpose reservoirs due to less water or less certainty. Examples include flood management and the ability of a reservoir to provide downstream environmental flows.
4. *Greater evaporation from reservoirs*
5. *Increased sedimentation*

Over the short term, higher rates of glacial melt will increase the availability of water for hydropower for projects below glaciers. For example, 2003 was one of the hottest and driest years in Switzerland in the past 500 years. Despite lower precipitation, Swiss hydropower production was only 0.8 per cent below the ten-year average due to significant glacial loss. However, over the long-term the decline of snowpack and glaciers will result in a decline in generation for projects that previously depended on meltwater. Climate change will also likely lead to increases in evaporation from reservoirs, which can be significant in arid areas, such as the Zambezi where, currently, approximately 16 per cent of the river's mean annual flow is lost to evaporation from reservoirs, resulting in hydropower being the single biggest consumer of water in the basin (SADC-WD and Authority 2008).

Growing expectations that runoff will increase in the North (Russia, Scandinavia, Canada, Alaska) and some other regions will provide incentives for increased hydropower investment, while in other regions the technical-financial hydropower potential will be reduced. Hamududu and Killingtveit (2012) conducted a global analysis of how climate change will affect hydropower, using runoff as the main determinant to generation. At the continental scale, the effect of climate-altered hydrology on hydropower was relatively small, in part due to offsetting increases and decreases. For example, within Africa, projections for hydropower in east Africa increased, while southern and northern decreased and the west stayed about the same. Much larger changes were forecast at the scale of individual countries, with hydropower generation in Uganda forecast to increase by 15 per cent and generation in Namibia forecast to decrease by 21 per cent.

Climate studies predict a 26 to 40 per cent decline in average annual runoff for the Zambezi River basin by 2050, compared to a 1960–90 baseline. Due to declining inflows, a World Bank study projected a 32 per cent decline in reliable energy capacity from the Zambezi hydropower system (from 30,000 GWh/year to 20,000 GWh/year) and a 21 per cent decline in average energy production (from 56,000 GWh/year to 44,000 GWh/year; Beilfuss 2012).

In general, run-of-river projects will be more vulnerable to hydrological changes than storage reservoirs, and projects that rely on snowmelt will also be more vulnerable. For example, Connel-Buck et al. (2011) found that low-elevation reservoirs in California that have storage will see declines of 4.5 per cent in hydropower generation. In comparison, Madani and Lund (2009) found that California's high-elevation hydropower system, which are generally run-of-river dams with limited storage and depending on the snowpack, would experience a 14 per cent decline in generation.

Some parts of the world have already experienced hydrological patterns that are considerably different than historical averages, causing hydropower companies to adjust expectations of generation. Hydro Tasmania has experienced a series of drought years and they have downgraded their estimate of reliable annual yield from 10,000 gigawatt-hours to 8,700 gigawatt-hours, a 13 per cent decline (Hydro Tasmania n.d.).

Countries that have a high reliance for hydropower for their electricity supply can be vulnerable to droughts, and this vulnerability may increase due to climate change (Ebinger and Vergara 2011;Matthews et al. 2011). Regionally, southern Africa is 60 per cent dependent on hydropower for their electricity, and this region is already vulnerable to droughts and is forecasted to become drier with a greater frequency of droughts (Beilfuss 2012).

The changes in temperature and precipitation will not only affect hydropower. Within river basins, hydropower interacts with a number of other sectors and resources, and climate change will also affect these resources, and potentially increase competition between various uses for water within multipurpose reservoirs and water-management systems (Lee et al. 2009). Reduced discharge and increased evaporation will reduce supply for irrigation while, through increased rates of evapotranspiration, higher temperatures will increase demand for irrigation water. Flood managers will seek greater storage allocation to flood management due to increased flood risk while water-supply managers will seek increased allocation to storage to manage for increased drought risk. For example, in the Zambezi River, water managers will confront trade-offs between maximising head for hydropower (with high reservoir levels) and leaving empty storage to attenuate floods (Beilfuss 2012).

Reservoir management will also see competition between storage for firm hydropower and the release of environmental flows, such as for salmon in the Columbia River (Payne et al. 2004). Climate change could degrade habitat for temperature-sensitive species like salmon (Yates et al. 2008) and declines in population of species could result in new requirements for mitigation to avoid extinction. Temperature-sensitive species will generally migrate to cooler portions of the river network, but dams may act as barriers to this movement (Palmer et al. 2008). Interestingly, for some temperature-sensitive species, flow management from dams may promote regional persistence in areas undergoing climate change. For example, Yates et al. (2008) forecasted that Shasta Dam can maintain a 'cold pool' of water to support flows of the appropriate temperature to support spawning and rearing habitat for salmon in the Sacramento River, assuming a mid-century warming of 2 degrees Celsius, although the reservoir may not be able to maintain a cold pool if warming reaches 4 degrees Celsius.

Changes to other water-management sectors, along with impacts on species' viability, will affect the economics of hydropower projects. Increased demand for other uses of water may reduce the financial viability of hydropower within multipurpose projects, or environmental requirements may reduce the financial viability of single purpose hydropower dams.

Management responses: Incorporating climate change in hydropower planning and operations

Until recently, the practice of water management has operated on the principle of stationarity – that future hydrology could be predicted based on observed past hydrology. However, in much of the world, precipitation trends have shifted sufficiently to suggest that stationarity can no longer be assumed, and forecasts predicting continued shifts and increased variability further erode the assumption (Milly et al. 2008). Shifting precipitation and temperature patterns have already reduced the effectiveness of water-management infrastructure in some regions, and scientists now strongly recommend that planning for water-management infrastructure fully consider the potential effects of climate change and climate variability. Matthews, Wickel and Freeman (2011) suggest that infrastructure development which doesn't account for climate change will be more vulnerable to technical, economic or ecological failure, exacerbating risks to the reliability of energy supplies, ecosystem services and misspending limited capital.

However, to date, hydropower planners and managers have generally not planned for climate change (with some exceptions; see Hydro Tasmania in previous section). Beilfuss (2012) reports that hydropower dam planning in the Zambezi basin is based on current hydrology, despite the projected major changes to river discharge reviewed above. In the United States, hydropower operators have generally not incorporated climate change into the analyses required during the relicensing process (Aspen Environmental Group 2005; Viers 2011).

Hydropower planners may not be incorporating climate change in part because of the relatively high uncertainty in precipitation projections and the perception that the effects will occur over time periods longer than economic planning periods (Rydgren et al. 2007). Although planners generally seek more reliable runoff predictions specific for their sites, and efforts are on-going to provide downscaled hydro-meteorological information, it appears likely that planners will have to settle for less reliable hydrological information than previously expected, even as modelling continues to improve.

Increasing storage capacity is a common recommendation to deal with this uncertainty (Rydgren et al. 2007), although increasing storage for hydropower is costly and may lead to conflicts with other water users and river resources (Hamududu and Killingtveit 2012). We synthesise several recommendations for how hydropower planners and managers can account for climate change (Arnell and Hulme 2000; Matthews et al. 2011; Pittock and Hartmann 2011; Beilfuss 2012):

1. Plan and operate hydropower at the basin or regional scale, integrated with other sectors and resources.
2. When planning individual projects or systems, use scenario analysis to incorporate potential future hydrological changes. To deal with the loss of stationarity, scientists recommend that hydropower planners use scenario analysis; Matthews et al. (2011) and Arnell and Hulme (2000) recommend that planners use a range of future climate projections instead of a single 'best guess' scenario and generate risk probabilities based on ensemble model simulations.
3. Diversify the regional power pool to reduce dependency on hydropower that could be vulnerable to droughts.
4. Prioritise improvements to existing capacity over new infrastructure.
5. Invest in infrastructure that is reversible and/or resilient/adaptable. Ensure operational flexibility through design and operation rules, such as a broader range of turbine sizes (Balciunas and Zdankus 2007).
6. Build in larger safety margins (eg increased spillway capacity for potentially larger floods).
7. Incorporate green infrastructure into system planning and operation.
8. Develop strong institutional capacity to implement adaptive management.

The first of these recommendations – plan and operate hydropower within a basin-wide context – is the necessary foundation for most of the other recommendations, and the remainder of the chapter focusses on this topic.

Sustainable, climate-resilient hydropower at the basin scale

Although great improvements to environmental performance can be made through dam design and environmental flow releases (Krchnak et al. 2009), a number of important impacts from hydropower dams cannot be mitigated effectively at the dam site. Particularly for large dams, effective passage of fish and sediment past the dam can be very difficult or impossible. While the release of environmental flows can mitigate some impacts from hydrological alteration, storage dams can be constrained in the type of flows they can release, given the operational demands for storage and release that they were built to meet. Also, the purpose of releasing environmental flows in one segment of a river may be defeated if other dams upstream or downstream do not do the same. Thus, balancing hydropower and other objectives can generally be achieved most effectively at larger spatial scales.

It is now broadly recognised – by conservation organisations (WWF-International 2006), industry (eg the 'early stage' of the Hydropower Sustainability Assessment Protocol, described below; IHA 2010) and funders (Ledec and Quintero 2003; ADB et al. 2010) – that achieving balance between hydropower and other objectives and values can best be achieved by planning at basin to regional scales: the spatial scale recommended for Integrated River Basin Management or Integrated Water Resources Management. Building resiliency to climate change into energy-water systems only underscores the need for this spatial scale.

Achieving this balance between hydropower and other water-management as well as river objectives, and doing it within context of changing climate, will require evolution of hydropower planning and operations away from a single-project focus, which predominates today, and towards basin scale. In this section, we review existing frameworks, policies and examples that either explicitly move hydropower towards a basin-scale approach or can serve as the building blocks of a basin-scale approach, and then describe two applications of these concepts.

Existing frameworks for sustainable hydropower

World Commission on Dams

The World Commission on Dams (WCD) report, *Dams and Development* (2000), received a great deal of attention (Postel and Richter 2003); for example, the IHA Hydropower Sustainability Protocol (2010), described in the next section, was initiated in response to the WCD. Throughout the report, the WCD emphasised comprehensive planning, earlier review of projects to eliminate inappropriate projects and integration across sectors. The report describes five key decision points that if applied according to their guidelines, should result in more sustainable projects being developed. These decision points include an initial needs assessment that determines the accurate need for water and energy services followed by the selection of alternatives for meeting those needs. The report recommends that a Strategic Environmental Assessment (SEA) be conducted at this stage (see the next section for more on SEAs). Recognising the values derived from undammed rivers, the WCD recommended maintaining some portions of a river or basin in a 'natural state'.

Regulatory and planning tools for hydropower

Governments have a range of tools for planning, licensing and regulating hydropower. These tools have great potential to move hydropower towards a basin-scale approach that fits well with other sectors and resources and is more resilient to climate change. However, these tools are not yet universally used, especially in a system-scale approach.

These governmental planning tools include environmental impact assessment, energy master plans, options assessments, strategic environmental assessment, river basin planning (basin master plans) and licensing.

Environmental Impact Assessment. Governmental tools are most developed and most widely applied at the scale of individual project. Most countries now require Environmental Impact Assessments (EIA), which review the environmental, and often social, impacts from a project or various alternatives to a project.

Project-level environmental review has significant limitations that have been well documented (Brismar 2004). The primary weakness of project-level review is that it generally takes place after many decisions about the project have already been made, significant investments in the project have already occurred, and momentum and support for the project are well developed. These conditions place significant pressure on the review process to not reject the project or move its location. Indeed, reviews of EIAs have found that they rarely result in the rejection of a project (Sadler et al. 2000). Thus, the EIA process provides little hope of rejecting inappropriate projects, except in unusual cases, and instead the review process generally results in minor modifications and a set of mitigation strategies. Finally, project-level reviews frequently include inadequate consideration of cumulative effects over time or the aggregate effects of multiple projects. The WCD advocated "cutting the direct link between the feasibility study and project approval" in favour of a review process that occurred early enough that it could realistically reject or relocate inappropriate projects (World Commission on Dams 2000).

Energy master plans and options assessments include realistic assessments of current and forecast energy needs and an options assessment of how best to meet those needs. This stage of planning is not focused just on hydropower, but instead considers all ways to meet energy needs, including demand-side management (conservation and efficiency) and other sources of generation. This type of planning process has the potential to provide a credible range of estimates of the role that hydropower should play in a nation or region's energy future and thus can serve as a key input into regional and basin-scale hydropower planning. Utilities can conduct an Integrated Resource Plan (IRP) to forecast future loads and plan the mix of options, considering both supply and demand, to meet that load. The contribution of hydropower can be determined within an IRP (Wilson and Biewald, 2013), which may include a preliminary ranking of hydropower projects according to their technical and economic feasibility.

Following an energy master plan and options assessment, the decision-making focus can switch to the river basin or region, employing policy and planning tools such as *Strategic Environmental Assessments* (SEA) and *river basin plans.*

Strategic Environmental Assessments address the environmental effects of programs, plans and policies, rather than individual projects. Although SEA's haven't yet lived up to their potential, the number of countries requiring or otherwise undertaking SEAs is on the rise, and SEAs still have the potential to improve how projects are selected and developed (Hussey and Pittock 2012). SEAs can be conducted at various levels, from a river basin to a sector to a country. An SEA for the Vu Gia Thu Bon basin in Vietnam considered cumulative effects and multiple values and resources, and produced a plan to meet a hydropower generation target while maintaining two 'intact' river corridors (Figure 6.4).

Licensing. Globally, only a few jurisdictions have time-bound licenses, for which licensees have to periodically apply for extensions and projects are subjected to a review of their performance. Original licenses for individual projects can be linked to broader planning efforts (ADB et al. 2010), such as those described above. Opperman et al. (2011a) include recommendations of how to incorporate basin-scale planning, operations and opportunities (eg for restoration) in a licensing framework that focusses on single projects.

The main benefit of a relicensing scheme, such as under the United States' Federal Energy Regulatory Commission (FERC), is that shifts in societal expectations and increases in environmental and social knowledge can be incorporated into updated license conditions (Pittock and Hartmann 2011).

Integrating planning and regulatory tools

The WCD posited that state governments have a responsibility for ensuring that this comprehensive planning approach is adopted. However, the full use of the tools described above – project-level review and licensing that occurs within a basin-scale planning framework, informed by credible energy master plans and options assessments – rarely occurs.

Governmental involvement in hydropower planning, management and regulation ranges from comprehensive to limited. In Mexico, the Constitution reserves energy generation to the government, and a quasi-governmental agency, the Federal Commission for Electricity (CFE) plans, builds and operates all large hydropower dams, along with all other major forms of energy generation and distribution. CFE does engage in energy master plans, national-scale hydropower planning and basin-scale plans, in addition to project-level EIAs. CFE is currently working with The Nature Conservancy to integrate conservation-planning tools into their basin-planning process.

In most countries, the government engages in limited amounts of large-scale planning for energy and water generally, or hydropower specifically, and instead responds to project proposals from dam developers through environmental review and licensing. Policy reforms and innovation required to move hydropower planning and operation to the basin scale will vary across this broad spectrum of governmental involvement.

There are a few examples of governments that have supported comprehensive planning for hydropower to identify where new hydropower should be built and which rivers should be protected to support other values and benefits. In the 1980s, the government of Norway conducted a range of planning processes to determine on which rivers future hydropower

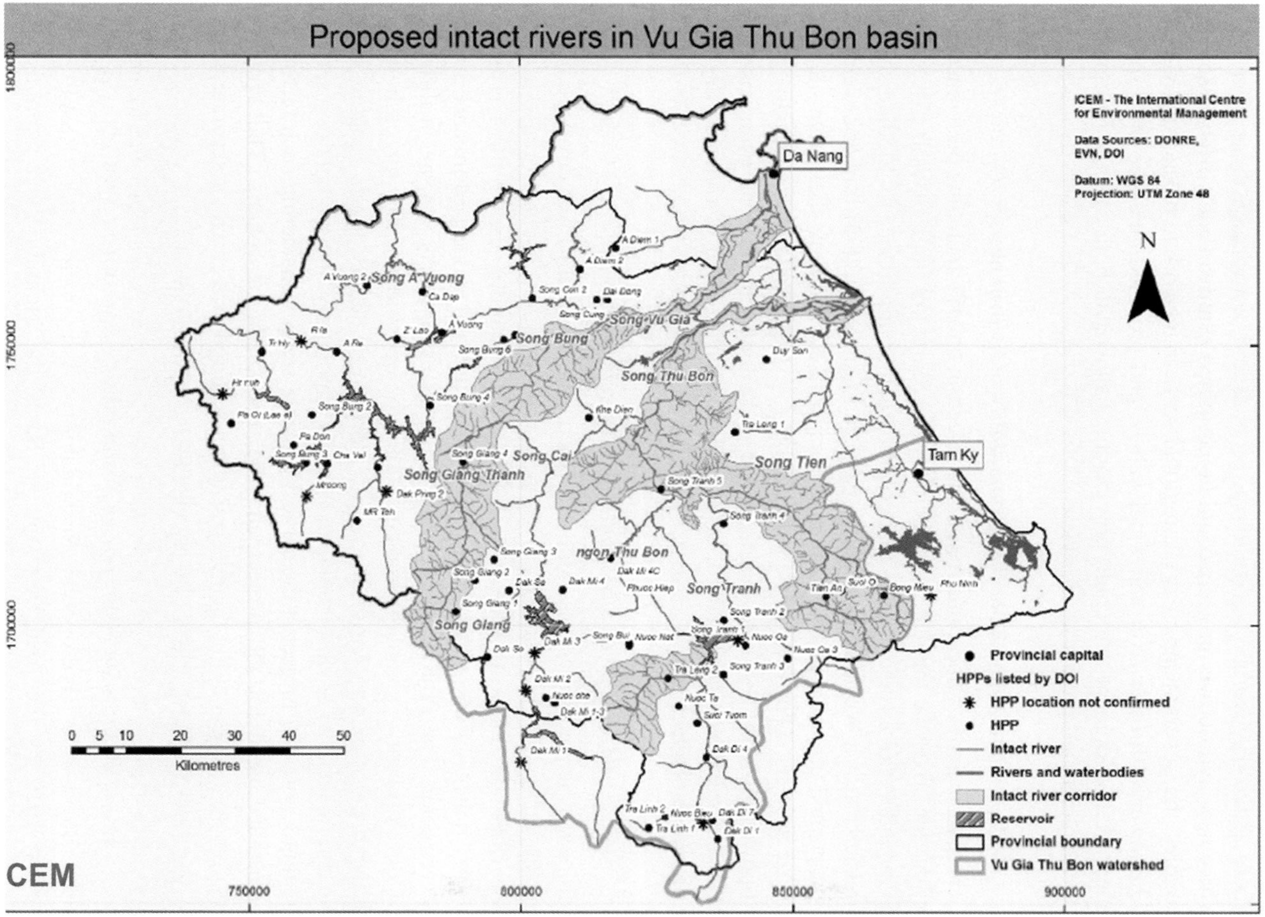

Figure 6.4 Proposed intact rivers for the Vu Gia Thu Bon basin in Vietnam (blue shading) that were designated through a Strategic Environmental Assessment of hydropower expansion in the basin (from ICEM 2008).

should occur and where it shouldn't (Huse 1987). Also in the 1980s, the state government of Maine in the United States developed the Maine Rivers Policy, in anticipation of increased demand for new hydropower licenses. The Rivers Policy identified eighteen rivers where additional hydropower could not be built and, conversely, those river segments where new hydropower projects could be permitted (Sullivan and Giffen 1985). In both examples, the governments created a regional scale framework to balance the values from hydropower with other river and water values, and provided guidance on a spatial arrangement of development that could provide this balance. In both cases, this regional approach complemented project-level review.

Tools to evaluate the sustainability of hydropower

The WCD report includes some guidance on site selection criteria, including preference for tributary locations over the main stem (principal watercourse) of large rivers for issues of connectivity and the avoidance of areas that are of particular importance for biodiversity or cultural resources. In a World Bank report, Ledec and Quintero (2003) further developed those concepts and proposed quantitative indicators to assess and rank alternative sites for hydroelectric projects. They offer thirteen indicators to compare and rank alternative sites for hydropower projects, and note that this type of ranking can be done even absent a detailed SEA. Examples include length of river impounded, length of river left dry and number of people displaced.

Hydropower Sustainability Assessment Protocol

The Hydropower Sustainability Assessment Protocol (the 'Protocol') was developed by a multi-stakeholder forum (International Hydropower Association 2011). Because it does not define what constitutes an acceptable level of sustainability, the Protocol is not a standard but, rather, an assessment tool used to measure and guide performance in the hydropower sector. The Protocol can be applied to assess projects that are in different stages of development – planning, implementation or operation – and, for each stage, scores relative performance across twenty or more sustainability topics ranging from economic and financial to environmental and social. Importantly, it also contains a section for evaluating a potential project at the earliest possible stage, considering such factors as the demonstrated need for the project and the assessment of options. The Protocol considers climate a cross-cutting theme, so it is not featured as a single or separate topic, and provides a list of topics where climate change is most relevant.

Rapid Basin-wide Hydropower Sustainability Assessment Tool

The Rapid Basin-wide Hydropower Sustainability Assessment Tool (RSAT) was developed in 2010 by the Asia Development Bank, the Mekong River Commission and the World Wide Fund for Nature (WWF). The objective of the RSAT is to "assess hydropower sustainability within an IWRM based framework" (ADB et al. 2010). Similar to the Protocol, the RSAT is a tool for measuring sustainability, although the RSAT places greater emphasis on river-basin planning and can be used to evaluate a system of hydropower projects. While developed for the Mekong basin, the RSAT developers state that it can be adapted for use in other regions. The RSAT is composed of ten topics and twenty-seven sub-topics. Climate change is addressed under topic 7 ("flows and reservoir management") and sub-topic 7.1 ("multiple water use optimisation and efficiency").

Financial safeguards

A special group of guidelines are those issued by banks. Since hydropower projects are highly capital intensive, access to loans from commercial and development banks is a major consideration and banks can exercise influence over project choice and design. Banks protect themselves from material and reputational risks, and ensure that projects they fund are in line with their sustainability policies through lending guidelines or 'safeguards'. These are usually generic across sectors, though in some cases banks have issued specific guidelines, such as the World Bank on dam safety and HSBC on freshwater infrastructure lending. Sustainability commitments by banks either reference their own standards or other international guidance. The Equator Principles require compliance with the Performance Standards of the International Finance Corporation (IFC) and have been adopted by close to eighty banks and cover more than 70 per cent of international Project Finance debt in emerging markets (Equator Principles 2013).

Integrated basin management for climate resiliency

The policies, guidelines and tools described above can move the planning, development and management of hydropower towards larger spatial scales – such as river basins or regions – which allow for the greatest opportunities for achieving balanced outcomes between hydropower and other river- and water-based values and benefits. This chapter concludes by describing two applications of these concepts, including a proposed framework for achieving more balanced hydropower development at the river-basins scale ('hydropower by design') and the integration of flood-risk management with hydropower management to increase resiliency of water-energy systems to climate change.

Hydropower by design

This chapter has emphasised that system-scale approaches for planning and operating water-management infrastructure offer the best opportunity for development that can provide the benefits of regulated rivers while maintaining the benefits provided by naturally functioning rivers. The expansion of hydropower and other dams is unlikely to achieve this balance if dams are planned, reviewed and licensed with a framework focused on individual sites or projects. Rather, sustainable dam development requires a framework focused on river basins or regions.

The Nature Conservancy has proposed a framework called 'hydropower by design' to guide decisions on hydropower development. The framework integrates hydropower planning with spatial information on environmental and social resources to identify alternatives that provide the greatest balance between energy and other values (Figure 6.5). This process is underlain by the mitigation hierarchy and follows the concepts of 'development by design' (see Kiesecker et al. 2010). The steps of the mitigation hierarchy, with their hydropower application, are:

Avoid. New development should avoid rivers or sections of rivers that have the most valuable resources and for which other mitigation will not be effective (eg a river that contains a unique biological, social or cultural resource).
Minimise. For dams that are built, impacts should be minimised to the extent possible during design, construction and operation. For example, where appropriate, dam design should include the potential for temperature control, fish passage and the ability to release environmental flows.
Restore. Dam operations should strive to restore and maintain key river resources, such as through the release of environmental flows to maintain downstream fisheries.
Offset. Not all impacts can be mitigated through site-specific actions to minimise and restore impacts. The 'residual' impact can be offset by making conservation investments elsewhere in the river basin to protect or restore similar resources.

This landscape-scale approach moves mitigation away from piecemeal efforts and towards an overall conservation plan with the best opportunity for protecting areas from inappropriate dam development, both because 'avoid' areas can be identified as early in the development process as possible and because an integrated plan can more effectively direct mitigation funds towards protection and management of these important areas. For government agencies – ranging from those charged with promoting energy development to those charged with protecting environmental and social resources – this approach offers the greatest potential for achieving balance across public-policy objectives. The landscape-scale, comprehensive planning framework can also increase the efficiency of individual project review.

Projects that emerge through hydropower by design may face lower controversy because they were identified through a process that specifically focused on ensuring meaningful protection of environmental and social resources. Less controversy, a straightforward review process and clarity on mitigation requirements should lead to greater certainty surrounding project construction, completion and operation, potentially improving the acquisition and terms of financing for dam projects.

In summary, hydropower by design strives to identify more balanced outcomes by comparing alternatives when the greatest flexibility for change remains: early in the dam siting process and with large spatial scales (Figure 6.6). The Nature Conservancy is currently working on hydropower by design projects in Mexico, Colombia and Brazil.

Integrating floodplain management with hydropower for climate resiliency

Large multipurpose reservoirs often have competing objectives of flood control, water supply and hydropower, and these various objectives can compete with each other (Kundzewicz et al. 2008). In addition to managing flood risk in reservoirs through storage and attenuation of flood pulses, flood risk can also be managed in the floodplain through levees and land-use policies that reduce exposure of people and property to floods. In addition, floodplains can

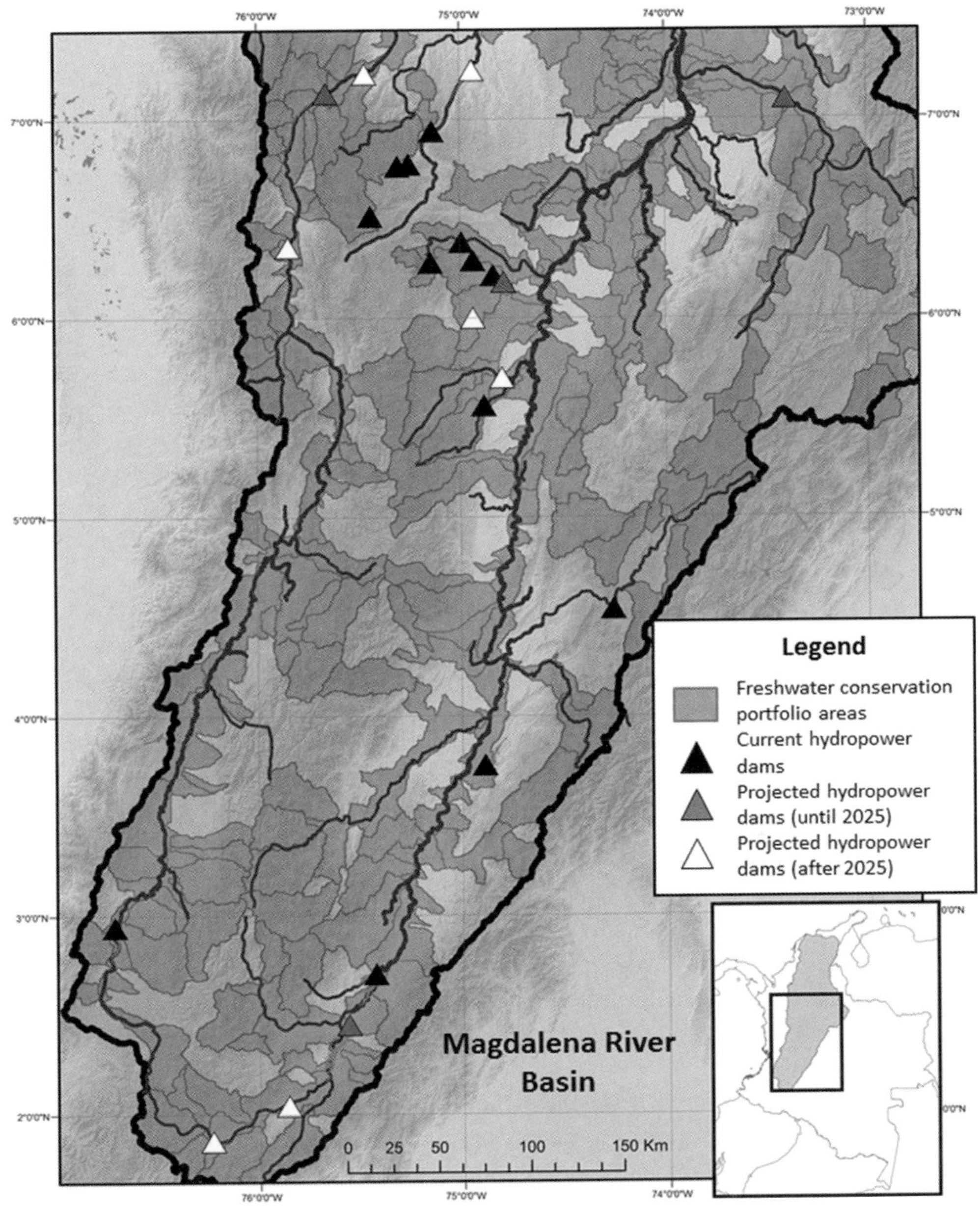

Figure 6.5 On the left, a conservation blueprint for the Magdalena River (Colombia) and, on the right, current and proposed hydropower dams in the basin.

provide storage and conveyance of floodwaters, either in the form of natural floodplains or in managed flood bypasses (Opperman et al. 2009; Opperman et al. 2013).

Closer co-ordination of flood risk between reservoirs and the floodplain can produce a different allocation of benefits and potentially improve the resiliency of energy-water-management systems to climate change. Specifically, transferring flood-management

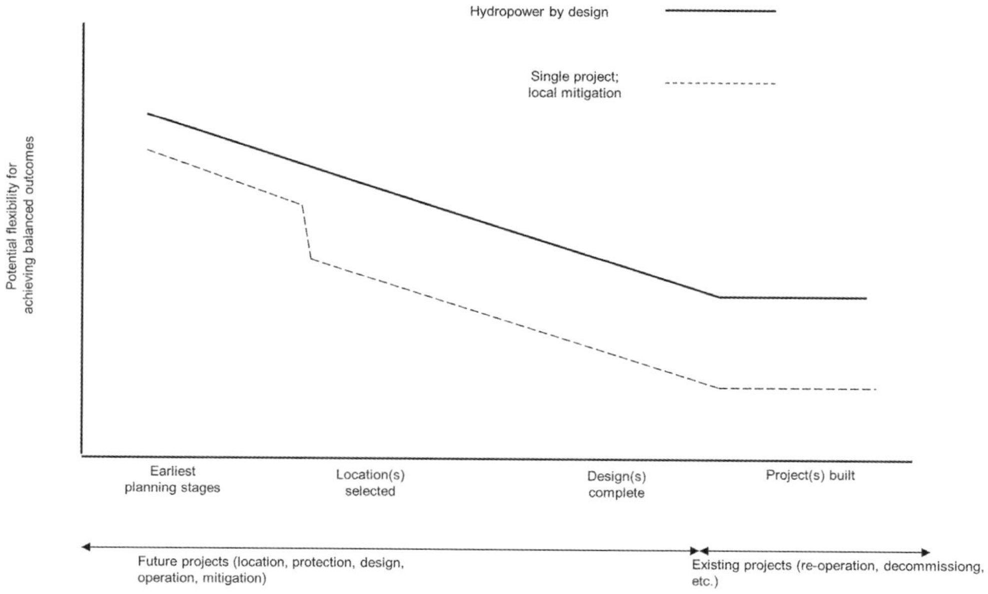

Figure 6.6 A conceptual diagram for improving the sustainability of hydropower projects. The y-axis represents the potential incremental improvement in environmental sustainability for a given energy generation target. The flexibility to achieve a balanced outcome declines as the planning process proceeds, and a hydropower by design approach (regional or basin-scale) provides a greater set of potential solutions, and thus greater potential sustainability, than a single-project approach.

functions from the reservoir to the downstream floodplain, such as through large-scale floodplain reconnection below dams, can enable a reduction of reservoir flood-storage space. By liberating a portion of a reservoir's flood-control storage, the reservoir can provide greater benefits in the form of increased hydropower, enhanced water-supply reliability and environmental flows.

This concept has been explored in a set of feasibility studies examining co-ordination of reservoir and floodplain management on the Yangtze River (China) and two rivers in the United States, the Mokelumne (California) and Savannah (Georgia and South Carolina) (The Nature Conservancy 2012). In each of these systems, researchers modelled various flood-storage volumes within reservoirs (ranging from 100 per cent of current storage volume to complete elimination of reservoir flood storage) and then estimated the resulting changes to downstream flood inundation patterns and the production of benefits from the reservoir in terms of hydropower, water supply, recreation and the provision of environmental flows.

For the Yangtze, the study estimated that eliminating the flood-storage volume in a cascade of hydropower dams could increase annual hydropower revenue by approximately US$400 million per year (Opperman et al. 2011b). The Nature Conservancy has proposed that a portion of increased hydropower revenues could be placed into a 'sustainability fund' to invest in basin-scale conservation efforts and in downstream flood-risk mitigation in the floodplain. The feasibility studies suggest that these

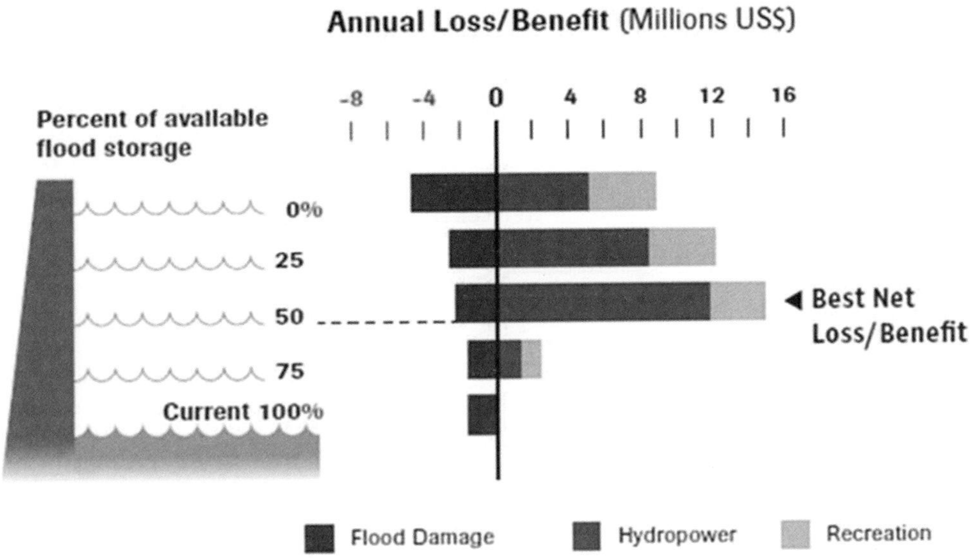

Figure 6.7 Estimated net losses and benefits from reallocation of reservoir flood storage towards other benefits, including hydropower and recreation. The greatest net increase in benefits compared to status quo operations ("current 100%") was forecasted to occur with 50% of the flood pool dedicated to other purposes (ie raising the level of the reservoir).

floodplain investments could be far more effective in reducing floodplain residents' flood risk than using the reservoirs for flood management.

For the United States rivers, the researchers compared the costs of the reservoir reallocation with the benefits (increased hydropower or water supply). Costs were summarised as either increased flood damages or the costs of changes in land use to mitigate for increased flood risk. For both rivers, they found that reducing flood storage to 50 per cent of current volume would result in considerably higher total benefits from the system. For example, with this reallocation of reservoir storage, annual hydropower revenue would increase by approximately US$12 million per year for the Savannah River (Figure 6.7). For the Mokelumne, the benefits were primarily for water supply. In addition, the change in reservoir storage patterns in the Mokelumne would allow the reservoir to meet environmental flow releases for salmon four times more frequently (The Nature Conservancy 2012).

Conclusion

While hydropower is a globally important source of generally low-carbon energy and provides other important energy services, this form of energy generation interacts with a broad array of other water sectors and the multiple benefits and values derived from rivers. Finding balanced solutions for this portion of the energy-water nexus will require integrated planning and management at the scale of whole systems. By affecting both the supply of and demand for energy and other water services, as well as the vitality of river systems, climate change will generally increase the stress in energy-water systems. System-scale planning offers the best opportunity to find climate-resilient solutions that balance across the benefits derived from managing water and those benefits provided by free-flowing rivers.

References

National Research Council (NRC) (1996). *Upstream: Salmon and Society in the Pacific Northwest*. Washington, DC: National Academy Press.

Abril, G., Guerin, F., Richard, S., et al. (2005). Carbon Dioxide and Methane Emissions and the Carbon Budget of a 10-year Old Tropical Reservoir (Petit Saut, French Guiana). *Global Biogeochemical Cycles* 19.

ADB, MRC and WWF (2010). Rapid Basin-Wide Hydropower Sustainability Assessment Tool (RSAT), http://www.mrcmekong.org/assets/Publications/Reports/RSAT-Revision-3-for-printingOCT-3-2010-Corrected-FINAL.PDF/.

Akhmatov, V., Rasmussen, C., Eriksen, P. B. and Pedersen, J. (2007). Technical Aspects of Status and Expected Future Trends for Wind Power in Denmark. *Wind Energy* 10: 31–49.

Arnell, N. and Hulme, M. (2000). Implications of Climate Change for Large Dams and Their Management. Cape Town, South Africa: World Commission on Dams.

Aspen Environmental Group (2005). Potential Changes in Hydropower Production from Global Climate Change in California and the Western United States.

Balciunas, P. and Zdankus, N. (2007). Harmonization of Hydropower Plant with the Environment. *Renewable & Sustainable Energy Reviews* 11: 1260–74.

Barnett, T. P., Adam, J. C. and Lettenmaier, D. P. (2005). Potential Impacts of a Warming Climate on Water Availability in Snow-Dominated Regions. *Nature* 438: 303–9.

Barros, N., Cole, J. J., Tranvik, L. J., et al. (2011). Carbon Emission from Hydroelectric Reservoirs Linked to Reservoir Age and Latitude. *Nature Geoscience* 4: 593–6.

Beilfuss, R. (2012). A Risky Climate for Southern African Hydro. *International Rivers*.

Brismar, A. (2004). Attention to Impact Pathways in EISs of Large Dam Projects. *Environmental Impact Assessment Review* 24: 59–87.

California Energy Commission (2005). *California's Water-Energy Relationship*. Sacramento, California.

Clean Energy Council (2011). Clean Energy Australia Report 2011.

Connell-Buck, C. R., Medellin-Azuara, J., Lund, J. R. and Madani, K. (2011). Adapting California's Water System to Warm Vs. Dry Climates. *Climatic Change* 109: 133–49.

dos Santos, M. A., Rosa, L. P., Sikar, B., Sikar, E. and dos Santos, E. O. (2006). Gross Greenhouse Gas Fluxes from Hydro-Power Reservoir Compared to Thermo-Power Plants. *Energy Policy* 34: 481–8.

Durack, P. J., Wijffels, S. E. and Matear, R. J. (2012). Ocean Salinities Reveal Strong Global Water Cycle Intensification During 1950 to 2000. *Science* 336: 455–8.

Ebinger, J. and Vergara, W. (2011). *Climate Impacts on Energy Systems: Key Issues for Energy Sector Adaptation*. Washington, DC: The World Bank.

Ethiopian Electric Power Corporation (EEPCO) (2008). *Gibe III Hydroelectric Project – Environmental and Social Impact Assessment*. Addis Ababa, Ethiopia.

Equator Principles (2013). *About the equator principles* [Online]. www.equator-principles.com/, accessed 16 October 2013.

Fearnside, P. M. (2005). Brazil's Samuel Dam: Lessons for Hydroelectric Development Policy and the Environment in Amazonia. *Environmental Management* 35: 1–19.

Freeman, M. C., Bowen, Z. H., Bovee, K. D. and Irwin, E. R. (2001). Flow and Habitat Effects on Juvenile Fish Abundance in Natural and Altered Flow Regimes. *Ecological Applications* 11: 179–90.

Gillilan, D. M. and Brown, T. C. (1997). *Instream Flow Protection: Seeking a Balance in Western Water Use*. Washington, DC: Island Press.

Gleick, P. H. and Chalecki, E. L. (1999). The Impacts of Climatic Changes for Water Resources of the Colorado and Sacramento-San Joaquin River Basins. *Journal of the American Water Resources Association* 35: 1429–41.

Hamududu, B. and Killingtveit, A. (2012). Assessing Climate Change Impacts on Global Hydropower. *Energies* 5: 302–22.

Heggelund, G. (2006). Resettlement Programmes and Environmental Capacity in the Three Gorges Dam Project. *Development and Change* 37: 179–99.

Huse, S. (1987). The Norwegian River Protection Scheme – A Remarkable Achievement of Environmental Conservation. *Ambio* 16: 304–8.

Hussey, K. and Pittock, J. (2012). The Energy-Water Nexus: Managing the Links between Energy and Water for a Sustainable Future. *Ecology and Society* 17.

Hydro Tasmania (N.d.). Climate Change Impacts on Hydro Tasmania.

International Centre for Environmental Management (ICEM) (2008). *Strategic Environmental Assessment of the Quang Nam Province hydropower plan for the Vu Gia – Thu Bon River basin*. Prepared for the ADB, MONRE, MOITT and EVN. Hanoi, Vietnam: ICEM.

International Energy Agency (IEA) (2011). *World Energy Outlook*. Paris.

International Hydropower Association (2011). *Hydropower Sustainablity Assessment Protocol*. London: International Hydropower Assocation.

International Rivers (2007). Failed Mechanism: How the CDM is Subsidizing Hydro Developers and Harming the Kyoto Protocol.

Intergovernmental Panel on Climate Change (IPCC) (2007). *Climate Change 2007: Mitigation Contribution of Working Group III to the Fourth Assessment Report of the Intergovernmental Panel on Climate Change*. In Metz, B., Davidson, O., Bosch, P., Dave, R. and Meyer, L. (eds.). Cambridge University Press.

Jacobs, A. (2013). Plans to Harness Chinese River's Power Threaten a Region. *The New York Times*, 5 May 2013, A6.

Kiesecker, J. M., Copeland, H., Pocewicz, A. and McKenney, B. (2010). Development by Design: Blending Landscape-Level Planning with the Mitigation Hierarchy. *Frontiers in Ecology and the Environment* 8: 261–6.

Krchnak, K., Richter, B. and Thomas, G. (2009). Integrating environmental flows into hydropower dam planning, design and operations. *Water Working Notes*. Washington, DC: The World Bank.

Kumar, E. A. (2011). Hydropower. In *IPCC Special Report on Renewable Energy Sources and Climate Change Mitigation*, eds. Morejon, C. B. and Krug, T. Cambridge University Press.

Kundzewicz, Z. W., Mata, L. J., Arnell, N. W., et al. (2008). The Implications of Projected Climate Change for Freshwater Resources and Their Management. *Hydrological Sciences Journal-Journal Des Sciences Hydrologiques* 53: 3–10.

Ledec, G. and Quintero, J. D. (2003). Good Dams and Bad Dams: Environmental Criteria for Site Selection of Hydroelectric Projects. *Latin America and Caribbean Regional Sustainable Development Working Paper 16*. The World Bank.

Lee, S. Y., Hamlet, A. F., Fitzgerald, C. J. and Burges, S. J. (2009). Optimized Flood Control in the Columbia River Basin for a Global Warming Scenario. *Journal of Water Resources Planning and Management-Asce* 135: 440–50.

Ligon, F. K., Dietrich, W. E. and Trush, W. J. (1995). Downstream Ecological Effects of Dams. *BioScience* 45: 183–91.

Madani, K. and Lund, J. R. (2009). Modeling California's High-Elevation Hydropower Systems in Energy Units. *Water Resources Research* 45.

March, J. G., Benstead, J. P., Pringle, C. M. and Scatena, F. N. (2003). Damming Tropical Island Streams: Problems, Solutions, and Alternatives. *Bioscience* 53: 1069–78.

Matthews, J. H., Wickel, B. A. J. and Freeman, S. (2011). Converging Currents in Climate-Relevant Conservation: Water, Infrastructure, and Institutions. *PLOS Biology* 9.

Mekonnen, M. M. and Hoekstra, A. Y. (2012). The Blue Water Footprint of Electricity from Hydropower. *Hydrology and Earth System Sciences* 16: 179–87.

Milly, P. C. D., Betancourt, J., Falkenmark, M., et al. (2008). Stationarity is Dead: Whither Water Management? *Science* 319: 573–4.

Montero, F. P. and Perez, J. J. (2009). Pumped Storage to Support Wind. *Hydro Review Worldwide* 17(3): 12–23.

Nilsson, C., Reidy, C. A., Dynesius, M. and Revenga, C. (2005). Fragmentation and Flow Regulation of the World's Large River Systems. *Science* 308: 405–8.

Noonan, M. J., Grant, J. W. A. and Jackson, C. D. (2012). A Quantitative Assessment of Fish Passage Efficiency. *Fish and Fisheries* 13: 450–64.

Opperman, J. J., Galloway, G. E. and Duvail, S. (2013). The Multiple Benefits of River-Floodplain Connectivity for People and Biodiversity. In *Encyclopedia of Biodiversity, Second Edition, Volume 7*, ed. Levin, S. Waltham, MA: Academic Press.

Opperman, J. J., Galloway, G. E., Fargione, J., et al. (2009). Sustainable Floodplains through Large-Scale Reconnection to Rivers. *Science* 326: 1487–8.

Opperman, J. J., Royte, J., Banks, J., Rose Day, L. and Apse, C. D. (2011a). The Penobscot River (Maine, USA): A Basin-Scale Approach to Balancing Power Generation and Ecosystem Restoration. *Ecology and Society*.

Opperman, J. J., Warner, A., Girvetz, E. H., Harrison, D. and Fry, T. (2011b). Integrated Reservoir-Floodplain Management as an Ecosystem-Based Adaptation Strategy to Climate Change. *Proceedings to American Water Resources Association 2011 Spring Specialty Conference on Climate Change and Water Resources*. Baltimore, MD: Amercan Water Resources Assocation.

Palmer, M. A., Liermann, C. A. R., Nilsson, C., et al. (2008). Climate Change and the World's River Basins: Anticipating Management Options. *Frontiers in Ecology and the Environment* 6: 81–9.

Payne, J. T., Wood, A. W., Hamlet, A. F., Palmer, R. N. and Lettenmaier, D. P. (2004). Mitigating the Effects of Climate Change on the Water Resources of the Columbia River Basin. *Climatic Change* 62: 233–56.

Pittock, J. (2010). Better Management of Hydropower in an Era of Climate Change. *Water Alternatives* 3: 444–52.

Pittock, J. and Hartmann, J. (2011). Taking a Second Look: Climate Change, Periodic Relicensing and Improved Management of Dams. *Marine and Freshwater Research* 62: 312–20.

Poff, N. L., Allan, J. D., Bain, M. B., et al. (1997). The Natural Flow Regime. *BioScience* 47: 769–84.

Postel, S. and Richter, B. (2003). *Rivers for Life: Managing Water for People and Nature*. Washington, DC: Island Press.

Richter, B. D., Postel, S., Scudder, T., et al. (2010). Lost in Development's Shadow: The Downstream Human Consequences of Dams. *Water Alternatives* 3: 14–42.

Rydgren, B., Graham, P., Basson, M. and Wisaeus, D. (2007). *Addressing Climate Change-Driven Increased Hydrological Variability in Environmental Assessments for Hydropower Projects – a Scoping Study*. Stockholm, Sweden: Vattenfall Power Consultant AB.

SADC-WD and Authority, Z. R. (2008). Integrated Water Resources Management Strategy and Implementation Plan for the Zambezi River Basin.

Sadler, B., Verocai, I. and Vanclay, F. (2000). Environmental and Social Impact Assessments for Large Dams. Cape Town, South Africa: The World Commission on Dams.

Saunders, S., Findlay, D., Easley, T. and Spencer, T. (2012). Doubled Trouble: More Midwestern Extreme Storms. Louisville, CO and New York: The Rocky Mountain Climate Organization and the Natural Resources Defense Council.

Scudder, T. (2005). *The Future of Large Dams: Dealing with Social, Environmental, Instituional and Political Costs*. London: Earthscan.

Sherman, B., Todd, C. R., Koehn, J. D. and Ryan, T. (2007). Modelling the Impact and Potential Mitigation of Cold Water Pollution on Murray Cod Populations Downstream of Hume Dam, Australia. *River Research and Applications* 23: 377–89.

Sullivan, M. and Giffen, R. A. (1985). The Maine Rivers Policy. In *Proceedings of the Symposium on Small Hydropower and Fisheries, 1985*, ed. Olson, F. W., White, R. G. and Hamre, R. H., American Fisheries Society, pp. 315–20.

Tharme, R. E. (2003). A Global Perspective on Environmental Flow Assessment: Emerging Trends in the Development and Application of Environmental Flow Methodologies for Rivers. *River Research and Applications* 19: 397–441.

The Nature Conservancy (2012). Increasing Social, Economic, and Environmental Benefits through Integrated Reservoir and Floodplain Management: Improving Return on Investment in Water Infrastructure.

Turner, W. R., Brandon, K., Brooks, T. M., et al. (2007). Global Conservation of Biodiversity and Ecosystem Services. *Bioscience* 57: 868–73.

van Vliet, M. T. H., Yearsley, J. R., Ludwig, F., et al. (2012). Vulnerability of US and European Electricity Supply to Climate Change. *Nature Climate Change* 2: 676–81.

Viers, J. H. (2011). Hydropower Relicensing and Climate Change. *Journal of the American Water Resources Association* 47: 655–61.

Wilson, R. and Biewald, B. (2013). Best Practices in Electric Utility Integrated Resource Planning. Synapse Energy Economics, Inc.

World Commission on Dams (2000). *Dams and Development: A New Framework for Decision-Making*. London: Earthscan Publications.

WWF-International (2006). Free-Flowing Rivers: Economic Luxury or Ecological Necessity?.

Yates, D., Galbraith, H., Purkey, D., et al. (2008). Climate Warming, Water Storage, and Chinook salmon in California's Sacramento Valley. *Climatic Change* 91: 335–50.

Ziv, G., Baran, E., Nam, S., Rodriguez-Iturbe, I. and Levin, S. A. (2012). Trading-Off Fish Biodiversity, Food Security, and Hydropower in the Mekong River Basin. *Proceedings of the National Academy of Sciences of the United States of America* 109: 5609–14.

7

Water and biofuels

ANNA DALLA MARTA, FRANCESCA ORLANDO, MARCO MANCINI AND SIMONE ORLANDINI

Introduction

The driving factor of the climate problem is the continuous increase of greenhouse gas (GHG) emissions from productive activities. Such emissions drive radiative forcing of climate and affect the radiation balance of the Earth (IPCC 2007) with global warming as a consequent effect. At a global level, GHG emissions originate from activities related to energy supply (electricity and heat generation) (41 per cent), industry (20 per cent), transport (22 per cent), residential and commercial sector (6 per cent) and 10 per cent other (forestry, agriculture, fishing, waste, etc.) (IEA 2012). The most problematic GHG, particularly due to released quantities, is carbon dioxide originated by fossil fuels combustion (60 per cent), which is mainly due to thermal and electric energy production (46 per cent) and transport (road, aviation, maritime, etc.; 23 per cent). Thus, fossil fuel is both the most widespread energy source and one of the biggest problems of modern society, due to its environmental impact and due to growing concern about the security of supply. All of these factors, together with a vision of development accompanied by energy independence, have led many countries (ie Europe, the United States, Brazil, India, China) to consider the production of energy from renewable sources, such as biomass, as an alternative to fossil fuels.

The term *biomass* encompasses a range of materials of heterogeneous nature with an organic matrix, able to be renewed and converted into energy. Biomass is the raw material for the production of biofuels for the transport sector, as well as of electricity and heat. In general, biomass can be considered a reservoir of solar energy captured and forfeited through the photosynthetic and metabolic processes of living organisms. Although also true for fossil fuels (coal, oil), which embed solar energy, they are relics of the geological past and cannot be renewed. In this context, energy crops should play an important role in the short-to-medium term in the replacement of fossil fuels, with global expansion of cultivated areas.

Biofuels can be divided into three main categories according to their physical state:

- Solid: a matrix consisting mainly of wood products intended for combustion. They are used mainly for the production of heat and electricity and have the potential to be converted into biofuel for transport with the development of second-generation technology.

- Liquid: products derived mostly from dedicated crops through extraction (pure vegetable oil) and esterification (biodiesel) of vegetable oils or fermentation and distillation of sugar materials (bioethanol).
- Gaseous: products derived from processes of anaerobic fermentation of wet biomass organic components such as manure, civil or agro-industrial bacterial colonies in the absence of oxygen to produce gas (mainly methane).

Among these, liquid biofuels are considered the most interesting due to a number of factors. First, they are ideal for powering internal combustion engines (diesel and Otto cycle), so their use has been traditionally connected to the transport sector, which is one of the main sources of CO_2 emissions. Secondly, liquid biofuels can be obtained by widespread and diversified types of dedicated crops (sugar cane, sugar beet, oilseed rape, sunflower, sweet sorghum, maize, etc.), which can be grown under a variety of environmental conditions.

On the basis of these characteristics and of the often-assumed, but questionable, assumption of a carbon-neutral balance (which often does not consider the energy spent for in-field agricultural processes), the use of biofuels from energy crops is considered a mitigation strategy for climate change, and national governments have translated this concept into practice by creating specific financial subsidies for supporting the production of bioenergy. The most common support measures are blending mandates, financial incentives and tax breaks.

However, there is vigorous debate in the scientific community about the real environmental benefit related to the production and large-scale replacement of fossil fuels with biofuels. One of the most debated problems concerns the energy balance of the biofuel production, particularly if based on annual crops. To reach a yield level capable of making the transformation into biofuels economically worthwhile, the cultivation phase requires high energy inputs such as nitrogen fertilizers and other chemical products, and involves an intensive-use fossil-fuel-based activity, such as pumping of irrigation water and mechanic operations (harrowing, ploughing, harvesting, etc.). As a result, the energy invested could be greater than the energy returned, thus compromising the sustainability of the entire production process (Dalla Marta et al. 2011; Spies and Dandy 2012).

Although many sustainability studies have been developed so far, they have mainly focused on the impact on air quality, land-use change, competition with food production, effect on food prices and biofuels efficiency. On the other hand, the same attention has not been paid to the complex relations between biofuel production and pressure on the water resource. Agriculture represents 70 per cent of global water use (FAO n.d.), and large-scale cultivation of biomass for fossil fuel substitution could modify the future water demand of agriculture (Berndes 2002). Plants use water for photosynthesis processes, and mainly through the transpiration system, both processes are directly related to biomass productivity. In addition to the direct water requirements of the crop, the type of canopy cover influences the rate of water loss due to evaporation from soil to atmosphere also. Evapotranspiration incorporates the sum of the water losses due to these two main water-using processes. The water used by crops to grow and produce has two origins; one is naturally present in the environment as rainfall stored into the soil. If rainfall is not sufficient to satisfy the crop needs, additional water can be supplied through irrigation. Of course, the amount of irrigation water needed for cultivation depends on the type of crop and on the environment. The same crop could be rain fed or irrigated depending upon the location, and it can have a very different impact on the water resource. Water withdrawn from surface or groundwater for irrigation may also be the main source of fresh water for human activities, including food production and drinking. One measure of quantifying the amount of water needed for crop cultivation is the concept of water footprint (WF), which is defined as the quantity of water used to produce a good. The WF of an agricultural product is the volume of water used during the crop growing period (Hoekstra et al. 2011) and can be classified as either green water, which is rain or soil humidity evapotranspirated by the crop (effective rainfall), or blue water, which is the amount supplied through irrigation (effective irrigation).

Beside the problem of water scarcity, which varies from country to country, another major problem is water pollution from agricultural production. Pollution is a particular issue in intensive productive systems where the maximisation of yield is paramount, fertilizers and pesticides are massively used and the pollutants (particularly nitrogen and phosphorous) can be easily leached through runoff which can recharge and accumulate in aquifers. Water pollution is a major concern, not only in relation to the safety of water quality for drinking purposes, but also for the aquatic ecosystem due to eutrophication and eco-toxicity. Similarly to the green and blue water, this issue can be defined through the concept of grey water, that is the volume of water required to dilute pollutants and to restore its quality standards (Hoekstra et al. 2011).

Policies to support supply chain development of biofuels

EU policy

The programs adopted by European Union (EU) for the renewable energy sector play a key role in the development of the biofuels' production chains and related markets.

The development of the EU strategy for biofuels is described in the documents: 'Biomass Action Plan' (COM 2005/0628 – SEC 2005/1573), 'An EU Strategy for Biofuels' (COM 2006/34 – SEC2006/142), 'Directive on the promotion of the use of biofuels or other renewable fuels for transport' (Directive 2003/30/EC, 08/05/2003) and 'Directive on the promotion of the use of energy from renewable sources' (Directive 2009/28/EC, 23/04/2009).

The first Biomass Action Plan provided an assessment of the potential biomass production for energy use in the EU, under the constraints not to increase the pressure on farmland and forest biodiversity and the impact on soil and water resources. The analysis compared the energy production observed in 2003, with the potential production estimated for the three decades after, for three main biomass sources (Figure 7.1).

The results showed that the forestry sector appears to have already reached its limit of growth, while in 2003, the sectors of waste and agriculture showed a great potential still

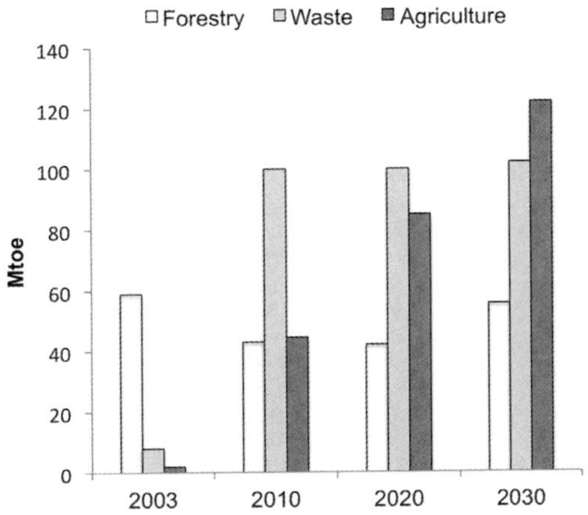

Figure 7.1 Energy production from biomass of three main productive sectors: forestry, waste and agriculture. The values in 2003 are those observed, while the values of the following years represent an assessment of potentiality for each biomass source (Elaboration of data shown in 'Biomass Action Plan' COM 2005/0628).

untapped. For this reason, the EU strategy has mainly targeted the promotion of liquid and gaseous biofuels from waste or energy crops, with the aim to stimulate both demand and supply and, at the same time, to ensure environmental benefits and the compliance with sustainability criteria. Since it has been estimated that the production of energy from waste could be developed more quickly than biomass, this latter was particularly promoted by EU policy, especially as alternative to the fossil fuel for the transport sector.

With the directive 2003/30/EC, the EU required each member state to ensure a minimum share of biofuels of 5.75 per cent within 2010 to replace fossil fuels in the transport sector. The general aim was to increase the share of renewable energy to more than 20 per cent of the European energy consumption by 2020. Biofuels for transport belong to four categories:

- Bioethanol and ETBE (Ethyl Tertiary Butyl Ether): originated by the fermentation of sugar/starch crops
- Biodiesel: produced from biomass or used frying oils
- Bio-oil: produced from biomass by pyrolysis
- Biogas and biomethanol: originated by the bacterial fermentation of organic matter

The directive 2009/28/EC repealed the directive 2003/30/EC, establishing a new target for the member states: increase the share of energy from renewable sources in the transport sector at least to 10 per cent by 2020. However, the main added value of this document is the constraint of environmental sustainability for biofuels production. Although biofuels production based on raw materials coming from outside the community is allowed, it is forbidden to use biomass from land with high biodiversity or with high-carbon stock. Further sustainability criteria are established in relationship to the greenhouse gas savings and agro-environmental practices. In particular, the agriculture activity is subjected to environmental requirements concerning the protection of soil, water and air. In regard to water resources, the EU highlights the importance of the conservation of groundwater and surface water quality and the avoidance of excessive water consumption in areas where water is scarce.

US policy

The development of biofuels policy in the United States is contained in a series of documents: the 'Energy Policy Act' of 1992, 2005 and 2011, the 'Energy Independence and Security Act of 2007', and the 'Food, Conservation, and Energy Act of 2008'. Since 1992, the United States developed a range of programs aimed at reducing the dependence on foreign oil through local biofuels production. As in the European Union, US action plans mainly focused on transport sector.

The Energy Policy Act of 1992 was directed to support the biofuels through the set up of facilities and incentives for the use and development of alternative fuel vehicles. In particular, programs were planned to accelerate the introduction of biofuels for the federal fleet, heavy-duty vehicles and vehicles for mass transit. The Energy Policy Act of 2005 required the blending of renewable fuels such as ethanol and biodiesel in transportation fuels. The government aimed to make biofuels competitive with fossil fuels, through technological improvements of the industrial stage of processing and through development of alternative engines rather than specific measures for feedstock production. Key goals included the improvement of hybrid and flexible vehicles and the production of bioethanol from cellulosic biomass through a second-generation process.

With the 'Energy Independence and Security Act' of 2007, the US goal was for renewable resources from agriculture and forestry to provide at least 25 per cent of national energy needs. The Act mandated an annual production of 56.8 billion litres of ethanol from corn by 2015 and an additional 60.6 billion litres from cellulosic crops by 2022, a total that represents 15 per cent of the petroleum-based fuels used in the United States in 2006.

Coinciding with this increase of production, concerns were raised about the socio-economic and environmental sustainability of biofuels. Concerns about the level of GHG emissions and energy consumption during production and combustion of biofuels, and the potential environmental impacts of the related agricultural activity have prompted policy makers to question the logic of production of biofuels with lifecycle emissions that exceed those of conventional fuels. Government policy also attempts to protect the market from adverse economic impacts caused by the conversion of agricultural production, from foodstuffs to energy products.

In this context, the 'Food, Conservation, and Energy Act' of 2008 highlighted the need to assess how ethanol production may be impacting farm economy, environment and food prices.

Notwithstanding concerns about food production, the 'Energy Act' of 2011 aims to ensure at least 50 per cent of cars and light trucks manufactured for sale in the United States have dual power systems, and CO_2 emissions are reduced by 25 per cent through the production and use of renewable fuels.

Impact of the policy guidelines

It is evident that US and EU economies are able to affect the direction of the global markets and the local productive sectors. On the basis of the policy guidelines for renewable energy in these countries, it is easy to understand how agriculture is playing an increasingly important role in the non-food sector.

As a result of energy policies, global biofuel production has risen more than 80 per cent in the last decade, from about 18 billion litres in 2000 to 110 billion litres in 2011 (US EIA n.d.), with 87 and 23 billion litres of ethanol and biodiesel respectively produced worldwide. The United States is the main producer, followed by Brazil, Europe, China and India (Figure 7.2). The United States is also the main consumer (52 billion litres), followed by Brazil (22 billion litres) and Europe (20 billion litres). Based on these data, while the United States is able to export some biofuels, Europe is not yet self-sufficient (US Energy Information Administration).

> Today, biofuels provide around 3% of total road transport fuel globally (on an energy basis) and considerably higher shares are achieved in certain countries. Brazil, for instance, met about 21% of its road transport fuel demand in 2008 with biofuels. In the United States, the share was 4% of road transport fuel and in the European Union (EU) around 3% in 2008. (International Energy Agency)

Government policies tend to promote all categories of biofuels, but bioethanol have been preferentially promoted in the United States. Research into second-generation biofuel technology, where bioethanol is derived from cellulosic biomass, is underway. This option would represent a huge improvement because it makes use of the whole plant, and thus leads to the optimisation of the agronomic production thanks to a higher harvest index (fraction of the total dry matter allocated in the harvested parts of the crop). Moreover, the ethanol production from cellulosic crops avoids the critical issue of competition with food production that affects energy crops, such as oil-seed (eg sunflower), sugar (eg sugar cane)

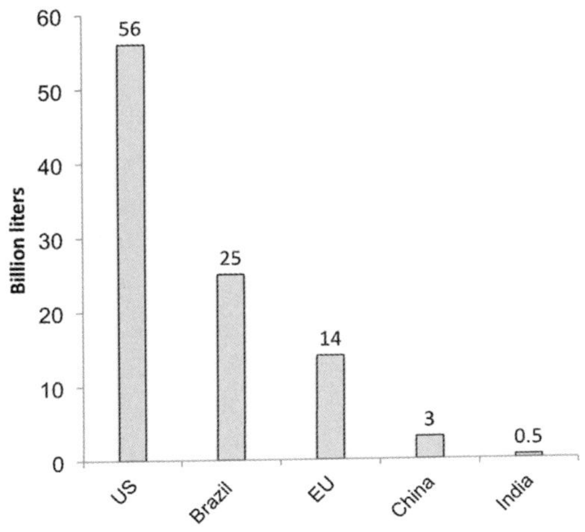

Figure 7.2 Global biofuel production in 2011. (Based on data from the US Energy Information Administration.)

and amylaceous species such as maize. Nevertheless, despite the research efforts made in recent years to improve the energy conversion process of ligno-cellulosic biomasses into biofuels, the technology has not yet been demonstrated to be economically feasible.

Biofuels development and impact on water resources

In most regions of Europe, as well as in the United States, India, Africa and other major energy crop producers, agricultural production systems are vulnerable to a potential decrease in water availability. The impacts of climate change (increasing temperatures, shifts of seasonal precipitations and decreasing summer rainfall) could cause water availability to reach critical limits in many areas. As the major user of water, agriculture influences the regional water resources, which are available for all key sectors. Moreover, the increasing world population – and, consequently, the increasing need for food, feed and bioenergy – will require more efficient agricultural production. Therefore, irrigation will probably become even more important. However, water scarcity due to climate change is likely to impose limits on irrigation expansion in many countries. Many attempts have been made to improve the effective use of water in agricultural crop production, but results are often only locally valid and cannot easily be extrapolated to other regions. Also, factors influencing water-use efficiency in crop production are manifold and vary with crops, crop management and the environment. The assessment of impacts of biofuel production on water resources plays a key role in its sustainability, since energy production from biomass consumes large quantities of water compared to fossil fuels (see Table 7.1; Gerbens-Leenes et al. 2008; Dominiguez-Faus et al. 2009).

Thus, water used for irrigation and the agrochemicals applied in cropping systems could adversely impact both the availability and quality of water.

The rapid expansion of biofuel production during the last decade has increased the area of arable land dedicated to biomass cultivation; however, to minimise competition with food and traditional crops, the energy crops have expanded into new land in marginal or not-yet-cultivated areas. The expansion of cultivated land increases the anthropogenic pressure on natural resource, especially in marginal lands where the climatic conditions and the poor soil fertility often require intense agronomic input to obtain satisfactory and economically sustainable harvests. Therefore, the development of biofuels chain is leading to a larger water demand, as well as to increased water pollution from fertilizers and pesticides.

Table 7.1 Water requirements per energy production by different feedstock (UNESCO-IHE 2008).

Process	Water requirement per energy production (m^3/GJ)
Oil refining	0.045
Corn ethanol irrigation	9.1–200
Soybean biodiesel irrigation	61.1–138

Although the refining, processing and transport of the product involve water costs, the feedstock production represents the most water-intensive stage of the biofuel chain and deserves special attention. The feedstock cultivation dominates the WF, and the other chain stages are relatively unimportant. Therefore, the water sustainability of the entire chain is mainly conditioned by the impact of the agricultural stage.

In many countries, water used for biomass production for food or fibre already competes with other consumptive users, such as urban supply and industrial activities (Gerbens-Leenes et al. 2009a). The crucial problem is that most energy crops require large quantities of water, and species considered drought-tolerant take advantage of available water to increase their productivity. Consequently, irrigation is required at almost every crop location (Gerbens-Leenes et al. 2009a).

Global evaluation of this issue highlights the strong nexus between water management and the policy targets for biofuels. The scenario for road transport projected for 2030 by IEA Alternative Policy Scenario involves both an increase of biofuels use and changes in global water use (Gerbens-Leenes et al. 2012). On this basis, the blue water footprint of biofuel is expected to rise from 0.5 per cent in 2005 to more than 5 per cent of the available blue water in 2030 (Gerbens-Leenes et al. 2012). Therefore, the expansion of the global biofuels trade has the potential to contribute to water scarcity.

Considering the situation in individual countries, it is clear that national programs in the renewable energy sector will increase the pressure on freshwater resources. Analysing the US situation, bioethanol production has increased water consumption in many states, and further expansion of dedicated corn cultivation will significantly impact the water sustainability (Chiu et al. 2009). Assuming the full implementation of the 'Energy Independence and Security Act' of 2007, the increased production of biofuel from soybean and corn could lead to an increase of 26 per cent of water withdrawal for crop irrigation (McDonald et al. 2012). In particular, the goal to reach a production of about 57 billion litres of ethanol from corn by 2015 involves an increased share of maize production of about 44 per cent compared to 2007 (Dominguez-Faus et al. 2009). The irrigation requirement to support this agricultural activity is assessed to be around 6 billion cubic metres per year. This amount of water represents about 3 per cent of the total irrigation water used in 2000 by the United States (Dominguez-Faus et al. 2009a).

Similar concerns have been raised on analysis of the efficiency of biofuels based on calculations of water intensity, defined as 'gallons of water per mile travelled'. For light duty vehicles (LDV), fossil fuels are less water intensive than fuel derived from biomass, particularly if based on irrigated crops (King and Webber 2008). In the United States, water intensities are calculated as 28 and 36 gallons of water per mile for corn ethanol (E85) for consumption and withdrawal, respectively. For soy-derived biodiesel, the average consumption and withdrawal rates are 8 and 10 gallons of water per mile. Water intensities are much lower for fossil fuels such as gasoline and diesel ($<$ 1 gal H_2O/mile) (King and Webber 2008).

The scenarios are not better in the European Union, especially in Mediterranean countries, which are typically affected by problems of water scarcity. The case of Spain provides an example of how the water footprint of the feedstock production is a constraint in reaching the national target for biofuels (Galan-del-Castillo and Velazquez 2010). The achievement of the goal for biofuels share in the transport sector would result in a notable increased pressure on national water resources, involving an increase of about 50 per cent of national WF (Galan-del-Castillo and Velazquez 2010). Similarly, in Italy, the production of 1 litre of bioethanol from irrigated maize would require more than 2,000 litres of water, 30 per cent of which supplied by irrigation. Furthermore, the situation is worsening due to the impact of climate variability that is causing a continuous increase of irrigation demand in Mediterranean environments due to a diminution of rainfall during the growing season over the last fifty years (Dalla Marta et al. 2010, 2012).

Similar to the European Union and the United States, India has also developed a strong biofuel policy with the aim of reducing the dependence from foreign sources, reducing GHG emissions from road transport, and creating new job opportunities and socio-economic improvement. However, this program has implications for water management, as the increase of biofuel production from agricultural biomass will lead to increased water withdrawals under a range of scenarios (Islam 2012). Sugar cane and jatropha are the main energy crops cultivated in India but, as sugar cane is an irrigated crop, additional cultivated land will require more irrigation. Although jatropha is a drought-tolerant crop, its yield could be increased considerably under irrigation (Islam 2012).

Similarly, in China, significant impacts on water resources are predicted as a consequence of increasing national biofuels production (De Fraiture et al. 2008). In contrast, expansion of sugar cane for biofuels using irrigation is not considered to be a problem for Brazil, because of its high rainfall and limited use of irrigation (Goldemberg et al. 2008). However, water pollution could be a major concern, as the deterioration of water quality would seriously reduce the availability of this resource for both agriculture and civil purposes.

Groundwater degradation and water eutrophication are problems often associated with energy crops. In particular, annual crops typically used as energy feedstock are grown under high-input cropping systems and are prone to discharge nitrogen from the field to surface water through run-off, sediment transport, tile drainage and subsurface flow (Dominguez-Faus et al. 2009). However, each crop has different needs for agrochemical input. In this context, corn and sorghum have higher nitrogen and pesticide requirements for producing ethanol compared to switchgrass, soybean and sugar cane (Dominguez-Faus et al. 2009).

Energy crops and water use

Before addressing this issue, it is necessary to introduce the concepts of water need and water-use efficiency of a crop. Crop water need is the amount of water used by a crop to meet the water loss through evapotranspiration and corresponds to the water needed for optimum growth. This mainly depends on the climate, crop type and its growing stage. On average, maize, which is the main energy crop used in the United States, requires between 500 and 800 mm of water during its growing season; sugar cane, mainly grown in Brazil, requires between 1,500 and 2,500 mm and soybean between 450 and 700 mm (Brouwer and Heibloem 1986). Another concept related to the productivity of water used by a crop is water-use efficiency (WUE), which corresponds to the amount of biomass or marketable yield produced per unit of water used in evapotranspiration. WUE is also affected by the environmental conditions, mainly climate and soil and crop type, but crop management and agro-techniques applied have an important role, especially related to water and nutrient supply. In the Mediterranean region, WUE from 0.22 to 2.16 kilograms per cubic metre has been reported for maize, from 0.39 to 0.77 kilograms per cubic metre for soybean and from 3.69 to 4.20 kilograms per cubic metre for sweet sorghum (Katerji et al. 2008). The WUE concept could be easily transferred to the biofuel itself, where the WUE corresponds to the amount of biofuel obtained per unit of water used. WUE in this case depends also on the efficiency of the conversion technology. In fact, under current technologies for energy conversion, ethanol derived from mischantus and maize have similar WUE, being about 1.2 and 1.1 litres of ethanol per cubic metre of water, respectively. Assuming an improvement in biomass-to-biofuel conversion efficiencies, for each cubic metre of water used, miscanthus could reach a potential biofuel production of 1.8 litres of ethanol per cubic metre (Hickman et al. 2010; Zhuang et al. 2013).

These studies suggest that biofuels production from ligno-cellulosic biomass have a lower water impact compared to the other feedstock. However, the benefits in water saving associated with cellulosic biomass compared to other feedstocks are not always achieved, as the WUE of these crops are influenced by many variables. The water need per unit of produced energy (cubic metres per gigajoule), computed on the basis of the ratio between water requirement (cubic metres per hectare) and biomass yield (tonne/hectare), shows wide variability among the countries, with low values of 9 cubic metres per gigajoule for maize and 20 cubic metres per gigajoule for miscanthus grown in the Netherlands and, increasing to 200 cubic metres per gigajoule for maize and 64 cubic metres per gigajoule for miscanthus grown in Zimbabwe (Gerbens-Leenes et al. 2009b).

However, despite bioethanol having a high impact on water demand when the production is based on sugar-starch crops, it still provides best prospects of development in the renewable energy sector, assuming it can be produced through a second generation process.

Oil-seed crops involve a higher water footprint compared to the sugar-starch crops in both electricity and biofuels production (see Figures 7.3 and 7.4). Soybean, rapeseed and jathropa have a total water footprint of 394, 409 and 574 cubic metres of water per gigajoule, respectively, for producing biodiesel. Among the sugar-starch species, only sorghum has such poor performance with a total water footprint of 419 cubic metres per gigajoule of produced bioethanol. The biomass for the bioethanol chain has a total water footprint far below 400 cubic metres per gigajoule, with half the impact on water resources, with values around 200 cubic metres per gigajoule or below. The situation is similar when biofuels are used in the production of electricity, indicating that the low performance of oil-seed species does not only depend on a different efficiency in the energy conversion process. In fact, the high water footprint is mainly due to the low energy content of raw material, which is low in proportion to the high crop water demand. Considering the ranges of irrigation requirements, an amount of 400 litres (cassava) to more than 4,000 litres (sorghum) is needed for producing 1 litre of bioethanol, while 7,000 to 8,000 litres are required for 1 litre of biodiesel (soybean and rapeseed) (Gerbens-Leenes et al. 2008).

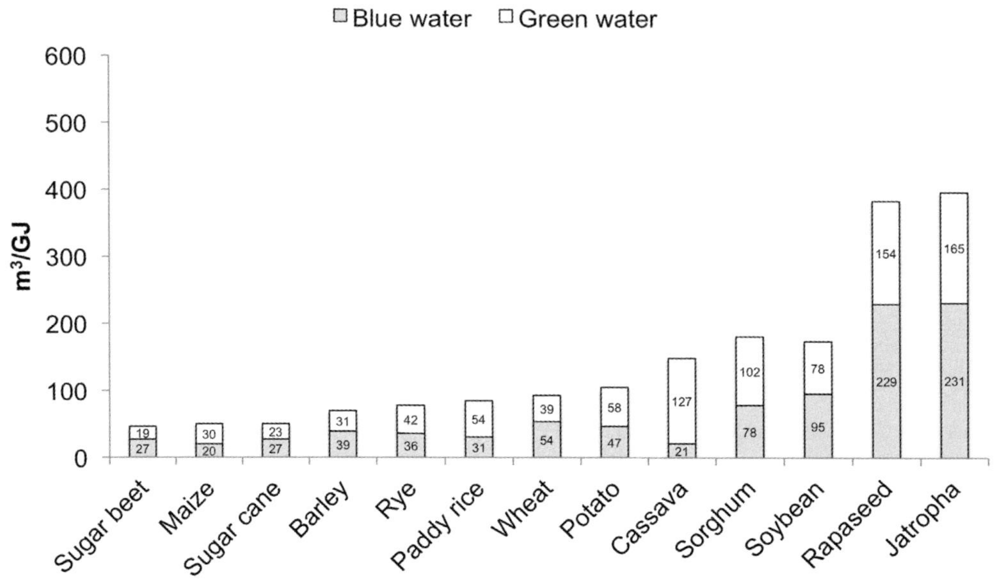

Figure 7.3 Total weighted-global average water footprint (green and blue water footprint) of thirteen energy crops, in m^3 of water per GJ of provided electricity, through the burning of the total biomass yield (Gerbens-Leenes et al. 2009a).

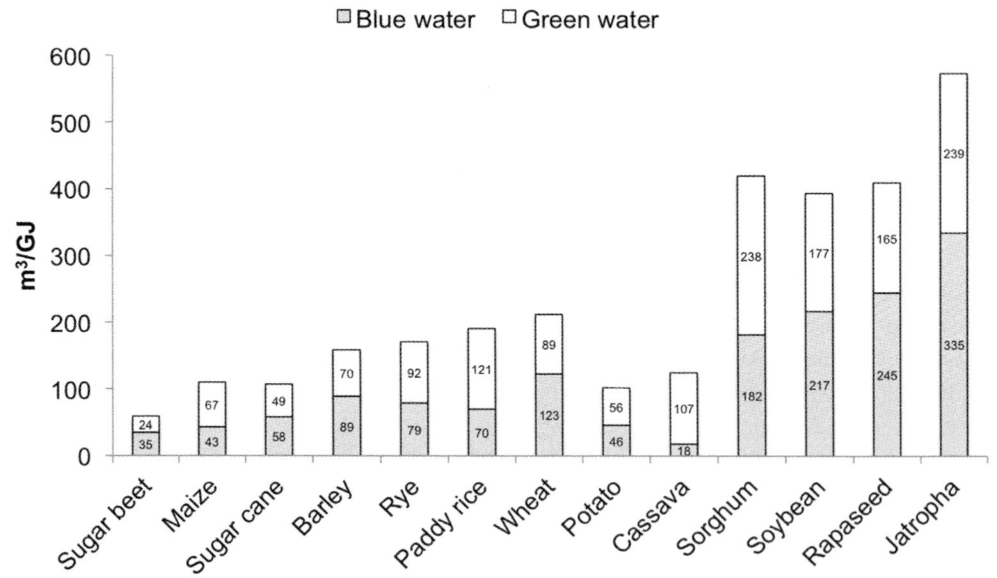

Figure 7.4 Total weighted-global average water footprint (Green and Blue water footprint) in m^3 of water per GJ of provided biofuel. Ten crops supply biomass for bioethanol chain and the last three crops (soybean, rapeseed and jathropa) supply biomass for biodiesel production (Gerbens-Leenes et al. 2009a).

The efficiency of the energy conversion process significantly affects the water footprint, as shown by comparing the combustion of the whole biomass to produce electricity directly with the same feedstock used to produce energy through biofuels.

The conversion of the energy content of biomass to biofuels improves the utilisation and efficiency of this renewable energy. However this conversion results in a loss of the initial energy content, which leads to a higher water footprint per unit of produced energy. The water footprint is an important consideration for first-generation biofuels, highlighting once again the importance to develop second-generation technologies to convert all the ligno-hemicellulose biomass of the plant into biofuel.

Climate

As mentioned in the previous sections, climate together with the agricultural system plays a key role in determining the amount of water necessary for producing the same quantity of biofuel. For example, 421 litres of irrigated water is required to produce 1 litre of bioethanol from maize in the United States, based on a total biofuel production of 12,907 million litres and irrigation withdrawals of 5.44 cubic kilometres (De Fraiture et al. 2008). The water requirement is six times higher in China, where 1 litre of bioethanol from maize in China requires 2,584 litres of irrigation water (total biofuel production of 3,649 million litres and an irrigation withdrawal of 9.43 km^3) (De Fraiture et al. 2008). The irrigation water needs to produce bioethanol from sugar cane varies widely from country to country (De Fraiture et al. 2008; Figure 7.5).

Within each country, the spatial variability in climate leads to different consequences in terms of water footprint of energy crops (Chiu and Wu 2012; Dalla Marta et al. 2012). In some regions of the United States, the rainfall can satisfy almost all the irrigation requirement of maize. In Iowa and Ohio, the cultivated surface with corn that require

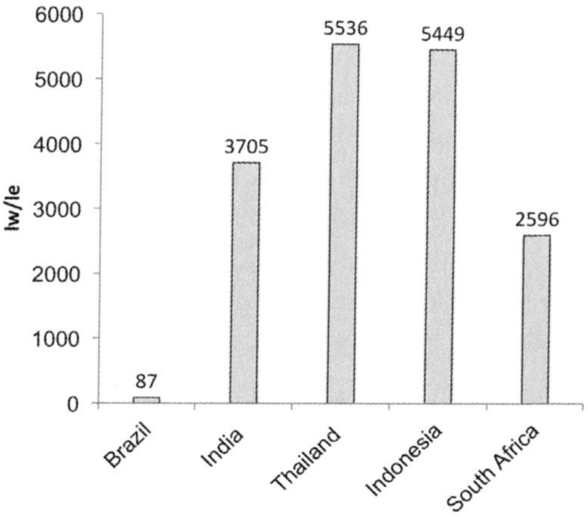

Fig. 7.5 Litres of irrigation water (lw) required to produce 1 litre of bioethanol (le) from sugar cane in different countries (computation performed on the basis of the total national biofuel production from sugar cane and the total irrigation withdrawals for that crop) (De Fraiture et al. 2008).

irrigation are 1 per cent and less than 1 per cent, respectively (Dominguez-Faus et al. 2009). Other states, such as Nebraska (where 61 per cent of maize acreage is irrigated), rely on irrigation for most if its corn production. These comparisons highlight the importance of choosing the most suitable species for each environment, since the same energy crop can result in different pressures on water reserves depending on the contribution of rainfall to plant growth.

The rainfall trend also affects the share of grey water in the water footprint. For example, in corn and soybean grown in rows, the average of nitrogen (N) discharged in surface water is between 24 and 36 per cent of the total N applied, but in years of extreme drought and flooding this percentage can reach 5 per cent and 80 per cent, respectively (Powers 2007).

The water footprint of energy biomass is the result of a complex interaction between the water and nutrient requirements that characterise each crop and the climate of the cultivation areas. Moreover, the same climate may minimise the impact on water quality but adversely impact water reserves. Management of this resource has different solutions from region to region, and the best measure must be able to minimise significantly both the blue and grey water footprint.

Conclusion

The global population is forecast to increase to 9.2 billion people by 2050 (UN 2007), and the subsequent demand for food, feed, and fibre and energy conflicts with the constraint of limited availability of water and land resource. In this context, the issue of water security is becoming more and more relevant.

Uncertainty in projections for future changes in rainfall patterns makes it difficult to predict how the increase of irrigation requirements will exacerbate the competition for water and creates local water shortages. The increasing demand for water use to grow energy crops for biofuels has major implications for sustainability. In particular, the environmental sustainability of a large-scale cultivation is closely connected with the issue of food security. The production of biomass may increase the cost of food, directly impact on food security and result in competition for the use of scarce water resources. Moreover, since irrigation usually requires energy for pumping, the increasing pressure on water resources may result in conflicts of sustainability in terms of energy.

The use of renewable energy derived from biomass represents an opportunity to protect climate because it is considered a promising strategy for the abatement of GHG emissions. In particular, the potential of biofuels to stabilise CO_2 emissions from the transport sector and reduce dependence on imported fuels, reduce price volatility and generate socio-economic development in the agricultural sector is an important step, but needs to be balanced by the need for food security. The sustainability of a large-scale substitution of fossil fuels with biofuels derived from dedicated energy crops requires an in-depth examination of the water-energy nexus in relation to the large amounts of water involved in the agricultural phase and its modification in a changing climate.

References

Brouwer, C. and Heibloem, M. (1986). *Irrigation water management Training Manual no. 3: Irrigation water needs* United Nations Food and Agriculture Organization. www.fao.org/docrep/S2022E/S2022E00.htm.

Berndes, G. (2002). Bioenergy and Water – The Implications of Large-Scale Bioenergy Production for Water Use and Supply. *Global Environmental Change* 12: 253–71.

Chiu, Y. W., Walseth, B. and Suh, S. (2009). Water Embodied in Bioethanol in the United States. *Environmental Science and Technology* 43(8): 2688–92.

Chiu, Y. W. and Wu, M. (2012). Assessing County-Level Water Footprints of Different Cellulosic-Biofuel Feedstock Pathways. *Environmental Science Technology* 46: 9155–62.

Dalla Marta, A., Mancini, M., Ferrise, R., Bindi, M. and Orlandini S. (2010). Energy Crops for Biofuel Production: Analysis of the Potential in Tuscany. *Biomass and Bioenergy* 34: 1041–52.

Dalla Marta, A., Mancini, M., Natali, F., Orlando, F. and Orlandini, S. (2012). From Water to Bioethanol: The Impact of Climate Variability on the Water Footprint. *Journal of Hydrology* 444–5: 180–6.

Dalla Marta, A., Natali, F., Mancini, M. et al. (2011). Energy and Water Use Related to the Cultivation of Energy Crops: A Case Study in the Tuscany Region. *Ecology and Society* 16(2) [online]: www.ecologyandsociety.org/vol16/iss2/art2/.

De Fraiture, C., Giordano, M. and Liao, Y. (2008). Biofuels and Implications for Agricultural Water Use: Blue Impacts of Green Energy. *Water Policy* 10(S1): 67–81.

Dominguez-Faus, R., Powers, S. E., Burken, J. G. and Alvarez, P. J. (2009). The Water Footprint of Biofuels: A Drink or Drive Issue? *Environmental Science and Technology* 43: 3005–20.

Falkenmark, M. and Chapman, T. (1989). *Comparative Hydrology: An Ecological Approach to Land and Water Resources*. Paris: UNESCO, 479.

Galan del Castillo, E. and Velazquez, E. (2010). From Energy to Water: The Virtual Water Content and Water Foot Print of Biofuel Consumption in Spain. *Energy Policy* 38: 1345–52.

Gerbens-Leenes, P. W., Hoekstra, A. Y. and van der Meer, T. H. (2008). *The Water Footprint of Bio-Energy and Other Primary Energy Carriers*. Value of Water Report Series, 29. Delft, The Netherlands: UNESCO-IHE.

Gerbens-Leenes, P. W., Hoekstra, A. Y. and van der Meer, T. H. (2009a). The Water Footprint of Bioenergy. *PNAS* 106: 10219–23.

Gerbens-Leenes, P. W., Hoekstra, A. Y. and van der Meer, T. H. (2009b). The Water Footprint of Energy from Biomass: A Quantitative Assessment and Consequences of an Increasing Share of Bio-energy in Energy Supply. *Ecological economics* 68: 1052–60.

Gerbens-Leenes, P. W., van Lienden, A. R., Hoekstra, A. Y. and van der Meer T. H. (2012). Biofuel Scenarios in a Water Perspective: The Global Blue and Green Water Footprint of Road Transport in 2030. *Global Environmental Change* 22: 764–75.

Goldemberg, J., Coelho, S. T. and Guardabassi, P. (2008). The Sustainability of Ethanol Production from Sugarcane. *Energy Policy* 36: 2086–97.

Hickman, G. C., Vanloocke, A., Dohleman, F. G. and Bernacchi, C. J. (2010). A Comparison of Canopy Evapotranspiration for Maize and Two Perennial Grasses Identified as Potential Bioenergy Crops. *GCB Bioenergy* 2: 157–68.

Hoekstra, A. Y., Chapagain, A. K., Aldaya, M. M. and Mekonnen, M. M. (2011). *The Water Footprint Assessment Manual: Setting the Global Standard*. London: Earthscan.

International Energy Agency (IEA) (2012). *CO_2 Emissions from Fuel Combustion – Highlights*. Eds. IEA Statistics, p. 9.
Intergovernmental Panel on Climate Change (IPCC) (2007). *Climate Change 2007, WG I: The Physical Science Basis*. Contribution of Working Group I to the Fourth Assessment Report of the Intergovernmental Panel on Climate Change, 2007, eds. Solomon, S., Qin, D., Manning, M., Chen, Z., Marquis, M., Averyt, K. B., Tignor, M. and Miller, H. L. Cambridge University Press.
Islam, S. (2012). Implications of Biofuel Policies for Water Management in India. *International Journal of Water Resources Development* 28: 601–13.
Katerji, N., Mastrorilli, M. and Rana, G. (2008). Water Use Efficiency of Crops Cultivated in the Mediterranean Region: Review and Analysis. *European Journal of Agronomy* 28: 493–507.
King, C. W. and Webber, M. E. (2008). Water Intensity of Transportation. *Environmental Science and Technology* 42: 7866–72.
McDonald, R. I., Olden, J. D., Opperman, J. J. et al. (2012). Energy, Water and Fish: Biodiversity Impacts of Energy-Sector Water Demand in the United States Depend on Efficiency and Policy Measures. *PLoS ONE* 7: Article n. 50219.
Powers, S. E. (2007). Nutrient Loads to Surface Water from Row Crop Production. *International Journal of Life Cycle Assessment* 12: 399–407.
Spies, B. and Dandy, G. (2012). Sustainable Water Management: Securing Australia's Future in a Green Economy. Report of a study by the Australian Academy of Technological Sciences and Engineering (ATSE).
United Nations (2007). *World Population Prospects: The 2006 Revision – Highlights*. Working Paper n. ESA/P/WP. 202, Department of Economic and Social Affairs, Population Division, New York: United Nations.
United Nations Food and Agricultural Organisation (FA0) (n.d.). AQUASTAT database, http://www.fao.org/nr/water/aquastat/water_use/index.stm.
United States Energy Information Administration (US EIA) (n.d.). International Energy Statistics. United States Energy Information Administration, www.eia.gov/.
Zhuang, Q., Qin, Z. and Chen, M. (2013). Biofuel, Land and Water: Maize, Switchgrass or Miscanthus? *Environmental Research Letters* 8: 015020, 6 pp.

8

Trade-offs and synergies between water and energy use in rural Australia

SHAHBAZ MUSHTAQ, TEK N. MARASENI AND KATHRYN REARDON-SMITH

Introduction

The Food and Agriculture Organization (FAO) estimates that, to address global issues of food security and rural poverty, the area of land under irrigated agriculture must be increased by about 34 per cent over the next few decades. However, there are major constraints on irrigation development and sustainable water resource management in many of the agricultural regions of the world. While the task of managing scarce water resources is complex, water scarcity, driven by high climatic variability as well as increasing demands from multiple water users, dictates the need for greater water use efficiency in irrigated agricultural systems.

In Australia, the rural irrigation industry contributes significantly to the national economy and has been the mainstay of increased agricultural production for over half a century. However, the sector also faces many challenges; over the past decade, these have included severe and prolonged drought, reduced water availability and on-going reform of government water policies. Climate change and increasing climate variability are likely to increase the uncertainty of water supply. As a result, rural irrigated agriculture is under considerable pressure to adopt best practices to increase efficiency in terms of water use and productivity.

Increased efficiency in terms of water use and productivity, and the security of long-term water supplies under climate change, are key foci of the Australian Government's *Water for the Future* program. A major component of this initiative is the AU$5.8 billion *Sustainable Rural Water Use and Infrastructure* program aimed at supporting the upgrading of irrigation infrastructure. However, it is likely that on-farm infrastructure investment will increase energy consumption and associated greenhouse gas (GHG) emissions, suggesting potential conflicts in terms of mitigation and adaptation policies.

This chapter reports on an integrated framework, based on carbon and water accounting and economic modelling, developed to analyse trade-offs between water savings, GHG emissions, and economic costs and benefits associated with irrigation technological change. The framework is applied to five national-scale crop-level irrigation transformation scenarios and five farm-level case studies, to enable exploration of the impact of technological change on the climate-energy-water nexus in this important sector.

BOX 8.1 Crop level water use

In 2010–11, pasture for grazing accounted for the greatest area of irrigated land (538,000 ha), representing 27 per cent of the national total (Figure 8.2). Cotton cropping used the largest volume of irrigation water (1,882 gigalitres, or GL), which was 28 per cent of the national irrigation total for the year (6,645 GL), while rice required the highest irrigation application rate of 10.1 ML/ha (Figure 8.3). However, horticultural crops (vegetable, tree fruits and grapes) represent the highest value added per volume of water used (Figure 8.4).

Climate change and water availability for agriculture

Managing water effectively is a critical challenge in many agricultural regions of the world (Tilman et al. 2002). Australia is the driest inhabited continent, and its main agricultural regions are highly vulnerable to climate change and climate variability: particularly, to risks associated with increasing water scarcity (Donohue et al. 2009). As such, Australia's experience in optimising agricultural production under water scarcity provides a valuable case study.

Agriculture is the largest consumer of water in Australia, responsible for, on average, around 70 per cent of Australia's total water consumption. Intensification of agricultural farming systems over the past three to four decades has seen a significant increase in Australia's agricultural productivity (AGO 2006), largely due to a shift from dryland to irrigated farming systems. The Australian irrigation industry is an important contributor to the Australian economy, providing approximately 50 per cent of the national value of agricultural production from less than 5 per cent of the productive land area. It is also the dominant agricultural water user (at 90 per cent of agricultural water use), with pasture production and cotton cropping (Box 8.1) accounting for nearly half (47 per cent) of all irrigation water applied in Australia in 2010–11 (ABS 2013). Given the multiple demands on scarce water resources in Australia, this industry is also a key political and operational target for reform in Australia.

Prolonged droughts in recent decades (eg 'Millennium Drought', 1997–2010), in conjunction with historical over-allocation of scarce water resources, have contributed to a widening gap between water supply and demand and increased the uncertainty of water supply in many regions in rural Australia. Growing awareness of the need to provide water for the environment to sustain ecological function and concern that climate change will result in increasing water stress in Australian catchments (Sanders et al. 2010) are also driving political agendas to move over-allocated systems back to more sustainable levels. The need to manage water resources to support both increased agricultural productivity and environmental security is a critical challenge facing Australian agriculture and places pressure on the irrigation industry to adopt best-practice methods to increase efficiency in terms of both water use and productivity.

Conventional irrigation practices (eg flood or furrow irrigation; hand shift, roll line or cannon sprinkler systems) characterised by low water-use efficiency (WUE) create the potential for significant water savings that could result in increased productivity and increased water availability for alternative uses, such as environmental flows to maintain essential ecosystem services. Modern water-efficient irrigation technologies are seen as a way to improve water security and food security through gains in crop yields and reductions in irrigation water use (Playán and Mateos 2006). The adoption of new pressurised irrigation systems (eg drip or trickle systems; lateral move or centre-pivot sprinkler systems) has been heralded as an integral way of increasing water-use efficiency and creating water savings in irrigation systems (Green et al. 1996; Zehnder et al. 2003; Lal 2004). Increased irrigation efficiency represents a robust adaptation to climate change, and investment[1] in new irrigation technologies plays a significant role in accommodating the impacts of climate variability, increasing the value of irrigated agriculture and reducing agricultural water use (Mackinnon et al. 2009; Jackson et al. 2010).

However, new irrigation technologies, such as pressurised drip and sprinkler systems, also change patterns of on-farm energy consumption and associated GHG emissions. Decisions to invest in new irrigation technology are predominantly determined by consideration of the water conservation benefits and the costs associated with implementing the technology change (Qureshi et al. 2001; Pratt Water 2004; Mackinnon et al. 2009); however, the economic consequences of increasing energy prices and potentially adverse environmental consequences are often ignored (Zillman et al. 2008; Jackson et al. 2010). This is apparent at both the farm and the national policy level (eg DSEWPaC 2011). Herein lies a potential climate–water–energy conflict, as water management policies and measures aimed at water resource use mitigation (effectively, climate change adaptation in a drying climate) can in fact drive higher GHG emissions, which drive positive feedbacks to the climate system (Barnett and O'Neill 2010).

Ideally, mitigation and adaptation policies and strategies should complement each other in order to address climate change risks (Maraseni et al. 2009a); however, there is potentially a reciprocal relationship in the water sector that has been largely ignored at the policy level (Khan et al. 2004). The expectation in the past has been that the economic and social benefits of irrigation development would exceed environmental costs. Consequently, there has been a tendency to promote the role of energy-intensive pressurised irrigation systems in achieving water savings and water conservation (Khan et al. 2004), while the impact of water management policies in driving increased energy consumption and GHG emissions has been largely ignored. However, considering the potential impact of climate change and the implementation of GHG reduction policies such as carbon pricing, the perception of net economic and social benefits from irrigation may no longer be valid.

This is no trivial matter. The Australian agricultural sector is the second largest contributor of GHG emissions, accounting for 15 per cent of national emissions (DCCEE 2012). As such, the sector is a critical determinant of Australia's capacity to reduce GHG emissions to the unconditional mitigation targets of 5 per cent and 80 per cent below 2000 levels by 2020 and 2050, respectively (Calford et al. 2010). There is an urgent need to critically evaluate energy-intensive responses to scarcity in the water sector in order to understand the interconnectivity between water use and GHG emissions (Pittock 2010).

[1] The Australian Government has established the *Water for the Future* initiative, of which a major component is the AU$5.8 billion Sustainable Rural Water Use and Infrastructure program, which subsidises technology upgrades in the irrigation industry to address the challenges of long-term water availability and climate change (DSEWPaC 2011).

Not to do so risks contradictory policy signals, which may contribute to maladaptation (Barnett and O'Neill 2010) with water management policies effectively driving increased energy consumption and GHG emissions and reinforcing climate change.

In a low-carbon water-constrained economy where GHG emissions may be penalised and water savings rewarded, failure to consider potential trade-offs between mitigation and adaptation policies may result in failure to properly account for overall impacts and to mitigate environmental degradation. Hence, analysis of trade-offs between climate change and water sector policies is critical to ensuring that the economic efficiency of agricultural production is maintained while environmental impacts are minimised.

Irrigation technology used in rural Australia

A variety of irrigation technologies are used in Australia (Table 8.1). The most commonly used in 2008–09 was drip irrigation, with 22.6 per cent of irrigated agricultural businesses reporting its use (ABS 2010). Surface irrigation methods (flood, furrow and border check) were reportedly used on 16.5 per cent of irrigating establishments and accounted for the largest proportion (44 per cent) of the total irrigated land area (ABS 2010).

Table 8.1 Number of irrigation establishments and area using irrigation technologies in Australia, 2002–03 to 2008–09 (ABS 2010).

	2002–03		2008–09		Change (2002–03 to 2008–09)			
	Est*	Area	Est*	Area	Est*		Area	
On farm technology	No.	('000 ha)	No.	('000 ha)	No.	%	('000 ha)	%
Surface	12,970	1,344	7,674	804	−5,296	−40.8	−540	−40.2
Drip or trickle (above ground)	9,632	180	10,515	217	883	9.2	37	20.6
Drip or trickle (subsurface)	1,156	23	886	26	−270	−23.4	3	13.0
Sprinkler – micro spray	6,469	80	5,915	85	−554	−8.6	5	6.3
Sprinkler – portable irrigators	6,231	123	3,243	81	−2,988	−48.0	−42	−34.1
Sprinkler – hose irrigators	8,122	289	5,515	214	−2,607	−32.1	−75	−26.0
Sprinkler – large mobile machines	2,730	209	2,653	253	−77	−2.8	44	21.1
Solid set	5,487	91	2,921	51	−2,566	−46.8	−40	−44.0
Other	848	14	7,188	95	6,340	747.6	81	578.6
Total	**53,645**	**2,353**	**46,510**	**1,825**	**−7,135**	**−13.3**	**−528**	**−22.4**

*Establishments

Increasing water scarcity, particularly during extended drought conditions, is a key driver of the conversion of flood irrigation systems to more efficient pressurised systems to increase WUE and create water savings in irrigation systems (Mackinnon et al. 2009). Australian Bureau of Statistics records indicate an overall reduction in irrigation area and a 40 per cent decrease in the area under surface irrigation from 2002–03 to 2008–09, mainly due to water shortages associated with prolonged drought conditions. Many establishments ceased irrigation altogether, but there is also evidence of adoption of more efficient irrigation technologies over this period (ABS 2010). Between 2002–03 and 2008–09, the number of establishments irrigated by above-ground drip irrigation systems increased (by 9.2 per cent), as did the area (by 20.5 per cent). There was also an overall decrease in the number of establishments and area under the older, less efficient portable and hose sprinkler irrigation systems, but a 21 per cent increase in the area under large water-efficient mobile sprinkler systems (Table 8.1).

Despite the effects of drought on farm profitability, new investments in on-farm irrigation infrastructure were also reported by Mackinnon et al. (2009), who found that around 7 per cent of irrigation farms in the Murray–Darling Basin purchased new irrigation systems during 2006–07. Investment patterns during this period were influenced by the extended drought conditions and widespread water scarcity, indicating that future climate change and on-going water and environmental reforms may continue to play a part in driving investment decisions on irrigation farms (Mackinnon et al. 2009).

Agriculture and GHG emissions

Crop water use under Australian conditions is well understood. However, there has to date been limited research into energy use and associated GHG emissions in agriculture (Jackson 2009). Preliminary studies indicate a general trend of increasing energy consumption with increasing water use, particularly in groundwater-dependent regions. A more complex picture arises with conversion to alternative irrigation systems: Energy use and GHG emissions increase with reduced water use where gravity-driven flood irrigation is replaced by pressurised water-use efficient systems in regions with surface water resources, while in groundwater dependent regions, overall energy consumption and GHG emissions may decline due to the reduction in the volume of water pumped (Jackson 2009).

Other on-farm reforms include the adoption of better management practices that contribute to increased agricultural WUEs (the crop yield per unit of irrigation water applied) and water productivity (the value of irrigated agricultural output per megalitre [ML] of water used) (Mackinnon et al. 2009). However, changed land-use practices may also involve increased use of fuel, farm machinery and agrochemicals, and hence increased GHG emissions (Graham and Williams 2005; Maraseni et al. 2009a, 2010a and 2010b; Jackson et al. 2010).

Of the total energy used in agriculture globally, 51 per cent is expended in farm machinery manufacturing and 45 per cent in the production of chemical fertilizers (Helsel 1992). In addition, increased fertilizer use contributes significantly to emissions. Between 1987 and 2000, nitrogen (N) fertilizer use in Australia increased by 325 per cent (Dalal et al. 2003), with more than 50 per cent of the applied N either lost through leaching into the soil or released into the atmosphere as nitrous oxide (N_2O) (Verge et al. 2007), a GHG which has 298 times more global warming potential than CO_2 (IPCC 2007). Significant amounts of GHG emissions also occur due to the production, packaging, storage, transportation and use of farm inputs, which is poorly researched.

Chen and Baillie (2007) estimated operational energy usages from different cotton production processes, including fallow, planting, in-crop, irrigation, harvesting and post harvest. They reported that the total energy inputs for these farms were significantly influenced by the management and operation methods adopted, and ranged from 3.7 to 15.2 GJ/ha of primary energy, resulting in 275 to 1,404 kgCO_2e/ha of GHG emissions. Irrigation water energy use was found to be the highest contributor and was typically 40 to 60 per cent of total energy costs wherever water was pumped.

Integrated modelling framework

Potential trade-offs between water security and climate security associated with the adoption of more water efficient but energy-intensive irrigation technologies have been evaluated using an integrated economic framework (Maraseni et al. 2012a and 2012b; Mushtaq et al. 2013). This framework uses robust approaches, including hydrological modelling, energy and GHG modelling, and cost and benefits analysis to estimate water savings, GHG implications and economic outcomes under a range of carbon prices (Figure 8.1). Integrating these factors through an economic framework enabled conflicts between energy consumption (and GHG emissions) and water savings at various levels of investment to be readily identified.

The framework has been applied: (1) at the crop level to investigate possible irrigation transformation scenarios, which government policy could potentially drive at national

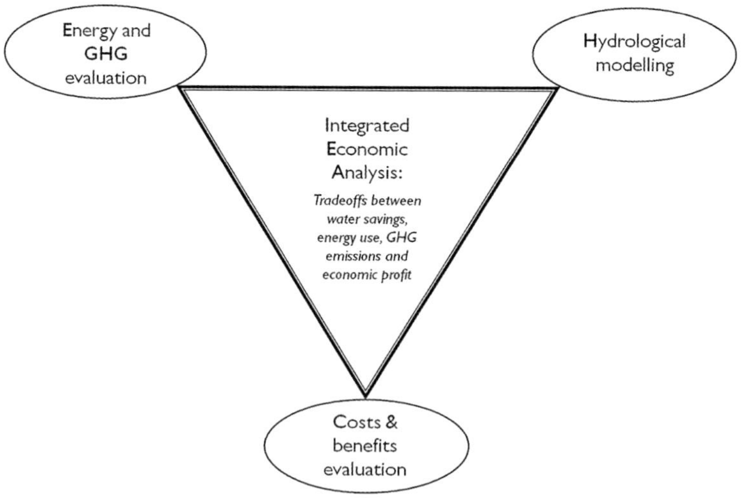

Figure 8.1 Integrated framework for assessing trade-offs between water savings, economics, energy use and greenhouse gas (GHG) emissions (Mushtaq and Maraseni 2011).

scales; and (2) to a number of farm-level case studies in order to investigate factors influencing the outcomes of new irrigation technology adoption decisions.

National scale crop level analysis

The integrated assessment framework was used to investigate five national level irrigation technology adoption scenarios – three historical and two potential – in terms of industry-wide trade-offs between water savings, energy consumption (GHG emissions) and economic returns under current Australian Government water resource policies (Table 8.2) using national figures on irrigation water use for different crops (see Box 8.1; Figures

Table 8.2 National level irrigation technology transfer scenarios based on historical adoption rates (Scenarios 1–3) and elicited information from farmers on planned adoption rates (Scenarios 4–5) (Mushtaq et al. 2013).

Transformation scenarios		Description
Future irrigation technology scenarios based on historical trends		
1	Converting surface irrigation to drip and sprinkler irrigation systems	Reducing the total area of surface irrigation systems from 44% during 2008–09 to 25% and replacing the balance with drip irrigation (40%) and sprinkler irrigation systems (60%).
2	Converting portable and hose irrigator sprinkler systems with drip and sprinkler irrigation systems	Reducing the total irrigation area under 'portable sprinkler irrigation systems' and 'hose irrigators sprinkler systems' from 16% during 2008–09 to 8% and replacing the balance with drip irrigation (50%) and efficient sprinkler irrigation (50%) (lateral-move and centre-pivot) systems.
3	Replacing surface irrigation with drip irrigation technology on horticultural crops	Increasing the drip irrigation area on horticultural crops from 13.3% during 2008–09 to 20% of the total irrigated area:
Future irrigation technology scenarios based on adoption		
4	Replacing surface irrigation with drip and sprinkler irrigation systems	Of the 7,674 establishments that used surface irrigation system during 2008–09, 15% will adopt drip irrigation and sprinkler irrigation systems on 50% of the farming land.
5	Replacing portable and hose irrigator sprinkler systems with drip and sprinkler irrigation systems	Of the 8,758 establishments under 'portable sprinkler irrigation systems' and 'hose irrigators sprinkler systems' during 2008–09, 15% will adopt either drip irrigation or more efficient sprinkler irrigation systems on 50% of the farming land.

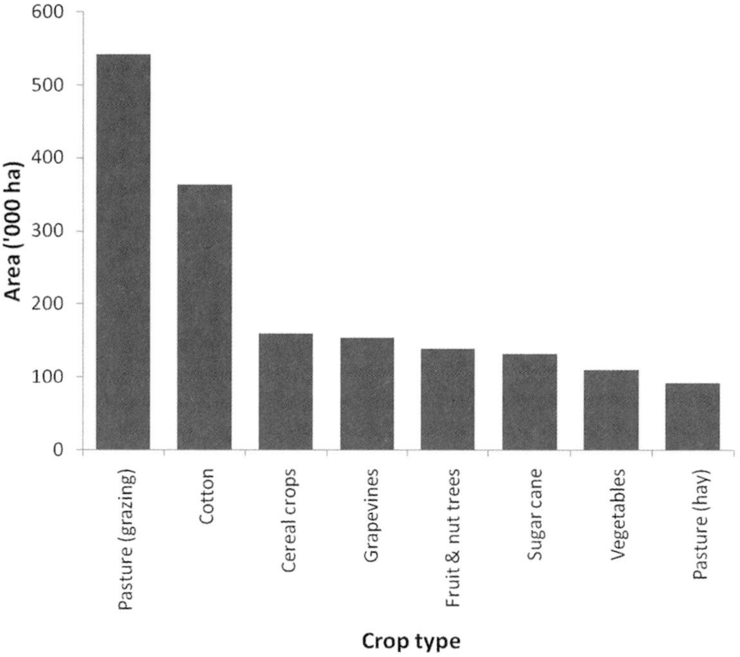

Figure 8.2 Irrigated cropping area by crop type, 2010–11 (ABS 2013).

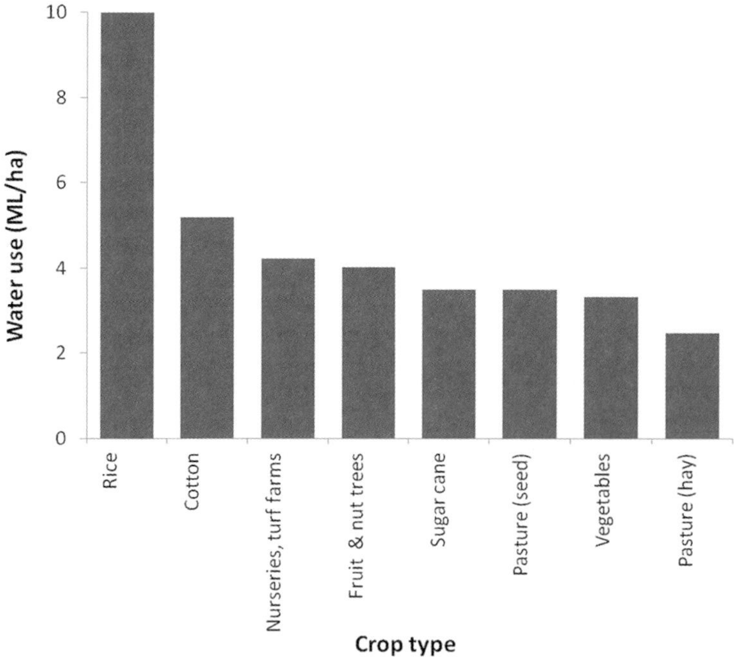

Figure 8.3 Water use per hectare by irrigated crop type, 2010–11 (ABS 2013).

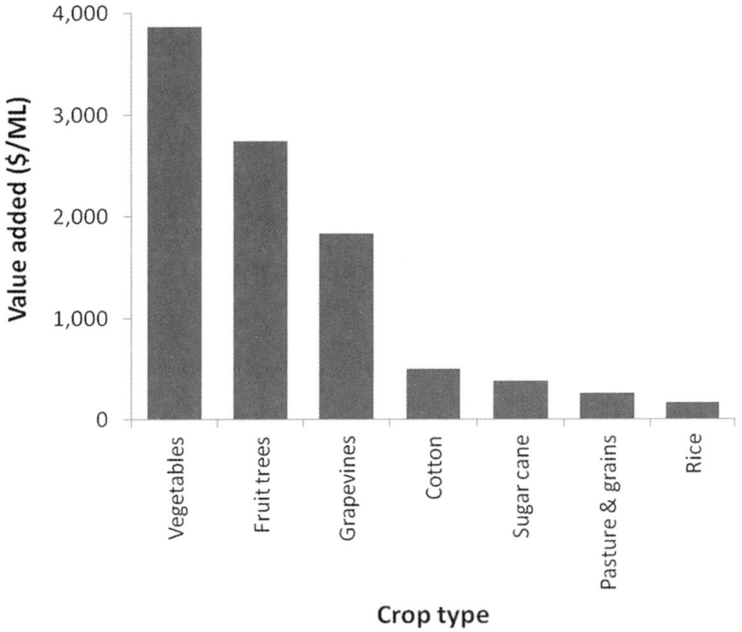

Figure 8.4 Value added per megalitre of water used by crop type, 2004–05 (ATSE 2012).

8.2–8.4). Economic assessments were conducted using four economic evaluation indexes – NPV, BCR, payback period and BE water savings – to assess the economic viability of new irrigation technology. NPV was calculated by measuring the total benefit of irrigation technology option (improvement in yield, quality, shifts in cropping rotation, reductions in input costs, labour savings, water savings and other benefits) and total costs (fixed and variable costs). The economics of technology adoption were evaluated using *Waterwork* (Khan et al. 2010), a farm-level irrigation technology modernisation model, simulated for twenty-five years with a 5 per cent interest rate. Detailed methods are described in full in Mushtaq et al. (2013), and the results are presented in Box 8.2.

Three of the transformation scenarios tested showed trade-offs between water savings and GHG emissions, with conversion of irrigation systems saving water but increasing both energy consumption and GHG emissions (Box 8.2). For example, 120 gigalitres per year of water savings achieved through drip irrigation adoption for cotton cropping would increase Australia's national energy consumption by 889 terajoules per year and GHG emissions by 250,000 tonnes of carbon dioxide equivalent per year, while an assumed carbon price[2] of

[2] Recent political changes (September 2013) mean that the current carbon tax ($24.15/t.$CO_{2e}$) could be scrapped in Australia. But if this did not occur, and a floating price were introduced (matched to the price on the European Union [EU] carbon market), the carbon price could drop significantly as Australian companies have unlimited access to Kyoto-eligible international markets. Currently, the Clean Development Mechanism (CDM) carbon offset price is less than a $1/tCO2e and the EU ETS price is <$5/tCO2e due to shrinking demand for and increasing supply of carbon credits. Causes include: (1) countries such as Japan refusing to extend their emission reduction targets under the Kyoto Protocol; (2) some developed countries, including the United States, with no mandatory emission reductions target; and (3) increase availability of funds to draw up Nationally Appropriate Mitigation Actions (NAMAs).

Box 8.2 Transformation scenario analyses (Mushtaq et al. 2013).

Table 8.3 Transformation Scenario 1: Converting surface irrigation to drip and sprinkler irrigation systems (estimated projected values based on historical adoption rates).

Crop	Area replaced ('000 ha)	New irrigation technology	Water savings (GL/yr)	Net present value (AUD millions/yr)	Energy use (TJ/yr)	GHG emissions ('000 tCO_2-e/yr)	Value of GHG emissions* ('000 AUD/yr)
Cereal crops for grain or seed	73	Sprinkler	62	−20	883	248	4,960
Cereal crops for hay	6	Sprinkler	5	−2	73	20	408
Cereal crops not for grain, seed or hay	6	Sprinkler	5	−2	73	20	408
Cotton	80	Drip	120	162	889	250	4,994
Pasture for grazing	105	Sprinkler	126	115	2663	748	14,968
Pasture for hay	59	Drip	207	401	1132	318	6,362
Pasture for seed production	10	Sprinkler	15	17	250	70	1,406
Pasture for silage	9	Sprinkler	14	18	225	63	1,266
Total	**348**		**553**	**690**	**6187**	**1739**	**34,771**

*based on an AU$20/t CO_2-e carbon price

AU$20 per tonne of carbon dioxide equivalent would result in additional costs nationally of about AU$5 million per year (Box 8.2; Table 8.3). However, the study finds that significant benefit in terms of water savings and GHG reduction can be achieved when replacing older inefficient and energy-intensive systems, such as hand shift and roll-line sprinkler systems, especially when they are replaced with drip irrigation systems (Box 8.2; Tables 8.4 and 8.7). For example, historical adoption rates for drip irrigation systems (Transformation Scenario 2, Tables 8.2 and 8.4) indicate significant potential to save water, energy and GHG emissions in both large-scale irrigated cropping (cotton and lucerne) and horticultural (grapes and tomatoes) production systems. These analyses indicate that on-farm irrigation infrastructure investment policies should prioritise investments in this area.

Table 8.4 Transformation Scenario 2: Converting portable and hose irrigator sprinkler systems to drip and sprinkler irrigation systems (estimated projected values based on historical adoption rates).

Crop	Area replaced ('000 ha)	New irrigation technology	Water savings (GL/yr)	Net present value (AUD millions/yr)	Energy use (TJ/ yr)	GHG emissions ('000 tCO_2-e/yr)	Value of GHG emissions* ('000 AUD/yr)
Cotton	21	Sprinkler	25	40	−11	−3	−50
Cotton	21	Drip	36	38	−204	−57	−1134
Grapes	52	Drip	73	399	−474	−132	−2,631
Lucerne	15	Sprinkler	15	77	35	11	213
Lucerne	15	Drip	33	92	−220	−61	−1,220
Tomato	30	Drip	45	235	−233	−64	−1,289
Total	**154**		**226**	**880**	**−1,108**	**−306**	**−6,112**

*using an AU$20/t CO_2-e carbon price

Table 8.5 Transformation Scenario 3: Replacing surface irrigation with drip irrigation technology on horticultural crops (estimated projected values based on historical adoption rates).

Crop	Area replaced ('000 ha)	New irrigation technology	Water savings (GL/yr)	Net present value (AUD millions/yr)	Energy use (TJ/ yr)	GHG emissions ('000 tCO_2-e/yr)	Value of GHG emissions* ('000 AUD/yr)
Citrus	38	Drip	57	248	691	194	3,882
Grapes	52	Drip	114	443	609	171	3,423
Tomato	30	Drip	45	261	400	112	2,247
Total	**120**		**216**	**952**	**1,700**	**478**	**9,552**

*using an AU$20/t CO_2-e carbon price

Farm level case studies

Maraseni et al. (2012b) also applied the integrated assessment framework described above to explore trade-offs associated with the adoption of more water efficient but energy-intensive irrigation technologies on individual irrigated cropping establishments in southern Queensland. Five different case studies were conducted, encompassing a range of irrigation technologies and cropping systems. Enterprises assessed included three cotton farms on the Darling Downs, a vegetable (lettuce) farm in the Lockyer Valley and a pasture-cropping (lucerne) farm on the southern Downs. Irrigation technology transitions investigated were from flood (furrow) to overhead sprinkler (lateral-move and centre-pivot sprinkler systems), from flood (furrow) to trickle irrigation, from overhead sprinkler (hand-

Table 8.6 Transformation Scenario 4: Converting surface irrigation to drip and sprinkler irrigation systems (estimated projected values based on elicited information from farmers on planned adoption rates).

Crop	Area replaced ('000 ha)	New irrigation technology	Water savings (GL/yr)	Net present value (AUD millions/ yr)	Energy use (TJ/ yr)	GHG emissions ('000 tCO_2-e/ yr)	Value of GHG emissions* ('000 AUD/yr)
Cereal crops for grain or seed	36.8	sprinkler	31	−10	445	125	2,500
Cereal crops for hay	2.8	sprinkler	2	−1	34	10	190
Cereal crops not for grain, seed or hay	1.9	sprinkler	2	−1	23	6	129
Cotton	6	drip	9	12	67	19	375
Pasture for grazing	67	sprinkler	80	73	1,692	475	9,510
Pasture for hay	2.5	drip	9	17	48	13	270
Pasture for seed production	1.9	sprinkler	3	3	48	13	267
Pasture for silage	1	sprinkler	2	2	25	7	141
Total	**120**		**137**	**96**	**2,381**	**669**	**13,382**

*using an AU$20/t CO_2-e carbon price

Table 8.7 Transformation Scenario 5: Converting portable and hose irrigator sprinkler systems to drip and sprinkler irrigation systems (estimated projected values based on elicited information from farmers on planned adoption rates).

Crop	Area replaced ('000 ha)	New irrigation technology	Water savings (GL/yr)	Net present value (AUD millions/yr)	Energy use (TJ/ yr)	GHG emissions ('000 tCO_2-e/yr)	Value of GHG emissions* ('000 AUD/yr)
Cotton	3	Sprinkler	4	6	−2	0	−7
Cotton	3	Drip	5	5	−29	−8	−162
Grapes	0.5	Drip	1	4	−5	−1	−25
Lucerne	8	Sprinkler	8	41	19	6	113
Lucerne	8	Drip	18	49	−117	−33	−651
Tomato	0.3	Drip	0	2	−2	−1	−13
Total	**23**		**35**	**107**	**−136**	**−37**	**−745**

*using an AU$20/t CO_2-e carbon price

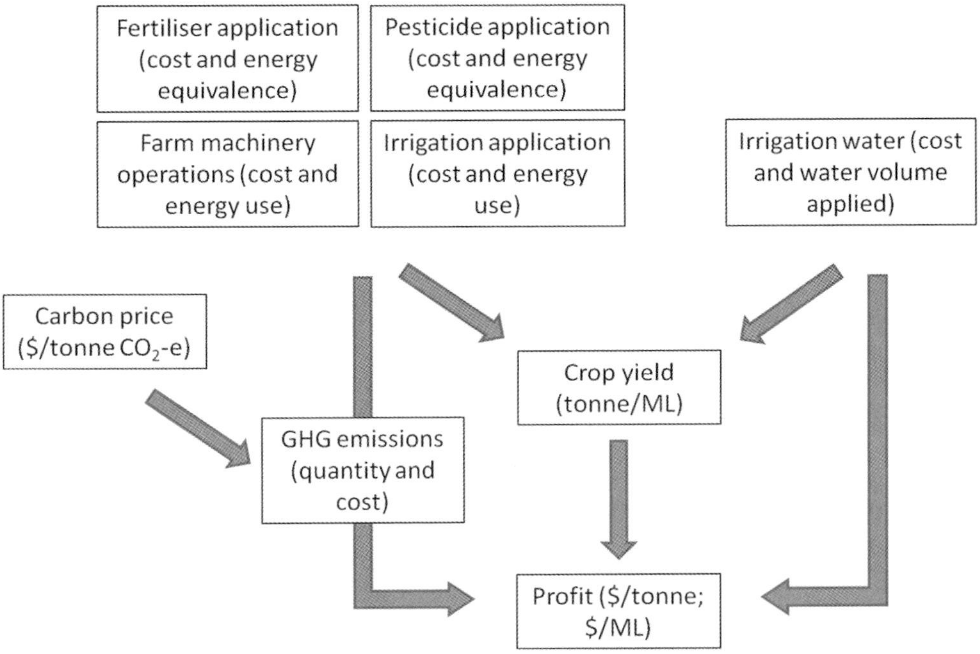

Figure 8.5. A conceptual diagram of the integrated analysis conducted on irrigated cropping systems adopting new irrigation technologies in southern Queensland (Maraseni et al. 2012b).

shift) to trickle irrigation and from older overhead sprinkler (roll-line) to improved overhead sprinkler (centre-pivot) systems. Economic analysis of these investment options was conducted on the costs of water and energy use and included the costs of the irrigation infrastructure upgrades as well as associated variable and operating costs (Figure 8.5). Detailed methodology for the study is available in Mushtaq and Maraseni (2011) and Maraseni et al. (2012a and 2012b).

Results from the five case studies indicated large variations in water and energy use at the field level between different cropping systems (Table 8.8). While the new irrigation technologies used less water per hectare of crop, analysis of the five case studies revealed trade-offs between water savings and energy use and irrigation-related GHG-related emissions with the adoption of new pressurised irrigation systems, except in cases where older water-inefficient and energy-intensive hand-shift and roll-line sprinkler irrigation systems were replaced by water use efficient drip or sprinkler irrigation technologies. In these two instances, results indicated significant water savings as well as significantly reduced irrigation-related GHG emissions (Table 8.8). This is an obvious win–win situation where water savings, energy use and irrigation-related GHG reductions can be achieved as a direct result of technology adoption and confirms the indication from the national scale analyses (described above) that priority should be given, in implementing on-farm infrastructure investment policy, to replacing inefficient and energy-intensive sprinkler irrigation systems such as hand-shift and roll-line.

Table 8.8. Trade-offs matrix for conversion to new pressurized drip and sprinkler irrigation systems based on five case studies in southern Queensland.

Cropping system	Irrigation technology conversion	Water savings (ML/ha/yr)	Emissions (kg CO_2e/ha/yr)*		Estimated Net Present Value (AUD/ha/yr)
			Irrigation-related	Total	
1. Cotton	Flood (furrow) to sprinkler (lateral-move)	2.0	491.8	−1,072.2	582
2. Cotton	Flood (furrow) to sprinkler (centre-pivot)	1.0	183.0	−488.4	176
3. Cotton	Flood (furrow) to drip	1.0	4,047.0	3,998.6	95
4. Lettuce	Sprinkler (hand-shift) to drip	2.0	−1,288.5	−489.1	9,065
5. Lucerne (alfalfa)	Sprinkler (roll-line) to sprinkler (centre-pivot)	8.0	−7,509.2	−7,509.2	1,657

*Difference in emissions between the new and old irrigation systems

The results also indicate that changes in farm-level inputs associated with new irrigation technology adoption, such as the use of agricultural machinery and agrochemicals, may also be important. Although new irrigation technologies require additional energy (and associated GHG emissions), these costs may be offset by reductions in the levels of farm machinery use and crop inputs; for example, crop nutrients may be applied through the irrigation system (by fertigation) rather than by conventional means. In four of the five case studies, there was a reduction in total GHG emissions when such associated changes were also considered (Table 8.8). In one study only (Case Study 3, a cotton production system converted from flood to drip irrigation), the new pressurised irrigation system resulted in irrigation-related GHG emissions that were 1.8 times higher per kilogram of crop yield due to the extent to which electricity use increased compared to the old gravity-fed flood irrigation system. While the case study analyses generally found that total GHG emissions due to the use of farm machinery operation and agrochemicals declined with conversion to new irrigation technology, Maraseni et al. (2012b) suggest that the observed reduction may not be due solely to the change in irrigation system, but in part to growing farmer experience with precision agriculture and awareness of environmental concerns. Nevertheless, the results signify that a win–win situation with both water savings and GHG reductions can be achieved through a combination of technology adoption and changed management practices.

Economic efficiency is a major consideration in both policy and farm management decisions. At the farm level, initial capital costs associated with conversion potentially limit the rate of adoption of new technologies; similarly, significant costs associated with energy requirements to operate and maintain particular systems can mean that the overall financial position of farmers is compromised. Both crop-level and farm-level economic modelling conducted in this study indicated that the conversion to irrigation technology may be economically viable, especially where higher-end water savings can be achieved, although the level of economic gain may be undermined by costs associated with increased energy use and associated GHG emissions when a carbon price is applied. However, the farm-level studies showed that these may be offset by reductions in GHG emissions where there are associated changes in farm machinery use and inputs (Mushtaq and Maraseni 2011; Maraseni et al. 2012b). The integrated economic analysis indicated that all five case studies generated positive economic returns, largely due to the significant water savings[3] and increased productivity derived from the adoption of the new irrigation systems (Table 8.8). The absence of national-level data that links crop management practices and irrigation technologies across a range of irrigated farming systems makes similar detailed GHG accounting at the national scale difficult. Further exploration of the complexities is required.

The study successfully highlights the complexity involved in evaluating the broader outcomes of investment in new irrigation systems primarily designed to deliver on-farm water-use efficiency. The trade-off analysis raises the critical point that mitigation and adaptation policies should be evaluated against each other in order to optimise economic investments in irrigation technologies while managing climate risk. While it is apparent that the use of pressurised systems can result in significant benefits in terms of savings and economic returns, the costs associated with increasing energy consumption and GHG emissions should be carefully considered. A more targeted approach that achieves balance between improvement in water use and the potential increase in energy consumption is required, particularly at the policy level. Without this, a policy focus on improving irrigation efficiencies to create water savings could result in unintended outcomes at the national level while subjecting irrigation enterprises to escalating costs under a carbon pricing mechanism.

Conclusion

While a primary goal of crop production systems is the pursuit of economic efficiency, agricultural development also brings considerable environmental change. Given current understanding of climate change, past expectations that economic and social benefits would exceed environmental costs may be no longer valid. There is growing awareness also that it is no longer sufficient for a farm to produce food and fibre as economic goods, but that it must also function to produce or protect environmental services (van der Werf et al. 2007; Jackson 2009). The importance of this role is critical given that agriculture is a significant contributor to GHG emissions, and thus climate change.

[3] Sensitivity analysis conducted with regard to water savings, water trading, carbon price, yield benefits, technology life and interest rates indicated that NPV was highly sensitive to 'water savings' but relatively less so to GHG emissions (Mushtaq and Maraseni 2011; Maraseni et al. 2012b).

The water sector presents major challenges, globally, in what is predicted to be a water-energy constrained future. Irrigated production systems have significant potential to deliver increased productivity while optimising water use through improved management and the adoption of new technologies. However, there is a risk that interventions at the farm-level could be counter-productive when evaluated in terms of climate change mitigation due to increased energy consumption and higher GHG emissions; hence possible conflicts (ie trade-offs) might arise as a result of increased levels of conversion to new irrigation technologies. On the other hand, synergies may also be apparent when analysis of trade-offs between adaptation and mitigation is extended to include associated changes in farming inputs.

The significant consequences of climate change, particularly for the water sector, emphasise the need for consistent and effective policy responses to achieve win–win synergies between mitigation and adaptation strategies and avoid maladaptation and unintended consequences that significantly increase the vulnerability of irrigators and rural communities. Better understanding through integrated assessment frameworks is required to enable simultaneous evaluation of both mitigation and adaptation approaches to guide effective and equitable evidence-based policy responses.

The trade-offs analysis illustrates the critical point that mitigation and adaptation policies must be co-evaluated to optimise economic investments and avoid unintended adverse outcomes due to the climate–water–energy nexus. The Australian case studies presented show positive synergies between water savings and GHG reduction; however, a careful assessment of different irrigation technologies across a range of cropping systems is needed to clarify the energy (and GHG) consequences of water policies in other countries with intensive irrigation. The challenge is to adopt new irrigation technology and management systems to enhance future food and water security without having unacceptable negative environmental and social impacts.

References

Australian Academy of Technological Sciences and Engineering (ATSE) (2012). *Sustainable Water Management: Securing Australia's Future in a Green Economy*. Melbourne: Australian Academy of Technological Sciences and Engineering.

Australian Bureau of Statistics (ABS) (2010). *Water Use on Australian Farms 2008–09*. Canberra: Commonwealth of Australia.

Australian Bureau of Statistics (ABS) (2013). *Water Use on Australian farms 2010–11*. Canberra: Commonwealth of Australia.

Australian Greenhouse Office (AGO) (2006). *Reducing GHG Emissions from Australian Agriculture: The Role of Benchmarking in Driving Best Management Practice*. Canberra: Australian Greenhouse Office.

Barnett, J. and O'Neill, S. (2010). Maladaptation. *Global Environmental Change* 20: 211–13.

Calford, E. M., Gurney, A., Heyhoe, E. and Ahammad, H. (2010). *The Effects of an Emissions Offsets Scheme on Australian Agriculture*. Canberra: Australian Bureau of Agricultural and Resource Economics (ABARE).

Chen, G. and Baillie, C. (2007). *Development of Energy/Calc – A Tool to Assess Cotton Onfarm Energy Uses*. Toowoomba: National Centre for Engineering in Agriculture (NCEA).

Dalal, R. C., Wang, W., Robertson, P. and Parton, W. J. (2003). Nitrous Oxide Emission from Australian Agriculture Lands and Mitigation Options: A Review. *Australian Journal of Soil Research* 41: 165–95.

Department of Climate Change and Energy Efficiency (DCCEE). (2012). *Australia's Emissions Projections 2012*. Canberra: Department of Climate Change and Energy Efficiency.

Donohue, R. J., McVicar, T. R. and Roderick, M. L. (2009). Climate-Related Trends in Australian Vegetation Cover as Inferred from Satellite Observations, 1981–2006. *Global Change Biology* 15(4): 1025–39.

Department of Sustainability, Environment, Water, Population and Communities (DSEW-PaC) (2011). *Sustainable Rural Water Use and Infrastructure*. Canberra: Department of Sustainability, Environment, Water, Population and Communities, www.environment.gov.au/water/programs/srwui/index.

Graham, P. W. and Williams, D. J. (2005). Optimal Technological Choices in Meeting Australian Energy Policy Goals. *Energy Economics* 25: 691–712.

Green, G., Sunding, D. and Zilberman, D. (1996). Explaining Irrigation Technology Choices: A Microparameter Approach. *American Journal of Agricultural Economics* 78: 1064–72.

Helsel, Z. (1992). Energy and Alternatives for Fertiliser and Pesticides Use. In Fluck, R.C. (ed.) *Energy in World Agriculture*. Amsterdam: Elsevier.

Intergovernmental Panel on Climate Change (IPCC) (2007). *Contribution of Working Group I to the Fourth Assessment Report of the Intergovernmental Panel on Climate Change, 2007*. Cambridge University Press.

Jackson, T. (2009). *An Appraisal of the Onfarm Water and Energy Nexus in Irrigated Agriculture*. PhD Thesis, Wagga Wagga: Charles Sturt University.

Jackson, T. M., Khan, S. and Hafeez, M. (2010). A Comparative Analysis of Water Application and Energy Consumption at the Irrigated Field Level. *Agricultural Water Management* 97: 1477–85.

Khan, S., Akbar, S., Rana, T., Abbas, A., Robinson, D., Dassanayke, D., Hirsi, I., Blackwell, J., Xevi, E. and Carmichael, A. (2004). *Hydrologic Economic Ranking of Water Saving Options Murrumbidgee Valley. A Study Conducted for the Pratt Water Murrumbidgee Valley Water Efficiency Feasibility Project*. Canberra: CSIRO. www.napswq.gov.au/publications/books/pratt-water/working-papers/pubs/hydrologic-main.pdf.

Khan, S, Mushtaq, S and Chen, C (2010). A Decision Support Tool for Irrigation Infrastructure Investments. *Irrigation and Drainage* 59: 404–18.

Lal, R. (2004). Carbon Emissions from Farm Operation. *Environment International* 30: 981–90.

Mackinnon, D., Mallawaarachchi, T. and Ashton, D. (2009). *Irrigation in the Murray–Darling Basin: Investment in On-Farm Irrigation Infrastructure, 2006–07*. Canberra: Australian Bureau of Agricultural and Resource Economics (ABARE).

Maraseni, T. N., Cockfield, G. and Maroulis, J. (2010a). An Assessment of Greenhouse Gas Emissions: Implications for the Australian Cotton Industry. *Journal of Agricultural Science* 148: 501–10.

Maraseni, T. N., Cockfield, G. and Maroulis, J. (2009a). An Analysis of Australia's Carbon Pollution Reduction Scheme. *International Journal of Environmental Studies* 66: 591–603.

Maraseni, T. N., Cockfield, G. and Maroulis, J. (2010b). An Assessment of Greenhouse Gas Emissions from the Australian Vegetables Industry. *Journal of Environmental Science and Health Part B* 45: 578–88.

Maraseni, T. N., Mushtaq, S. and Maroulis, J. (2009b). Greenhouse Gas Emissions from Rice Farming Inputs: A Cross Country Assessment. *Journal of Agricultural Science* 147: 117–26.

Maraseni, T. N., Mushtaq, S. and Reardon-Smith, K. (2012a). Integrated Analysis for a Carbon- and Water-Constrained Future: An Assessment of Drip Irrigation in a Lettuce Production System in Eastern Australia. *Journal of Environmental Management* 111: 220–6.

Maraseni, T. N., Mushtaq, S. and Reardon-Smith, K. (2012b). Climate Change, Water Security and the Need for Integrated Policy Development: The Case of On-Farm Infrastructure Investment in the Australian Irrigation Sector. *Environmental Research Letters* 7(3): 1–12.

Mushtaq, S. and Maraseni, T. N. (2011). *Technological Change in the Australian Irrigation Industry: Implications for Future Resource Management and Policy Development.* Waterlines Report Series No. 53, National Water Commission Canberra: NWC.

Mushtaq, S., Maraseni, T. N. and Reardon-Smith, K. (2013). Climate Change and Water Security: Estimating the Greenhouse Gas Costs of Achieving Water Security through Investments in Modern Irrigation Technology. *Agricultural Systems* 117: 78–89.

Pittock, J. (2010). A Pale Reflection of Political Reality: Integration of Global Climate, Wetland, and Biodiversity Agreements. *Climate Law* 1: 343–73.

Playán, E. and Mateos, L. (2006). Modernization and Optimization of Irrigation Systems to Increase Water Productivity. *Agricultural Water Management* 80: 100–16.

Pratt Water (2004). *The Business of Saving Water: The Report of the Murrumbidgee Valley Water Efficiency Feasibility Project.* Campbellfield: Pratt Water.

Qureshi, M. E., Wegener, M. K., Harrison, S. R. and Bristow, K. L. (2001). Economic Evaluation of Alternative Irrigation Systems for Sugarcane in the Burdekin Delta in North Queensland, Australia. In *Water Resource Management*, eds. Brebbia, C. A., Anagnostopoulos, K., Katsifarakis, K. and Cheng, A. H. D. Boston: WIT Press.

Sanders, O., Goesch, T. and Hughes, N. (2010). *Adapting to Water Scarcity.* ABARE Issues and Insights 10.5. Canberra: Australian Bureau of Agricultural and Resource Economics.

Tilman, D., Cassman, K. G., Matson, P. A., Naylor, R. and Polasky, S. (2002). Agricultural Sustainability and Intensive Production Practices. *Nature* 418(6898): 671–7.

van der Werf, H. M. G., Tzilivakis, J., Lewis, K. and Basset-Mens, C. (2007). Environmental Impacts of Farm Scenarios According to Five Assessment Methods. *Agriculture Ecosystems & Environment* 118: 327–38.

Verge, X. P. C., Kimpe, C. D. and Desjardins, R. L. (2007). Agricultural Production, Greenhouse Gas Emissions and Mitigation Potential. *Agricultural and Forest Meteorology* 142: 255–69.

Zehnder, A. J. B., Yang, H. and Schertenlieb, R. (2003). Water Issues: The Need for Action at Different Levels. *Aquatic Science* 65: 1–20.

Zillman, N. D., Redgwell, C., Omorogbe, O. and Barrera-Hernandaz, K. L. (2008). *Beyond the Carbon Economy: The Energy Law in Transition.* Oxford University Press.

9

Management of the urban energy-water nexus

STEVEN KENWAY

Introduction

The influence of urban water management on energy use is substantial. Approximately 8 per cent of Australia's 2009 greenhouse gas emissions (24 t CO_2-e/person) are influenced by water policy in cities. Much of the influence is associated with the use of water, in homes and industry. This is, consequently, 'hidden' from the much smaller (although still large) energy balances assessed and managed by water utilities. Understanding and managing this wider pool of energy is vital in order to avoid perverse impacts such as increasing greenhouse gas (GHG) emissions during water conservation efforts.

The energy consumption of urban water in Australia is anticipated to grow 200 to 250 per cent by 2030, from 2007 levels, largely as a result of 'climate resilient' desalination and water reuse systems. This chapter argues that co-ordinated management of water and energy in cities is both a need and an opportunity to find efficient and low-cost solutions to water shortages and emissions targets. Such solutions will require integrated water and energy research, planning, funding, regulation, standards, data, targets and reporting. Collaboration across the industry sectors and government will be necessary to generate knowledge and policy in major gaps, including (a) quantification of links between water and energy in the industrial and commercial sectors, (b) understanding of the strong economic, social and institutional factors of relevance and (c) generate real understanding of the variability in the systems as opposed to the current over-reliance on 'averages'.

The significance of water-related energy in cities

The energy associated with urban water provision in Australia in 2006–07 accounted for 13 per cent of national electricity use, plus 18 per cent of natural gas consumption. It totalled some 6,800 gigawatt-hours of primary energy use per 1 million people (Table 9.1; Kenway et al. 2011a), representing 9 per cent of Australian primary energy use. In 2009, Australia emitted 539 Mt CO_2-e, around 24 tonnes per person (AGO 2010). At 1.9 tonnes per person per annum, water-related energy would account for 8 per cent of Australia's total GHG emissions in 2009.

Approximately 86 per cent of water-related primary energy in the average Australian city is associated with use of water in households and in the commercial and industrial

Table 9.1 Breakdown of water-related energy per 1,000,000 Australians (2006–07) (Kenway 2012).

Aspect	Electricity GWhe		Natural gas GWhth		GHG emissions COZ-e (1,000 T)		Primary energy GWhth	
			Provision					
Energy use – utilities – water	106	8%		0%	106	6%	317	5%
Energy use – utilities – wastewater	68	5%		0%	68	4%	205	3%
Rainwater tank pumps	1	0%		0%	1	0%	4	0%
Highrise basement pumps	1	0%		0%	1	0%	3	0%
Bottled water		0%	50	2%	10	1%	50	1%
Subtotal	**176**	**13%**	**50**	**2%**	**186**	**10%**	**579**	**8%**
			Use					
Residential use (hot water)	535	39%	535	20%	642	34%	2,139	31
Residential use (non-hot water)	204	15%		0%	204	11%	611	9%
Commercial water use	245	18%	215	8%	288	15%	951	14%
Industrial water use	177	13%	1,650	62%	507	26%	2,181	32%
Subtotal	**1,161**	**84%**	**2400**	**90%**	**1,641**	**86%**	**5,882**	**86%**
			Resources					
Carbon loss (not captured)		0%	164	6%	33	2%	164	2%
N, P, K loss (resynthesis)		0%	61	2%	12	1%	61	1%
Subtotal	**0**	**0%**	**225**	**8%**	**45**	**2%**	**225**	**3%**
Heat Island – water component	**42**	**3%**		**0%**	**42**	**2%**	**125**	**2%**
Other e.g. Displaced urban agriculture			*not estimated*				*not estimated*	
Total water-related energy	**1,379**	**100%**	**2,675**	**100%**	**1,914**	**100%**	**6,811**	**100%**

Converting 1 kWh electrical =3 Kwh thermal (Glekk 2009); Note, most of the thermal energy use is natural gas

1.0 kg CO_2-e/kWh electricity and 0.2 kg CO2-e/kWh natural gas

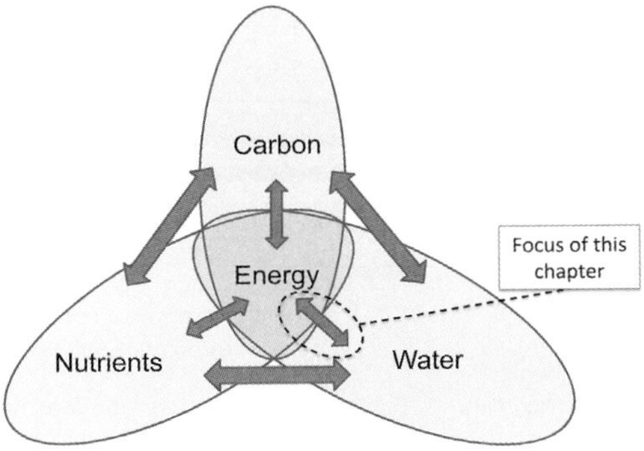

Figure 9.1 Water-related energy use in cities, and total energy use in Australia per 1,000,000 people (2006–07) (Kenway 2012).

sectors. Energy use by utilities (for water and wastewater services) represents about 8 per cent of total water-related energy. The remaining 5 per cent is comprised of energy use related to resources loss (such as Carbon, Nitrogen, Phosphorus and Potassium) and the 'water component' of the urban heat island effect.

While every city or nation is different, many modern cities could expect similar numbers. For example, water-related energy accounted for 19 per cent and 32 per cent of California's respective electricity and natural gas usage in 2005 (Klein et al. 2005). This is slightly higher than the numbers for Australia because the Californian estimates include energy related to agricultural water use and the Australian estimates do not. In colder climates, higher water-related energy use could be expected.

Figure 9.1 compares water-related energy in cities (shown below the dashed line) with the total direct energy use by the Australian economy (above the dashed line) (Kenway 2012). Water-related energy in cities is equivalent to approximately half the energy use of the Australian residential sector (excluding transport), or one quarter of the total energy use of the industrial sector including all manufacturing and mining, but excluding transport. It is more than five times the energy use of Australian agriculture. Water management has a large potential to affect significant quantities of energy.

What is the water-energy nexus?

Even though the water-energy nexus is widely discussed, it has rarely been defined. One definition argues that the nexus represents the "interconnections" or "cause-and-effect" relationships between water and energy (Kenway 2012). When there is a connection, a change in one leads to a change in another. For example, more water requires more energy, which requires more water. Alternatively, a change in a third factor could cause a change in both water and energy use. For example, increased food production or consumption typically means increases to both water and energy use.

There is a frequently described relationship between water, energy, carbon and climate: the 'WECC nexus' (eg Proust et al. 2007). The classic example, the Australian drought

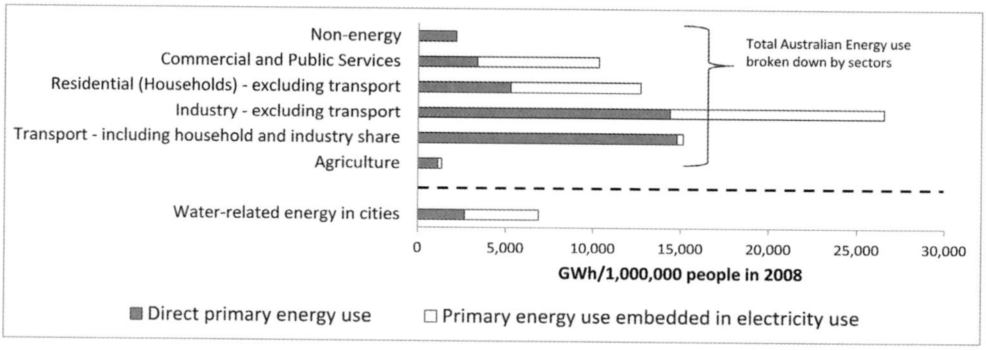

Figure 9.2 The energy, water, carbon and nutrient inter-relationship (Kenway 2012).

situation from 2002–07, demonstrates that a dry climate reduces available water supplies and increases water demand. This can lead to an increase in energy required to provide the water, for example, for use in cooling (McMahon and Price 2011). The additional energy use, in turn, contributes to further GHG emissions.

A less regularly articulated connection also exists between water, energy and nutrients. For example, industrial processes consume energy to manufacture nutrient fertiliser. Food production in agriculture requires water, nutrients and energy (eg for pumping water). Energy is also often invested to remove nutrients from wastewater in order to protect receiving environments. A conceptual outline of the energy-water-carbon-nutrient nexus is presented in Figure 9.2. While there are a wide range of interconnections between climate, carbon, water, food, nutrients and energy, this chapter is primarily focussed on the links between water and energy within cities.

Direct and indirect water-energy connections

In order to understand water-related energy in cities, it is helpful to consider both *direct* and *indirect* impacts. *Direct* energy impacts occur, for water utilities, when 'mains' water or wastewater is pumped or treated. *Indirect* energy impacts (for utilities) occur when water is used, for example, in heating water. Indirect impacts also occur for alternative water sources such as the use of pumps for rainwater tanks and transporting bottled water. Water is also consumed in evaporative air conditioning. Other indirect connections exist, for example, via flows of carbon and nitrogen in wastewater or in the production of food.

An emerging problem

In 2005, the 'Millennium' water crisis in Australia spawned a significant urban water infrastructure investment surge. Headlines warned of 'Armageddon'. We had to "Desalinate or Die" (Stolz 2007; ABC News 2007). Climate change was widely held as responsible (Howe et al. 2005; Bernstein et al. 2007; WSAA 2005). In 2005–06, a then-record AU$2.4 billion was spent in major cities for water (and wastewater) infrastructure. Most was aimed at increasing the proportion of 'climate-independent' sources such as desalination or wastewater reuse (WSAA 2008). This investment grew to over AU$7billion for 2008–09. Expenditure for 2009–10 was forecast over AU$14billion (WSAA 2009), but stabilised around AU$7billion, excluding the Melbourne Desalination Plant (National Water Commission 2011)[1]. Given the scale of this investment, it is important to know whether our solutions are solving problems or simply shifting them around.

Since 2008, increased rainfall has alleviated water shortages in eastern and southern Australia and reduced the pressure for action. However, a new challenge has emerged. The energy demand for water in Australia's major cities is anticipated to grow 200 to 250 per cent of 2007 levels by 2030 (Kenway et al. 2008b; Kenway and Lant 2011).

In December 2007, Australia ratified the Kyoto Protocol, and in 2012 established a long-term goal to reduce GHG emissions to 20 per cent of 2000 levels by 2050 (Australian Government Department of Climate Change and Energy Efficiency 2012). This means that if the water sector was to contribute proportionately to the abatement goal, then energy use for urban water must be reduced by more than 90 per cent from the projected 2030 levels. This is a massive challenge.

Lack of co-ordination across water and energy can lead to problem shifting or perverse outcomes. For example, during the millennium drought, water-efficient washing machines were promoted as a response to water shortages attributed to climate change. However, installing a water-efficient clothes washer can actually increase greenhouse gas emissions when the plumbing configuration shifts the water-heating energy source from a (common) natural gas hot water system (emitting 0.2 kgCO_2/kWh) to internal electrical heating within the washing machine (emitting 1 kg CO_2/kWh of delivered energy). A water problem was solved, but a GHG problem was likely exacerbated. Similarly, typical urban development creates enormous volumes of stormwater runoff, which are typically discharged, often with deleterious effects, into the local environment (Gardner and Sharma 2005). Because it is not captured and reused locally, additional water for urban supply must be pumped from distant sources in order to meet local demand (PMSEIC 2007). This has energy implications. While these problems may appear simple, they are complex, interconnected, confusing, value-laden, involve multiple stakeholders and change over time.

How can cities help address this emerging problem?

Cities offer considerable scope for solving many problems of resource efficiency (O'Meara 1999). Large flows of water, energy and materials pass directly through cities. Larger quantities of water and energy also flow through cities 'embodied' in goods and services. Because cities are also often being redesigned and recreated, there is a large opportunity to influence these flows.

Many aspects of urban planning and management influence urban water use. For example, water policy, pricing and restrictions have immediate impacts. Building design codes, technology rebates and land-use planning have longer-term impacts.

[1] Estimated then at around AU$4 billion.

If the population of Australia grows to 62 million by 2101, (ABS high estimate) an extra 40 million people will need to be provided for. This would require building one city the size of Sydney every ten years. It would demand 10 new desalination plants, 68 universities and over 16,000 new schools (Weller and Bolleter 2012). Simultaneous with this new growth, much of the existing urban infrastructure will also be refurbished or replaced. This provides enormous opportunity to improve resource efficiency.

There is much current effort towards the development of sustainable cities internationally. To date, however, little work has characterised the influence of urban water policy on holistic energy and GHG outcomes. Even more significantly, quantitative performance measures for cities are typically poorly developed, even for relatively common metrics (Satterthwaite 2008) such as population and water balance (Kenway et al. 2011b), let alone more complex measures such as water-related energy. An effort to benchmark cities for such parameters would be a challenging and worthwhile exercise.

Why should business consider a wider boundary of their effects?

Government and industry are anticipated to increasingly take responsibility for upstream (supply-related) and downstream (product-related) impacts, in addition to their operational impacts (GRI 2005; Heemskerk et al. 2002). This is not only for environmental stewardship reasons but also to manage business risks. An unsustainable supply chain creates operational risks.

Many current sustainability assessment frameworks are designed to assist organisations in reducing their individual impact, such as their energy use. However, relatively little effort goes into considering the city or 'system' of which the organisation is a component. Many water utilities would actively manage their own energy impacts, yet be unaware of wider energy use influenced indirectly by their operations or decision making. However, government entities arguably exist to enhance the overall 'system' of prosperity and efficiency. Increasingly, this philosophy is being embedded in governance documentation. For example, the Statement of Obligations (Government 2013) for the Queensland Bulk Water Authority requires the authority to "collaborate with all regional bodies and authorities to achieve whole of region benefits . . . at least cost to the community."

The emerging urban water-energy landscape

Resilience, or the ability to recover after shocks, is increasingly identified as a necessary strategy for cities. According to PMSEIC (2010), "Resilient pathways will simultaneously reduce GHG emissions, lower overall water demand, maintain overall environmental quality and allow living standards to continue to improve."

In order to achieve the challenge of resilience, improved co-ordination between water and energy planning and management will be necessary. In cities, water and energy management is frequently shared across federal, state and local government. Historically, little consideration is given with regard to interaction of the components, overall efficiency or how performance could be co-ordinated. However, this is anticipated to change, as summarised in Table 9.2.

Table 9.2 The emerging landscape of water-energy linkages in cities (Kenway 2012).

Where are cities now? (2013)	Where is it heading? (2030)
Planning, funding, regulation and standards for water and energy are managed in isolation. Unplanned 'trading' of water and energy impacts occurs.	Comprehensive simultaneous consideration and management of water, energy and related cycles (including nutrients). Governance frameworks ensure 'systemic efficiency' through planning and regulatory mechanisms (eg regional planning and building approval processes). A shift towards regulation for a 'whole of community' benefit.
Methodologies for assessing water and energy address only narrow component elements (such as the energy of individual system components, such as water or wastewater treatment or transport). The system boundary is either not considered or inconsistently applied.	Better informed decisions, enabled by comparison of options using standardised methods and boundaries will influence the total material and energy fluxes through cities.
Water and energy data sets and monitoring are scattered widely. For example, separate water and energy accounts are published in Australia at national level. Combined water and energy statistics are almost impossible to find other than for particular, highly studied individual households or sites.	Co-ordinated and inter-locking data-sets and monitoring programs will enable diverse and powerful analysis and management of water and energy simultaneously. (This is the metabolic equivalent of a co-ordinated sensory system and single brain rather than scattered ganglia and sense-response systems.)
Unquantified indicators of urban performance or limited quantified performance indicators (such as ecological footprint), which give little policy guidance, particularly to the water sector; analysis frameworks inconsistently applied.	Quantified performance indicators, which simultaneously consider water, energy and related flows underpinned by a metabolic framework and mass balances.
Optimisation analysis of the urban space for overall water and energy efficiency is absent.	Optimisation studies will contribute to or guide blueprints for urban futures (eg Regional and City Plans). They will also strongly inform development assessment and building codes.
Targets for urban systems are missing or unquantified.	Quantified targets for the resource efficiency of cities. Government, industry and business function in a co-ordinated manner to achieve them.
Separate *management and reporting* of water and energy performance.	Co-ordinated reporting is used to evaluate the efficacy of urban design and management (including human well-being and ecosystem health).
Research undertaken in isolation. For example, very little energy-impact research is undertaken by water research agencies, and likewise for energy research agencies.	Co-ordinated or better integrated water-energy research (see also AWE and ACEEE 2011; Kenway et al. 2011c; Kenway et al. 2013a).

At what scale could changes be made?

Innovative cities could reduce water and energy use and greenhouse gas emissions by an order of magnitude. They could achieve this while simultaneously maintaining or improving ecosystem health and human well-being. Enabling this change is a major challenge for society. Such a future is likely to require increased efficiency of cities. It is likely that this will require increased recycling of water, materials and heat, at local and district or city levels. There are examples both within Australia and internationally where this is happening with regard to water and energy independently. However, there are relatively few well-documented examples where a combined energy and water 'win' have been achieved. In 2003, The Pacific Institute identified that 22 per cent of California's urban water use could be conserved without technological change. Savings were reported the following year (from reporting agencies) indicating 4 per cent of this potential (about 1 per cent of urban water use) had been conserved. This resulted in savings of some 27 billion gallons of water, but importantly, more than 234 million kilowatt-hours of electricity and over US$200 million in net present value of water-related energy (Klein et al. 2005). Other major energy and cost savings initiatives have been reported by water utilities influencing residential and commercial water use. This includes, for example, demand management initiatives such as shower-head replacement programs and new heating and cooling technologies for the food industries (Larabee and Ashktorab 2007).

Improvements to the co-ordinated management of water and energy could be initiated across a range of scales, because decisions are taken at different points in the land and water (and energy) planning cycle. Decisions can influence efficiencies and savings. Considering just the case of water, of the maximum savings available from the full sequence of decisions, potential savings as a result of decisions by developers and builders may be as high as 80 per cent of total available savings (PMSEIC 2010). As little as 20 per cent of total potential savings are gained as a result of decisions by owners and occupiers. This is represented by the right-hand end of the diagram in Figure 9.3. While demand

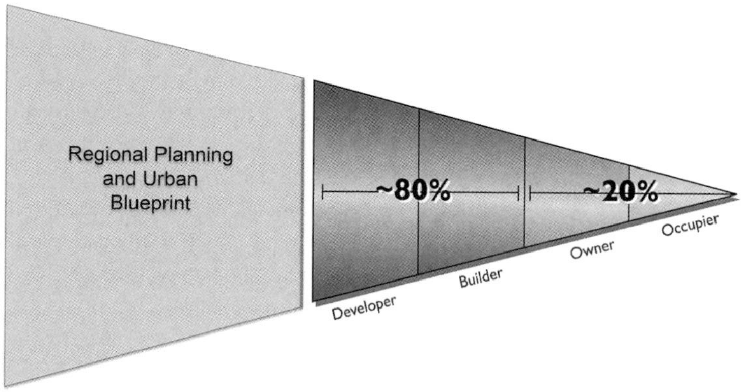

Figure 9.3 Water (and energy) savings possible at points in the development decision cycle (after PMSEIC 2007).

management at the owner or occupier stage of new housing is important, there is only so much that can be achieved at this late point in the decision cycle by relying on pricing signals to change behaviour as distinct from changing technology and infrastructure earlier in the planning cycle.

In addition to the stages of developer through to occupier (shown in Figure 9.3, based on PMSEIC [2007]), regional planning and urban blueprints can also have significant impact on urban water and energy balances. The shape and form of cities can have a high impact on water consumption (Kenway et al. 2008a) because it can affect the demand for water. It can also influence associated energy demands: for example, for water and wastewater pumping and treatment. Well-designed cities can act as water-supply catchments and provide local sources of rainwater and stormwater for suitable uses, potentially at low energy cost. These options are increasingly becoming components of urban water supply strategies.

What are the major knowledge gaps?

The majority of existing water-energy research has focussed on utility, household and technological aspects of the water-energy nexus (Kenway et al. 2011a). Water and energy links in the industrial and commercial sectors are, relatively, poorly publicly researched. In part, this is due to commercial sensitivity of data and process information. Similarly, the economic dimensions and implications of the water-energy nexus are not well documented. This is surprising, given that economic or financial factors have high influence on decision making. It is also surprising given that water demand response can be more sensitive to the price of energy ($/kWh) than water pricing because the energy cost of showering, bathing and using washing machines and dishwashers can be higher than the water costs (Hansen 1996; Kenway et al. 2012). There are even bigger gaps in the knowledge of social, political or legal dimensions of the nexus.

Further consideration of the influence of technologies is warranted because understanding these linkages will help us achieve desirable changes by design. However, many existing studies rely on average results rather than considering the range of possibilities (eg probability distributions). For example, the energy use of rainwater tanks is often considered to be around 1.5 kilowatt-hours per cubic metre, whereas systems range from 0.1 to 10 kilowatt-hours per cubic metre. It is necessary for the research to progressively understand the factors influencing individual systems and reduce our reliance on averages. Similarly, most studies addressing water-energy interrelationships rely on modelling. Very few studies have compared the modelled results with empirical data of performance (Kenway 2012).

Within the residential sector, city-scale knowledge of water-energy connections is needed in order to inform the design and implementation of successful water strategies aimed at reducing energy consumption. For city-scale analysis, there is a need to better understand a range of household types based on their water and energy performance and connections (Kenway et al. 2013b). Such knowledge would let policy be crafted to the applicable circumstance.

The influence of time needs further consideration with regard to the water-energy nexus. For example, how does peak water use (use between 6 and 8 PM) impact on peak electricity use and thereby contribute to elevated electricity distribution infrastructure costs? Also, how is water-related energy changing through decades, as supply and demand sources change and qualities of water and wastewater treatment also change?

Lack of a consistent theoretical framework and method (eg for quantifying the water-energy nexus) is a major knowledge gap. However, this is not to be unexpected, given the infancy of the overall research effort. Development, application and use of consistent framework could have a tremendous impact on our understanding of the relevant linkages. A well-developed theoretical framework may help to classify or group cities based on their relative performance. It could help differentiate urban performance and help advance our understanding from specific situations through to general principles. Such analysis would have numerous practical examples, including helping to quantify what comprises a water- and energy-sensitive city.

The use of performance indicators to a common boundary will be important in enabling this transition. There is a need for a framework to help us understand 'the city', including its flows of water, energy, GHG emissions, materials and waste. With such a framework, resultant methods and tools, we will be better able to systematically understand and manage the considerable influence of our cities. The concept of urban metabolism appears to have significant strengths in this domain (Wolman 1965; Kennedy et al. 2007; Sahely et al. 2003).

A benchmarking effort comparing a range of cities could be highly valuable. Analysis and comparison of cities over short- or long-term time periods could also be highly informative and help detect trends.

In order to close such gaps, collaboration and research across the water and energy sectors would be necessary. Such a need has been widely articulated by others, along with a need for increased collaboration between the (widely different) water and energy sectors (PMSEIC 2010; AWE and ACEEE 2011).

How could research and action in this complex domain be staged?

In 2011, a workshop was convened in California to identify the needs and priorities for a program of work aimed at improving management of water-related energy in cities (Kenway et al. 2013a).

The first step was to conceptualise a city successfully managing the water-energy nexus. Such a city had to be liveable, locally self-sustaining, and have diverse and affordable water and energy supply options. The city would have limited draw on the surrounding environment for inputs. Waste outputs would be small. Critically, it would operate within its local water and energy (carbon) budget. The efficiency paradigm is embraced for the 'whole system'. It fosters creativity, design innovation, business and economic success. The full detail of the work is described in Kenway et al. (2013a). The key points include:

1. Integrated standards, guidelines, funding and planning. It is vital that the planning and development processes in cities work together to optimise the entire system. Ideally, this involves simultaneously reducing water and energy flows through the city. There is a need for greater clarity of the vision, data needs, techniques, methods, appropriate

technologies, tools and development patterns. Barriers to progress here include fragmented jurisdictional boundaries, real physical differences between water and energy (and the associated system operation implications), short-term policy-making and the wider issue of agency 'entrapment'.
2. Water-energy education programs are needed at tertiary institutions, schools and within agencies and utilities. This is necessary to progress towards more interagency collaboration for planning and decision making.
3. Methods and then targets are needed for 'water-related energy' and associated GHG emissions. Methodologies are necessary to ensure that water-related energy is being accounted for consistently and accurately. Target setting would require cost-benefit analysis in order to find least-cost, system-wide implications.
4. Understanding, and then using, behavioural motivations to achieve urban resources efficiency. Due to the strong influence of customer behaviours, understanding motivational factors was viewed as vital to helping cities more rapidly progress towards efficiency modes.

In addition to these four broad priorities, other areas viewed as necessary included full cost charging, including for water and electricity rates; understanding the urban metabolism; aligning technology development and adoption with the need; integration with transport and waste; adaptive management; and blueprints for urban development. Improved tools and metrics were also identified as necessary, particularly regarding the economic implications of the water-energy nexus. Relationship building and trust across the myriad organisations is also important.

Conclusion

This chapter has demonstrated how urban water management has significant influence on energy use and related greenhouse gas emissions, and costs, particularly via the use of water. Managing this indirect or 'hidden' energy is important to meeting Australia's long-term GHG reduction goals. It is viewed as critical to finding least-cost water and energy solutions in cities.

The emerging landscape calls for far greater co-ordination across the water and energy sectors, including collaboration in areas of research, policy, planning, funding, regulation, standards, methods for assessing water-related energy, data management, quantifying performance, establishing targets and reporting.

Improved knowledge of water-energy interconnections would enable planners, managers and regulators to evaluate potential implications of a range of water strategies and options. Such knowledge could lead to major changes in the design and management of our cities in the future.

Acknowledgements

The support of The Australian Research Council, the Smart Water Fund (Victoria), Seqwater (Queensland) and the (former) Urban Water Security Research Alliance is acknowledged. The work and inputs of a number of people, including Professor Paul Lant, Professor Tony Priestley, Professor James McMahon, Professor Hans-Peter Bader, Ruth Scheidegger and others are generally referenced in this chapter and in the related specific citations. The support of the Swiss Federal Institute of Aquatic Science and Technology (Eawag) and the Australian-American Fulbright Commission enabled the work on which this chapter has built.

References

ABC News (2007). Beattie Scraps Water Poll Amid 'Armageddon' Situation. Brisbane.

Australian Greenhouse Office (AGO) (2010). *Australian National Greenhouse Accounts*. Canberra: Australian Department of Climate Change and Energy Efficiency.

Australian Government Department of Climate Change and Energy Efficiency (2012). National Targets. In *Think Change*. Canberra: Australian Government.

Alliance for Water Efficiency and American Council for an Energy Efficient Economy (AWE and ACEEE) (2011). *Addressing the Energy-Water Nexus: A Blueprint for Action and Policy Agenda*. Washington, DC: Alliance for Water Efficiency and American Council for an Energy Efficient Economy.

Bernstein, L., Bosch, P., Canziani, O., Chen, Z., Christ, R. and Davidson, O. (2007). Climate Change 2007: Synthesis Report. Summary for Policymakers. Lima: CONAM, 2006, p. 71 (Serie Indicadores Ambientales, 10).

Gardner, T. and Sharma, A. (2005). Water Resources for a Transit Orientated Sustainable Urban Development in Australia. In OzWater.

Government, Q. (2013). Statement of Obligations for Queensland Bulk Water Supply Authority Issued by the Treasurer and Minister for Trade, and Minister for Energy and Water Supply 2013.

Global Reporting Initiative (GRI) (2005). *Sector Supplement for Public Agencies: Pilot Version 1.0*. Boston: Global Reporting Initiative.

Hansen, L. G. (1996). Water and Energy Price Impacts on Residential Water Demand in Copenhagen. *Land Economics* 72(1): 14.

Heemskerk, B., Pistorio, P. and Scicluna, M. (2002). *Sustainable Development Reporting: Striking the Balance*. Geneva: World Business Council for Sustainable Development.

Howe, C., Jones, R., Maheepala, S. and Rhodes, B. (2005). *Melbourne Water Climate Change Study: Implications of Potential Climate Change for Melbourne's Water Resources* Canberra: CSIRO.

Kennedy, C., Cuddihy, J., and Engel-Yan, J. (2007). The Changing Metabolism of Cities. *Journal of Industrial Ecology* 11(2): 43–59.

Kenway, S., McMahon, J., Elmer, V., Conrad, S. and Rosenblum, J. (2013a). Managing Water-Related Energy in Future Cities – A Research and Policy Roadmap. *Journal of Water and Climate Change* 4(3): 161–75.

Kenway, S. J. (2012). *The Water-Energy Nexus and Urban Metabolism – Identification, Interpretation and Quantification of the Connections in Cities, thesis, School of Chemical Engineering*. Brisbane: The University of Queensland.

Kenway, S. J. and Lant, P. (2011). The Influence of Water on Urban Energy Use. In *Water Sensitive Cities*, ed. Howe, C. and Mitchell, C. Amsterdam: International Water Association and UNESCO-IHE.

Kenway, S. J., Lant, P., and Priestley, A. (2011a). Quantifying the Links between Water and Energy in Cities. *Journal of Water and Climate Change* 2(4): 247–59.

Kenway, S. J., Gregory, A., and McMahon, J. (2011b). Urban Water Mass Balance Analysis. *Journal of Industrial Ecology* 15(5): 693–706.

Kenway, S. J., Turner, G., Cook, S. and Baynes, T. (2008a). *Water-Energy Futures for Melbourne: The Effect of Water Strategies, Water Use and Urban Form.* 9780643095663. Canberra: CSIRO.

Kenway, S. J., Lant, P., Priestley, A. and Daniels, P. (2011c). The Connection between Water and Energy in Cities – A Review. *Water Science and Technology* 63(9): 1983–90.

Kenway, S. J., Scheidegger, R., Larsen, T. A., Lant, P. and Bader, H. P. (2012). Water-Related Energy in a Brisbane Household – Modelling Analysis and Reduction Potential. In *Urban Water Security Research Alliance Science Forum*, ed. Begbie, D. Brisbane: Urban Water Security Research Alliance.

Kenway, S. J., Scheidegger, R., Larsen, T. A., Lant, P. and Bader, H. P. (2013b). Water-Related Energy in Households: A Model Designed to Understand the Current State and Simulate Possible Measures. *Energy and Buildings* 58: 378–89.

Kenway, S. J., Priestley, A., Cook, S., Seo, S., Inman, M. and Gregory, A. (2008b). *Energy use in the Provision and Consumption of Urban Water in Australia and New Zealand.* 9780643096165. CSIRO and Water Services Association of Australia.

Klein, G., Krebs, M., Hall, V., O'Brien, T. and Blevins, B. (2005). *California's Water-Energy Relationship – Final Staff Report.* California Energy Commission.

Larabee, J. and Ashktorab, H. (2007). From Watts to Water. Climate Change Response through Saving Water, Saving Energy and Reducing Air Pollution.

McMahon, J. and Price, S. (2011). Water and Energy Interactions. *Annual Review of Environment Resources* 36(17): 1–29.

National Water Commission (NWC) (2011). *National Performance Report 2009–10 Part A – Comparative Analysis.* Canberra: National Water Commission.

O'Meara, M. (1999). *Reinventing Cities for People and the Planet.* Ed. Peterson, J. A., *Worldwatch Paper 147.*

Prime Minister's Science, Engineering and Innovation Council (PMSEIC) (2007). Water for Our Cities – A Report of the PMSEIC Working Group.

Prime Minister's Science, Engineering and Innovation Council (PMSEIC) (2010). *Challenges at Energy-Water-Carbon Intersections.* Canberra: Prime Minister's Science, Engineering and Innovation Council.

Proust, K., Dovers, S., Foran, B., Newell, B., Steffen, W. and Troy, P. (2007). *Climate, Energy and Water, Accounting for the links. Discussion Paper.* Canberra: The Fenner School of Environment and Society, Australian National University,.

Sahely, H. R., Dudding, S. and Kennedy, C. A. (2003). Estimating the Urban Metabolism of Canadian Cities: Greater Toronto Area Case Study. *Canadian Journal of Civil Engineering* 30(2): 468–83.

Satterthwaite, C. (2008). Cities' Contributions to Global Warming: Notes on the Allocation of Greenhouse Gas Emissions. *Environment and Urbanization* 20(2): 539–49.

Stolz, G. (2007). The Water Crisis – Its Desalinate or Die. *Courier Mail* [Brisbane], 1 February, section 1.

Weller, R. and Bolleter, J., eds. (2012). *Made in Australia: The Future of the World's Most Urbanised Nation.* Edited by U. of W. Australia. Perth: University of Western Australia.

Wolman, A. (1965). The Metabolism of Cities. *Scientific American* 213: 179–90.

Water Services Association of Australia (WSAA) (2005). *Testing the Water – Urban Water in Our Growing Cities: the Risks, Challenges, Innovation and Planning* Melbourne: Water Services Association of Australia.

Water Services Association of Australia (WSAA) (2008). *WSAA Report Card 2007–2008. Performance of the Australian Urban Water Industry and projections for the future.* Melbourne: Water Services Association of Australia.

Water Services Association of Australia (WSAA) (2009). *WSAA Report Card 2008–2009. Performance of the Australian Urban Water Industry and projections for the future.* Melbourne: Water Services Association of Australia.

10

Managing the electricity-water nexus in China, France, India and the United States

BENJAMIN K. SOVACOOL AND AJITH RAO

Introduction

Existing and planned reliance on electricity supply – generally from thermoelectric power plants, facilities that combust fuels or fission atoms – depends too heavily on assumptions of widespread, abundant water resources. For instance, the International Energy Agency recently estimated that global water withdrawals – water utilised by power plants but returned to the original source – for the electricity sector amounted to 585 billion cubic meters in 2010 – about 15 per cent of the world total, or an amount greater than the annual average discharge of the Ganges River in India. During this year, the world's thermoelectric power plants also consumed – used up water that did not return to the local hydrological cycle – 66 billion cubic metres of water. By 2035, the International Energy Agency (2012, p. 514) projected that withdrawals will increase by 35 per cent (in the Current Policies Scenario) to 790 billion cubic metres and consumption will more than double to 130 billion cubic metres.

The availability of water and downstream impacts on its quality is therefore a salient energy security concern. Affirming this point, one broad-ranging survey of 2,167 energy users in ten countries, including the United States, Japan, China, India and Saudi Arabia, found that 'enhancing the availability and quality of water' was the *highest* rated energy security dimension among sixteen options (Sovacool et al. 2012). This meant respondents rated it more highly than security of supply of fossil fuels and uranium and responding to climate change, among others.

To illustrate this importance, this chapter depicts how water may become a key constraint to thermoelectric power generation in four case studies covering northern China, France, India, and a part of the United States, based on current projections of water and power demand through 2040. Each case reveals something different about the electricity-water nexus: China, the sheer quantity of water needed to meet future electricity (and industrial) demand; France, the water-related vulnerability of nuclear power units; India, the water vulnerabilities of coal-fired power; and Texas, the significance of wind energy and natural gas in enabling power providers there to generate electricity in times of drought while reducing water dependencies. The chapter then discusses how an array of six different policy and technology options can enable planners to better manage the electricity-water nexus across our four cases and perhaps throughout the world at large.

Our primary method of data collection for this chapter is a database of existing and projected future electricity capacity, fuel mix and water resource needs (both by type of demand and overall volume) across our four case studies from 2010 to 2040. To build this database, we started by collecting data from the International Energy Agency and US Energy Information Administration but quickly identified a series of data holes. We 'filled' these holes through a complex process of contacting experts across our four case studies and verifying data with both officials and colleagues, a process explained in greater detail in a forthcoming report. We then supplemented our database with a number of secondary sources that we reference further.

Our chapter focusses on the *electricity-water* nexus rather than the broader concept of the *energy-water* nexus (or *water-energy* nexus) for multiple reasons. The electricity sector withdraws more significant amounts of water (primarily for cooling needs) than any other source of energy in most countries (Bogardi et al. 2012). The electricity sector is the largest contributor of greenhouse gas emissions and influencing climate change, which is already having major impacts on water resources around the globe (IPCC 2008). Demand for electricity is growing globally at a faster rate than for energy overall, due largely to major electrification efforts in developing countries (IEA 2012). And plans to electrify buildings and transportation, and potentially transition countries and regions away from direct use of largely fossil fuels with potentially cleaner heat-pump technologies and electric vehicles (including plug-in hybrids) and vehicle-to-grid systems, could dramatically increase our reliance on the electricity sector – ultimately increasing its water intensity as well (King and Webber 2008).

China: Fourfold increases in water demand

Our China case study focusses specifically on the North China Grid region, as it is not only home to major growth centres but also acutely water stressed. As Table 10.1 reveals, according to our projections overall, water demand within the North China Grid will jump by more than 460 per cent between 2010 and 2040 – a growth rate significantly higher than any of our other cases, even though it also presumes that water intensity will slightly improve due to enhancements in technology. The North China Grid serves the municipalities of Beijing and Tianjin as well as the provinces of Hebei, Shanxi, Shandong and parts of Inner Mongolia.[1] The North China Grid is one of the largest in the country, generating electricity for a population of nearly 250 million people and home to almost one-third of the total national installed capacity.

[1] According to the State Grid Corporation of China, the North China Grid consists of Beijing Electric Power Co. and Tianjin Electric Power Co. as two municipal branches; Hebei Electric Power Co., Shanxi Electric Power Co. and Shandong Power (Group) Corp. as three provincial subsidiaries; Qinhuangdao Power Generation Co., Ltd. and North China Electric Power Research Institute Co., Ltd. as two share-holding subsidiaries; and other twenty-three directly operated subsidiaries: Tianjin Dagang Power Plant, Shisanling Pumped Storage Power Plant and Panjiakou Pumped Storage Power Plant as three power generators; Tangshan Power Supply Co., Qinhuangdao Electric Power Co., Zhangjiakou Power Supply Co., Chengde Power Supply Co., Langfang Power Supply Co., Beijing EHV Power Transmission Co. and Datong EHV Power Transmission Co. as seven transmission and distribution subsidiaries; Beijing Electric Power Construction Co., Tianjin Electric Power Construction Co., Beijing Power Transmission and Transformation Co. and Beijing Power Equipment Group as four construction and manufacture subsidiaries; and other nine subsidiaries, such as hospitals, schools and sanatoriums.

Table 10.1 Total water demand (billion cubic meters), 2010–40 (Source: Authors).

	North China Grid	France	India	Electric Reliability Council of Texas (ERCOT)
2010	66.9	33.4	55.1	13.4
2040	311.1	40.4	76.5	16.4
% Change	465	21	39	22

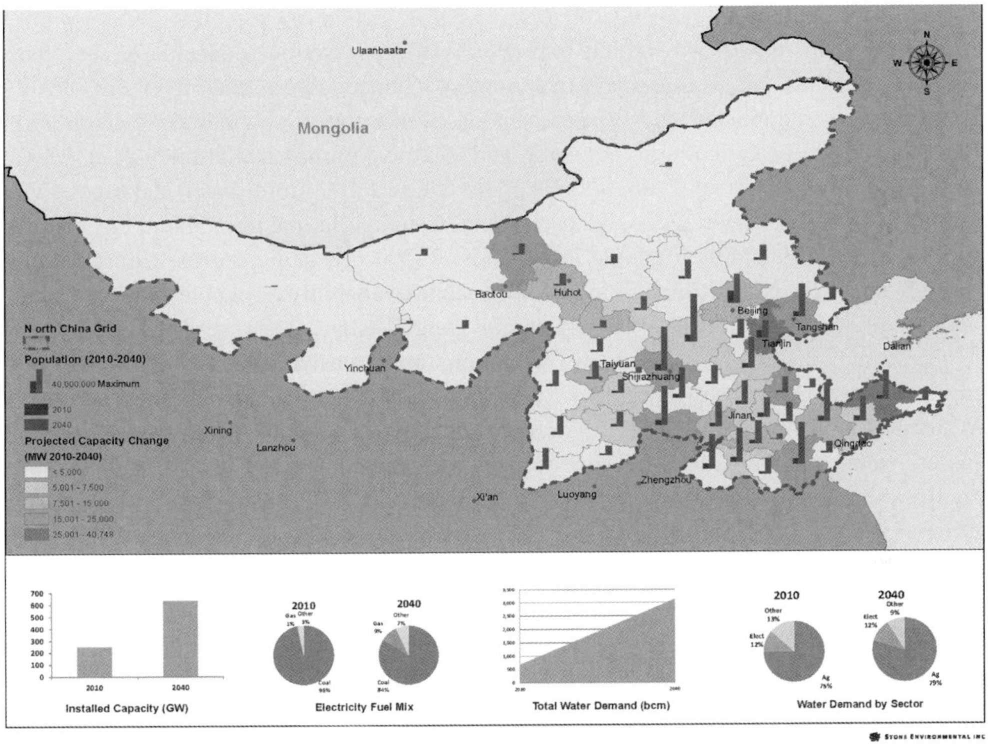

Figure 10.1 Population growth, electricity capacity additions, and water use in the North China Grid, 2010–2040.

According to the most recently available data presented in Figure 10.1, the North China Grid Company operated 247,840 megawatts of capacity in 2010, but this will rise to 631,444 megawatts of capacity by 2040. Most of that growth will occur in Shanxi, Shandong and Tianjin Provinces, as Figure 10.1 also shows. Rapid growth is not unique to the North China Grid; the International Energy Agency (2012) projects a near doubling of China's domestic electricity generation by 2035. The result will be more competition for increasingly scarce water for commercial, agricultural and industrial uses.

However, water security is of paramount importance for the North China Grid. Bloomberg New Energy Finance estimated in 2013 that the five largest electric utilities in China were planning to add new coal-fired capacity in the North China Grid – China Huaneng Group, China Datang Corp., China Huadian Corp., China Guodian Corp. and China Power Investment Corp. – but that 85 per cent of these additions will be located in water-scarce regions, placing pressure on both rivers and groundwater aquifers (Liu 2013). Similarly, HSBC Bank looked at the financial risks for new Chinese power plants and concluded that Hebei, Shanxi, Shandong, Henan and Jiangsu provinces, and three municipalities of Beijing, Shanghai and Tianjin, were "most at risk of water shortages" (Chan 2010). Most of these provinces and metropolitan areas fall within the North China Grid.

Therefore, increased competition for scarce water is a serious concern for the North China Grid, since an estimated 70 per cent of China's rivers and lakes are already significantly contaminated and 30 per cent of river water is so polluted that planners consider it unfit for agricultural, industrial and electrical purposes (Carmody et al. 2010). Nationwide, two-thirds of China's 660 largest cities suffer from water shortages, with 110 facing 'severe' shortages, and each year, water pollution throughout the country sickens at least 190 million people and causes 60,000 premature deaths. Consequently, those living in the North China Grid have per capita availability drinkable water resources about a quarter of the world average (Liu and Yang 2012).

Confirming the dire situation across the nation, one extensive study led by twenty-eight investigators, supported by more than 1,000 researchers and sponsored by the Ministry of Water Resources across all of China's 646 counties found that "heavy water and soil losses" were common with 82 per cent of the worst affected counties – those with the most significant shortages of water – residing in the Yangtze River and Yellow River valleys (Xinhua News 2009). A similar 2012 study led by the Chinese Academy of Sciences, Chinese Academy of Agricultural Sciences and a team of international researchers concluded that "China faces its own 'perfect storm' as rapid economic transition drives increasing per capita demand for water, food and energy with far reaching environmental consequences" (Wang et al. 2012).

Water shortages in China are thus contributing to both mammoth economic losses and social instability far beyond the North China Grid. A drought in August 2009 left 5 million people and 4.1 million livestock without drinking water, and it destroyed 8.7 million hectares of crop land, only to be followed by a rare winter drought in 2009, impacting 10 million hectares and leaving 4 million people and 2 million livestock without adequate water (Carmody et al. 2010). One particularly bad drought in the summer of 2010 affected 51 million Chinese and left more than 16 million people and 11 million livestock without water for days. The drought also 'paralysed' 90 per cent of hydroelectric power stations throughout Southwest China in Yunnan Province and the Guangxi Zhuang autonomous region, causing US$2.8 billion in direct economic losses (China Daily 2010). Collectively, researchers estimate that water shortages in China are now responsible for direct economic losses of at least US$35 billion per year, or 2.5 times the average annual losses due to flooding (Carmody et al. 2010).

France: The water intensity of nuclear power

In France, the electricity sector withdrawals a larger percentage of water than any other case study we investigated (and possibly more than any other industrialised country in the world). Table 10.2 shows that in 2010, the electricity industry withdrew and consumed 65 per cent of the nation's water, but that this figure will rise to 80 per cent by 2040. Électricité de France (EDF), the national grid operator, draws 19 billion cubic metres, or 670 billion cubic feet, of water for its nuclear operations from rivers and lakes every year, more than all the freshwater drawn from the environment and much more than used by agriculture and industry (Kanter 2007).

The reason is that French nuclear reactors are very water intensive – nuclear units withdraw more water per kilowatt-hour than any other commercial source of electricity. Our own modelling, shown in Figure 10.2, expects France to add 106,000 megawatts of capacity between 2010 and 2040, an amount almost equal to the entire grid today.

France's high reliance on nuclear generation makes it particularly vulnerable to water shortages and heat waves. During a hotter than expected summer in 2003, France lost 7 to 15 per cent of its nuclear electricity supply for five weeks in addition to 20 per cent of its hydroelectric capacity. The result was large-scale load shedding and shutoff transmission of exports to Italy (Hightower 2008). The reduction in nuclear generation was due to two reasons: river levels were not high enough to support power plant cooling cycles and water temperatures after the cooling process exceeded environmental safety levels (UNEP 2004). Électricité de France (EDF) had to purchase electricity from the European spot market and from importing countries, where demand drove the price per megawatt-hour for electricity as high as €1,000, more than ten *times* the 'normal' price of about €95 per megawatt-hour. In aggregate, EDF estimated that the heat wave cost them €300 million (Kanter 2007). In order to ensure adequate electricity supply, waivers were granted to allow reactors to exceed water discharge temperature requirements. This prompted criticism from environmental groups that argued the increased thermal discharges further threatened biodiversity already threatened by the summer's high temperatures (Gentleman 2003).

Similar situations occurred during heat waves in 2006 and 2009. During the heat waves of 2006, seventeen reactors had to shut down or limit their electrical output. In 2009, a power-workers strike and a drought combined to take as much as 20 gigawatts of France's nuclear-generating capacity offline (Pagnamenta 2009). In October 2009, for the first time in twenty-seven years, France became a net importer of electricity for an entire month (Réseau de Transport d'Électricité 2009). The heat waves also disrupted nuclear and thermal generation in Germany, the Netherlands and other neighbouring countries,

Table 10.2 Electricity's share of water use (%), 2010–40 (Source: Authors).

	North China Grid	France	India	ERCOT (Texas)
2010	12	65	2	6
2040	12	80	15	9

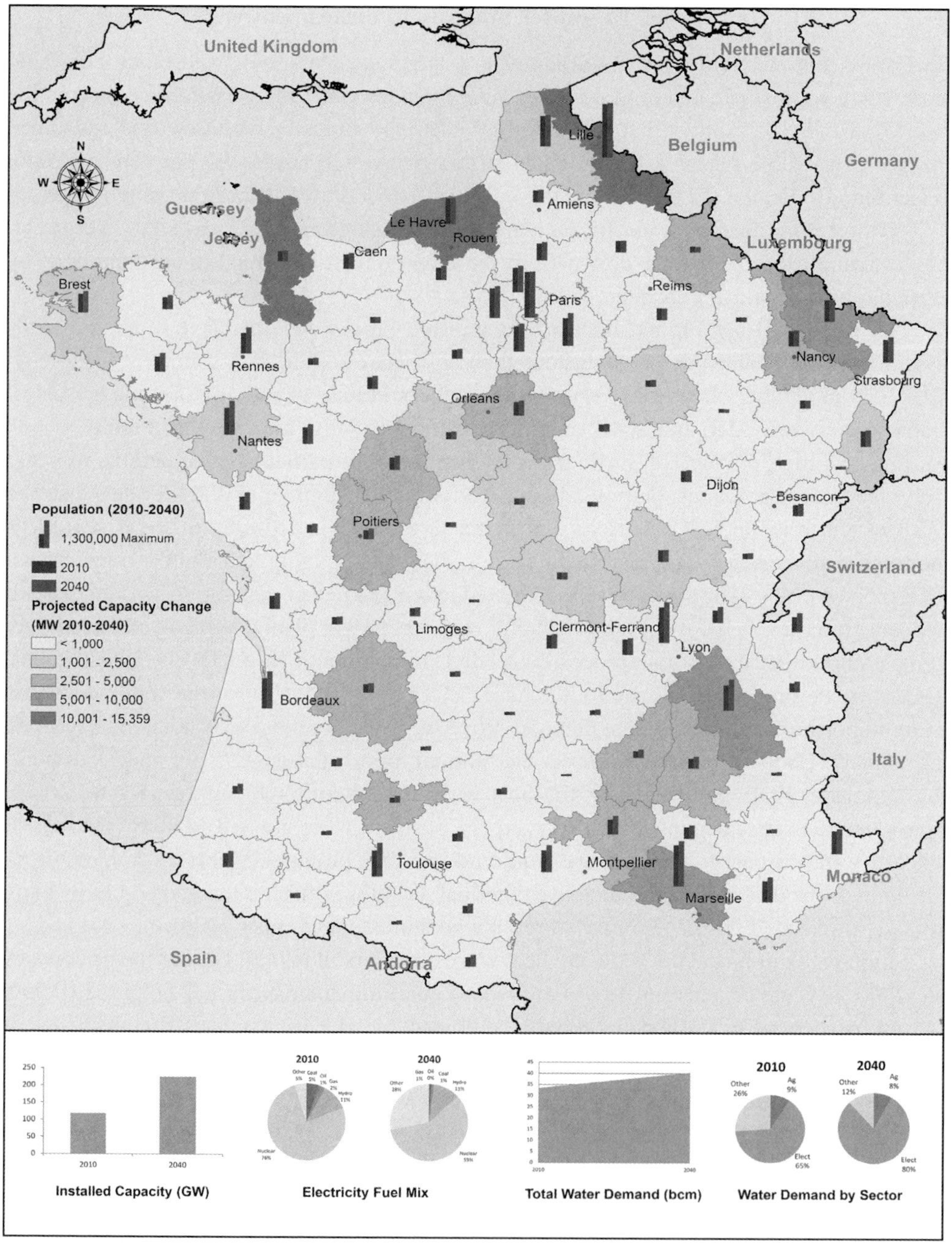

STONE ENVIRONMENTAL INC

Figure 10.2 Population growth, electricity capacity additions, and water use in France, 2010–2040.

threatening the stability of regional electricity supply and increasing the need for French electricity exports (Urban and Mitchell 2011).

India: Massive additions in coal-fired thermoelectric capacity

With an installed capacity of 189,000 megawatts in 2010, India consumes the sixth largest amount of electricity in the world. Coal-fired thermoelectric power plants produced about 80 per cent of the country's electricity, with nuclear, hydropower and other renewable sources of energy making up the remainder. In India, already experiencing a massive shortfall of power, planners expect to add a sobering 613,000 megawatts of thermoelectric capacity between 2010 and 2040 – more than the combined capacity of the *existing* grids for North China, France, India and ERCOT, as Table 10.3 shows. Figure 10.3 illustrates that India's growing economy and population are dramatically increasing demand for electricity in every state. India is both the third largest consumer and producer of coal in the world. India has thus planned truly momentous increases in capacity.

India's water situation may therefore be even more serious than North China. About 52 per cent of India's population lives in water-scarce regions, and 73 per cent of electricity capacity owned by the country's three large utilities – NTPC, Tata Power and Reliance Power – reside in water-scarce or stressed locations (Federation of Indian Chambers of Commerce and Industry and HSBC Bank 2011). India, heavily dependent on groundwater for irrigation and drinking, has lost 15 per cent of its aquifers due to contamination and overuse and is projected to lose 60 per cent by 2030 (Mukherji 2007).

Moreover, in 2012 the Raichur Thermal Power Station (RTPS) had to shut down four of its eight coal-fired units due to lack of water in the Krishna River. The loss of more than 800 megawatts of capacity left Kanataka State with erratic power supply. Most recently, in early 2013, extreme water shortages caused by a delayed monsoon forced the state of Maharashtra to curtail power generation and provide emergency water through the region by tankers. As one commentator astutely noted, it seems that for the moment "people in southern Maharashtra can enjoy either electricity or water, not both" (Umar 2013). The plight of Karnataka and Maharashtra may reveal broad problems facing India's thermal plants. Most of India will be water stressed by 2050, and the World Resources Institute (WRI) reports that the country's quantity and quality of freshwater is in decline due to climate change. India's plan for achieving 'Energy for All' could only worsen the situation,

Table 10.3 Installed thermoelectric capacity (MW), 2010–40 (Source: Authors).

	North China Grid	France	India	ERCOT (Texas)
2010	247,840	117,671	189,000	58,083
2040	631,444	224,354	802,000	78,688
Additions	383,604	106,683	613,000	20,605

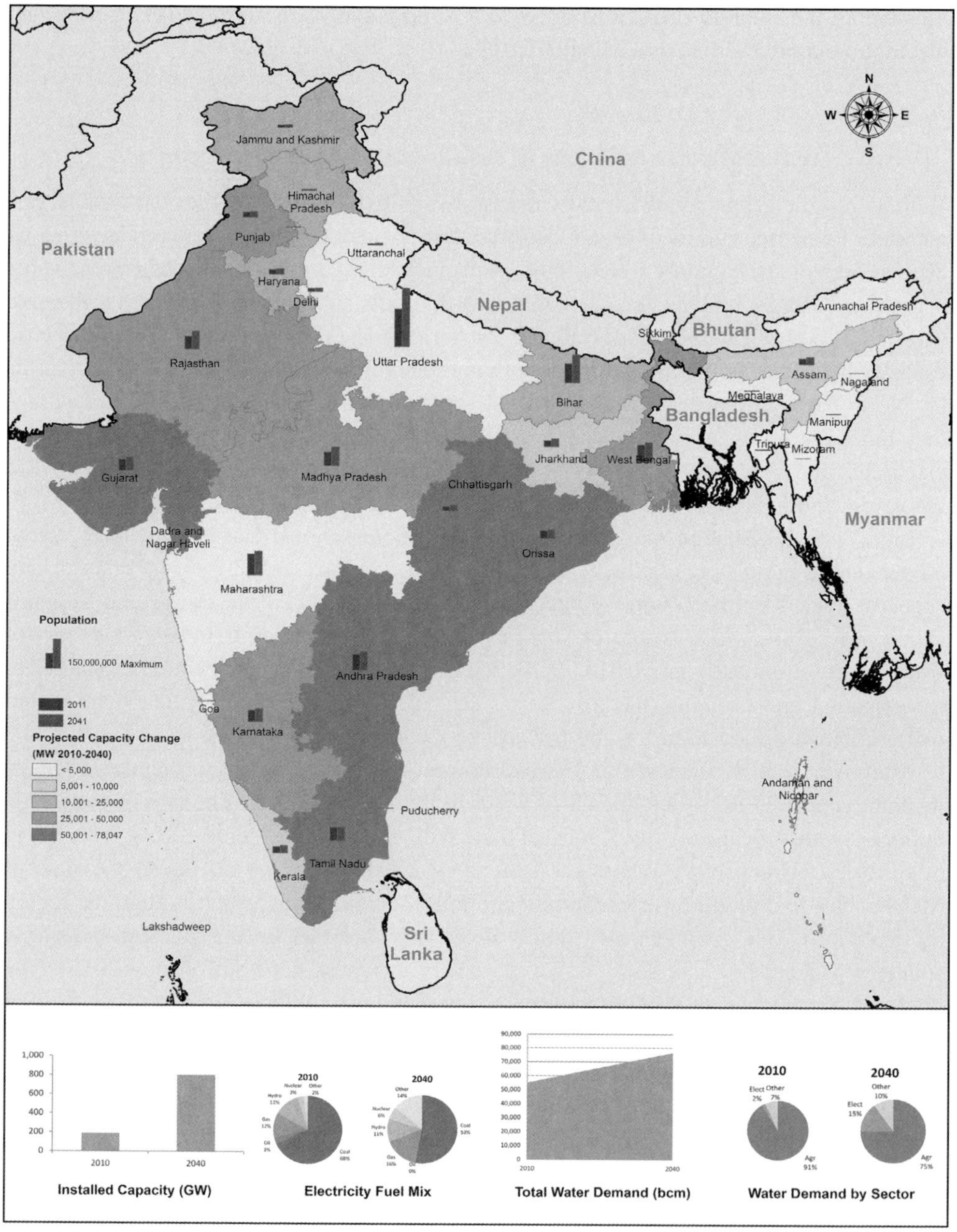

STONE ENVIRONMENTAL INC

Figure 10.3 Population growth, electricity capacity additions, and water use in India, 2010–2040.

as it calls for increasing capacity by 56 gigawatts, yet 79 per cent of these facilities are expected to be built in water-scarce or water-stressed areas (Sauer et al. 2010).

Texas: Dealing with drought through wind energy and gas

With a population of approximately 25 million people, projected to reach 55 million by 2050, and US$1.1 trillion in GDP, the state of Texas currently touts the second largest population and second largest economy of any state in the country, following California on both counts (US Census Bureau 2010). The Electric Reliability Council of Texas (ERCOT) is entrusted with managing the flow of electric power to 85 per cent of the state's electric load and schedules power on 405,000 miles of transmission lines and more than 550 generation units totalling about 58,000 megawatts of operational capacity, almost 6 per cent of the United States' total installed capacity of slightly more than 100,000 megawatts (ERCOT 2012). As Figure 10.4 shows, they expect to add about 20,000 megawatts – equivalent to one-third of today's installed capacity – by 2040.

The ERCOT region is prone to frequent droughts. During the summer of 2011, severe heat accompanied dry weather, and nine cities recorded at least fifty days above 100 degrees Fahrenheit. As air conditioners strained to cool buildings, demand for electricity broke records for days on end (Doggett 2012). Continuing drought conditions threatened to close 3,000 megawatts of thermoelectric capacity in 2012. In East Texas, other plant owners were forced to pipe in water from other rivers so plants could continue to operate and meet electricity demand (O'Grady 2011).

Researchers in Texas project that by 2015, unmet regional water needs would result in an approximately 13 per cent loss in personal income in *all* sectors of the economy. The Texas Water Development Board has estimated that if the state does not ensure it has enough water to meet projected need, the state will have 7.4 million fewer jobs, 13.8 million fewer people and 38 per cent less income by 2050 (US General Accounting Office 2003, p. 67). As one recent study blithely put it, "the free lunches of the original Texas water endowment have been consumed" (Griffin 2011, p. 238).

Despite these challenges, Texas – perhaps providing a ray of hope among our other bleak assessments – demonstrates how rapid deployment of wind turbines and natural gas combined cycle power plants (two of the least water-intensive sources of electricity, along with energy efficiency and solar panels) is helping alleviate the impacts of droughts on the power sector. As Table 10.4 shows, from 2010 to 2040 natural gas is expected to play a significant role in ERCOT's portfolio, along with wind energy (not shown in the table, but is expected to generate 3 to 4 per cent by 2040), which can reduce water needs for power generation. Indeed, a recent study found that continuing to build natural gas combined cycle plants would continue to reduce annual freshwater consumption in the state by an estimated 53 billion gallons per year, largely due to the higher efficiency of the natural gas plants (Grubert et al. 2011). Additionally, Texas led the United States in the installed capacity of wind-powered electricity, and it was the first state to reach 10,000 megawatts of wind capacity in 2010, though the intermittent nature of wind energy meant it supplied only about 1 per cent of ERCOT's actual power.

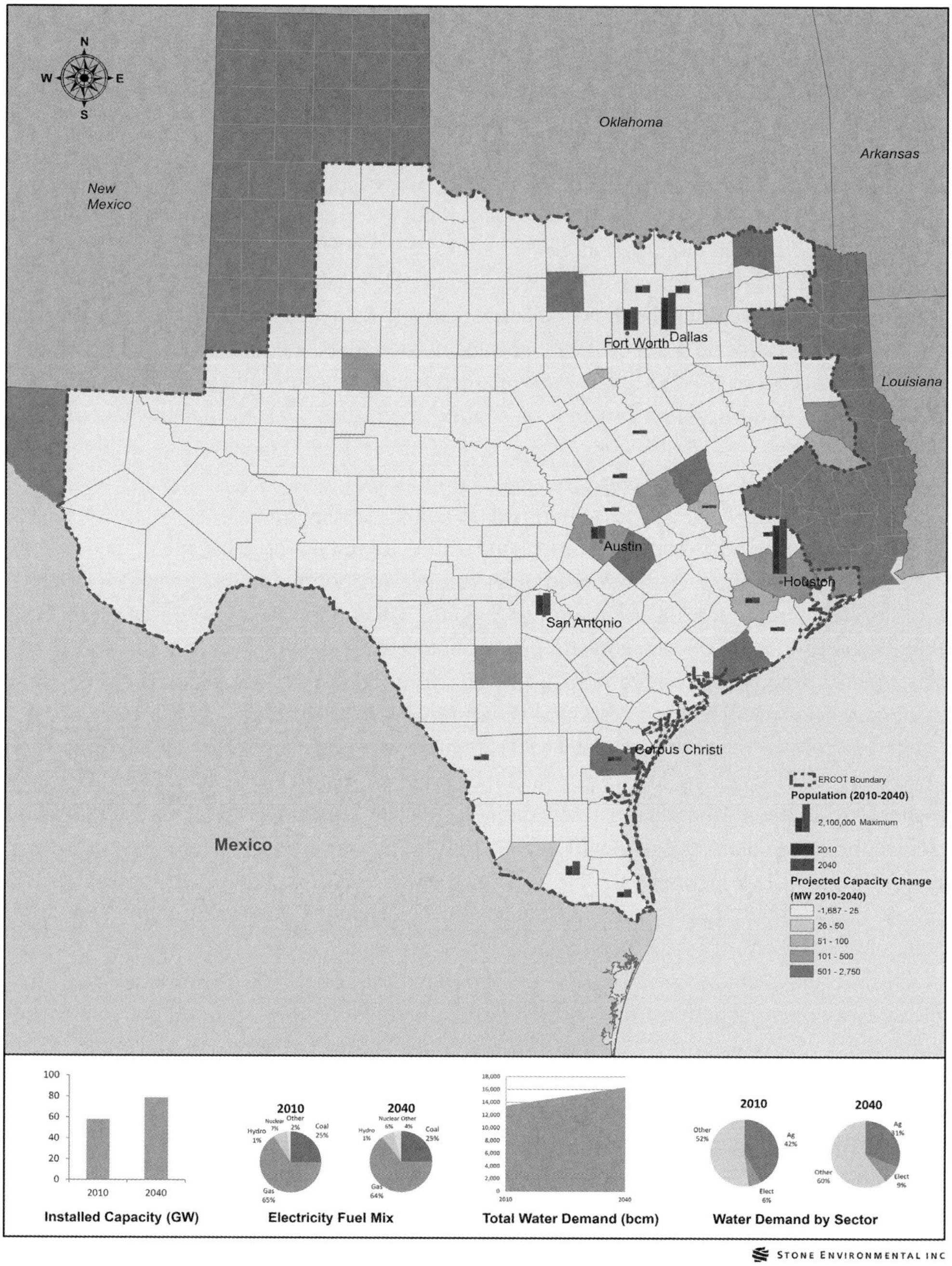

Figure 10.4 Population growth, electricity capacity additions, and water use in Texas (ERCOT), 2010–2040.

Table 10.4 Contribution of thermoelectric power plants (%), 2010–40 (Source: Authors).

	North China Grid	France	India	ERCOT (Texas)
2010	97 (mostly coal)	84 (mostly nuclear)	86 (mostly coal)	97 (mostly gas)
2040	93 (mostly coal)	61 (mostly nuclear)	75 (mostly coal)	96 (mostly gas)

Policy and technology options for avoiding electricity-water crises

Despite the seriousness of the electricity-water nexus challenges confronting different regions of the world, a variety of policy and technology options already exist for addressing it. We argue that an integrated set of six solutions would be most effective at avoiding future water shortages: (1) improving electricity-water data collection and monitoring, (2) increasing research and development funding to minimise water use by thermoelectric power plants, (3) changing permitting and licensing requirements to better consider water use, (4) placing a moratorium on thermoelectric power generation in water stressed regions, (5) promoting energy efficiency and water-efficient renewable electricity sources and (6) changing electricity prices and giving customers more feedback and information.

First, planners and policymakers need to sponsor better quality data on the electricity-water nexus. This solution is a prerequisite to achieving the other solutions below, given that they will depend on high-quality information concerning electricity generation and water usage. The existing quality and availability of data limit policy responses to the electricity-water nexus, even in the United States, which arguably has some of the best freely available data related to energy. The lack of better quality data was also highlighted in a recent study by the Union of Concerned Scientists (2012), which examined more than a decade's worth of water data related to electricity generation in the United States. It concluded, "collisions and near-misses between energy and water needs point to the importance of accurate, up-to-date information on power plant water demand." The analysis identified "a number of gaps and apparent inaccuracies in federal data." Similarly, a recent National Renewable Energy Laboratory report noted, "existing data [for the electricity-water nexus] collected from federal agencies are currently inconsistent and incomplete" (Macknick et al. 2011). Electricity-water data is even harder to come by for China, France and India. A corollary of this conclusion is the need for more co-ordination between water and energy regulators and authorities, interactions that could both generate new data and ensure that existing, reliable information is shared.

Second, a suite of new technologies can reduce thermoelectric power plant water vulnerabilities, and research and development of them should be accelerated. Alternative cooling systems, for example, reduce water use by adapting cooling systems to local water constraints. Non-traditional water sources include municipal wastewater, treated coalmine drainage and water recycled from plant processes. Power plants can also produce water by capturing water in flue gas, desalinating seawater using waste thermal heat and transforming water-intensive procedures to dry processes (Sovacool 2009). Though these options save water, they can be quite expensive. If the North China Grid Corporation were to retrofit all of its once-through cooling systems (more than 100,000 megawatts of coal-fired capacity), they would need to invest at least US$20 billion (Liu 2013), though it's possible they could also save utilities money by improving the electricity sector's resilience to water shortages and droughts

A third solution is incorporating water into integrated resource planning, and altering the permitting and licensing requirements for power plants so that they better incorporate water needs. In China, the National Energy Commission, or the provincial governments, can decline to issue permits unless water is expressly accounted for in Environmental Impact Assessments. In France, EDF is in charge of the power supply for the entire country, and it has the authority for licensing and re-licensing power plants. India's electricity market is regulated by the Central Electricity Regulatory Commission (CERC), a national agency under the umbrella of the Ministry of Power. CERC is the entity responsible for the development of the nation's grid, and is vested with licensing and permitting power for new intrastate generation projects. In the ERCOT region, the Texas Commission on Environmental Quality (TCEQ) may require a statement of environmental, social and economic impacts for any proposal. TCEQ is additionally responsible for water quality permitting and a variety of water conservation programs, making it a likely candidate for the authority to make water licensing a requirement.

Fourth, planners may want to consider a ban or moratorium on all new thermoelectric capacity additions in water-scarce or water-stressed regions. This solution acknowledges that the addition of new conventional power plants has two inherent water-related risks that suggest electric utilities should think carefully before constructing them: They are unable to withdrawal water needed for normal operation in times of scarcity, and new plants can increase water demands, worsening existing shortages.

Fifth, to offset the risks of placing a moratorium on future thermoelectric generators, electric utilities should rigorously implement energy efficiency and demand-side management (DSM) programs. Such actions would not only address impending electricity-related water shortages, but would also improve energy security, lower electricity and water prices, and enhance reliability. History suggests that energy efficiency, DSM and load management practices represent the most feasible and cost-effective way of responding to increases in electricity demand. As Jon Wellinghoff, the Commissioner of the FERC, put it, "the potential benefits from the incorporation of demand response into wholesale markets indicate that a considerable margin of gain is possible from accelerating such activity" (Wellinghoff and Morenoff 2007).

The US Department of Energy (2006) calculated the benefits of DSM and found that it lowers wholesale electricity prices, as costly power plants are displaced and total demand on the system decreases. Generating peak electricity is extremely expensive, often exceeding US$5,000 to US$10,000 per installed kilowatt (meaning a 100 megawatt plant can cost US$750 million to build and require US$75 million per year to operate), implying that DSM should be profitable for *all* utilities.

In situations where energy efficiency and DSM programs are unable to completely offset the need to construct new generation capacity, utilities could rely on wind turbines and solar photovoltaic panels to produce electricity. These two technologies use almost no water to generate electricity, and need only a very small amount for cleaning and maintenance. Cost is not a significant barrier, as wind and solar costs have been rapidly declining worldwide. The marginal levelised cost of building offshore and onshore wind turbines in 2009 – that is, the cost of constructing, operating, maintaining and fuelling a new facility – was between US$0.026 and 0.056 per kilowatt-hour, making them two of the six cheapest sources of power (Sovacool 2008). Solar PV is the most expensive at US$ .39 per kilowatt-hour, but it has already reached grid parity with many expensive peaking plants. The cost for solar PV has also decreased by an average of 7 per cent annually for the past thirty years (Naam 2011). Furthermore, the large pumped-storage hydropower components of Chinese, Indian, and US hydropower programs may also back up greater use of solar and wind and reduce vulnerability to water shortages. The United States already has more than 22,000 megawatts of pumped hydro storage with significant capacity operated by the Bonneville Power Administration in the Pacific Northwest and the 2,772-megawatt Bath County facility operated in Virginia (Erpenbeck 2011).

Sixth, and lastly, electric utilities, public service commissions and power providers should alter electricity and water prices and provide better feedback and information to electricity customers. Three changes would be most significant: cost-reflective electricity and water pricing, altered electricity billing practices and a utility-wide information program to educate consumers.

First, policymakers should consider putting a volumetric price on water so that it can go to the highest-value user. A price on water would also provide an economic incentive for power companies to upgrade facilities to reduce water use. In the same way, electric utilities can alter how they bill customers as well. Currently, consumers are generally unaware of hourly, daily, weekly and seasonal changes in electricity cost, and instead see only a monthly electricity bill. They thus consume electricity indiscriminately. Utility programs that reflect time-of-use through ‘real-time’, ‘interval metering’, ‘time-of-use’ or ‘seasonal’ rates could show customers how electricity production and consumption varies according to the time of day, week and month. Most electricity bills combine charges into a lump sum, making it difficult for consumers to tell how much of the bill results from the individual use of appliances or technologies, how much the bill could be decreased by using more efficient models or how much electricity use can be shifted to off-peak hours.

Because most people remain uninformed about the electricity-water nexus, a second form of feedback could be useful: making water usage associated with electricity generation ‘visible’ by including it in people’s electricity bills. California, for example, was the first state in the country to enact an Advance Recovery Fee on sales of some electronics. Whenever customers purchase new cell phones and televisions, a visible fee between six and ten dollars appears separately on receipts, showing consumers how much it will cost to collect and recycle some of the toxic elements inside their products (Sachs 2006). The same technique could be used in electricity billing, by showing consumers a separate line estimating the number of gallons per litres of water used (and/or its associated cost) to produce the electricity they used within their home that day, week or month. Environmentally minded utilities could also show the amount of greenhouse gas emissions and other desirable environmental attributes.

The last type of information that utilities can offer consumers is education. The US Department of Energy (2008) estimated in 2008 that 88 per cent of Americans would fail even a 'basic' electricity-literacy test. Another study in Kentucky found that 41 per cent of respondents identified coal, steel and oil as 'renewable resources' (Kentucky Environmental Educational Council 2005). Since they remain uninformed and misinformed about electricity, many consumers do not understand the scope and scale of the electricity industry's water needs. Utilities could fund and promote electricity information and education campaigns to teach the public about electricity and water use. These could include grade school classes on energy and the environment; public demonstrations and tours of renewable power facilities; free energy audits and training sessions for industrial, commercial and residential electricity customers; and utility-sponsored document 'clearing houses' that offer websites, free books and articles to help consumers gather and process information in order to make more informed choices about their electricity use.

Conclusion

To conclude, of all the commodities necessary for continued economic stability, water and electricity are two of the most vital. Yet existing global reliance on thermoelectric electricity and its projected growth, coupled with climate change effects, could contribute to potentially massive shortages of water in China, France, India and the United States, as well as in various other regions around the world. Climate change, in particular, is creating acute uncertainty about water resources in the future, thereby increasing the water related risks to the power sector. China and India even see their large-scale hydropower development programs as an explicit part of their national climate change policies, and China has pointed to its hydropower program as part of its Copenhagen commitment to reducing carbon intensity in its economy (Government of China 2007; PMCCC 2008; Pittock 2011) – creating a dangerous scenario where climate goals can trade-off with water conservation and quality goals. Mitigating these risks would require the electricity industry to develop foresight in anticipating how to meet generation needs in a water- and carbon-constrained future.

Fortunately, the technology and policy tools needed to lower the water intensity of electricity generation are widely available and include clean power technologies in the form of energy efficiency, wind farms and solar photovoltaic (PV) panels, as well as alternate cooling technologies, combined cycle natural gas power plants and some shrewd changes in pricing and policy. With this in mind, we advance three key conclusions.

First, the electricity-water nexus makes business as usual completely non-viable. Combined trends in population growth, increasing electricity consumption and increasing water usage indicate that we cannot, and should not, continue to generate electricity the way we do today. Because water is an essential part of the cooling process for conventional water-cooled thermoelectric power plants, they may become wholly unsuitable as the globe enters a new era of accelerated water stress and water scarcity.

Second, including water into electricity planning fundamentally 'changes the game' of how we make future projections about power plants and capacity additions. Accounting for water constraints in electricity planning will become a critical component of maintaining system reliability. Location-specific assessment of water resources can thus help guide new generation and decreasing electric water nexus vulnerabilities.

Third, our study is yet another reason, on top of climate change and energy security, to promote energy efficiency, wind turbines and solar energy. As we have shown, the electricity generation technologies that are beneficial for reducing greenhouse gas emissions correspond with those that require less water to operate, providing crucial benefits on both ends of the scale: minimising water use under a climate change worst-case scenario and mitigating the effects of climate change that will put additional stresses on water resources.

References

Bogardi, J. J. et al. (2012). Water Security for a Planet Under Pressure: Interconnected Challenges of a Changing World Call for Sustainable Solutions. *Current Opinion in Environmental Sustainability* 4: 35–43.

Carmody, L., Doré, D., Peiyuan, G., Nette, A-S., An, J., (2010). *Water in China: Issues for Responsible Investors.*

Chan, W-S. (2010). Does China Have Enough Water to Keep Building Three Power Stations a Week? *China Dialogue*, 5 May.

China Daily (2010). Drought Paralyzes Power Supply. 24 March. www.chinadaily.com.cn/china/2010-03/24/content_9631695.htm.

Doggett, T. and ERCOT (2012). ERCOT's Challenges & Opportunities. Presentation for the Texas Public Power Association, www.ercot.com/.

Electric Reliability Council of Texas (ERCOT) (2012). About ERCOT. www.ercot.com/about/.

Erpenbeck, D. (2011). View on Renewables: Rediscovering Pumped Storage. *Power Engineering Magazine* August: 6.

Federation of Indian Chambers of Commerce and Industry and HSBC Bank (2011). *Water Use and Efficiency in Thermal Power Plants*. New Delhi: FICC.

Gentleman, A. (2003). France Faces Nuclear Power Crisis. *The Guardian*. 13 August. www.guardian.co.uk/news/2003/aug/13/france.internationalnews.

Government of China. (2007). China's National Climate Change Programme (English version). Beijing: National Reform & Development Commission, Government of China.

Griffin, R. C., ed. (2011). *Water Policy in Texas: Responding to the Rise of Scarcity*. Washington, DC: RFF Press.

Grubert E. A., Beach, F. C. and Webber, M. E. (2012). Can Switching Fuels Save Water? A Lifecycle Quantification of Freshwater Consumption for Texas Coal- and Natural Gas-Fired Electricity, *Environmental Research Letters* 7(4).

Hightower, M. (2008). *Energy and Water*. Presentation to the Issues, Trends, and Challenges EPRI Workshop, 8 July, Sandia National Laboratories.

Intergovernmental Panel on Climate Change (IPCC). (2008) *Climate Change 2007: Synthesis Report*. Geneva: IPCC.

International Energy Agency (IEA) (2012). *World Energy Outlook 2012*. Paris: OECD.

Kanter, J. (2007). Climate Change Puts Nuclear Energy into Hot Water. *New York Times*, 20 May.

Kentucky Environmental Education Council (KEEC) (2005). *The 2004 Survey of Kentuckians' Environmental Knowledge, Attitudes and Behaviors*. Frankfort, KY: KEEC.

King, C. W. and Webber, M. (2008). The Water Intensity of the Plugged-In Automotive Economy. *Environmental Science & Technology* 42: 4305–11.

Liu, C. (2013). Water Demands of Coal-Fired Power Plants Are Drying Up Northern China, Report Says. E&E Asia, 25 March.

Liu, J. and Yang, W. (2012). Water Sustainability for China and Beyond. *Science* 337: 649–50.

Macknick, J., Newmark, R., Heath, G. and Hallett, K. C. (2011). *A Review of Operational Water Consumption and Withdrawal Factors for Electricity Generating Technologies*. Prepared under Task No. DOCC.1005.

Mukherji, A. (2007). The Energy-Irrigation Nexus and Its Impact on Groundwater Markets in Eastern Indo-Gangetic Basin: Evidence from West Bengal, India. *Energy Policy* 35: 6413–30.

Naam, R. (2011). Smaller, Cheaper, Faster: Does Moore's Law Apply to Solar Cells? 16 March, http://blogs.scientificamerican.com/guest-blog/2011/03/16/smaller-cheaper-faster-does-moores-law-apply-to-solar-cells/.

O'Grady, E. (2011). Drought Adds to 2012 Texas Power Supply Worry. *Reuters*.

Pagnamenta, R. (2009). France Imports UK Electricity as Plants Shut. *The Times*, 3 July.

Pittock, J. (2011). National Climate Change Policies and Sustainable Water Management: Conflicts and Synergies. *Ecology and Society* 16(2): 25.

Prime Minister's Council on Climate Change (PMCCC) (2008). *National Action Plan on Climate Change*. New Dehli: Government of India.

Réseau de Transport d'Électricité (2009). The French Electricity Report, www.rte-france.com/uploads/media/pdf_zip/publications-annuelles/rte-be09-en-02.pdf.

Sachs, N. (2009). Planning the Funeral at the Birth: Extended Producer Responsibility in the European Union and the United States. *Harvard Environmental Law Review* 30, 51–98.

Sauer, A., Klop, P. and Agrawal, S. (2010). *Overheating: Financial Risks from Water Constraints on Power Generation in Asia India, Malaysia, Philippines, Thailand, Vietnam*. Washington, DC: WRI.

Sovacool, B. K., Valentine, S. V., Bambawale, M. J., Brown, M. A., Cardoso, T. D. F., Nurbek, S., Suleimenova, G., Jinke, L., Yang, X., Jain, A., Alhajji, A. F., and Zubiri, A. (2012). Exploring Propositions about Perceptions of Energy Security: An International Survey. *Environmental Science & Policy* 16(1): 44–64.

Sovacool, B. K. (2008). Renewable Energy: Economically Sound, Politically Difficult. *The Electricity Journal* 21(5): 18–29.

Sovacool, B. K. (2009). Running on Empty: The Electricity-Water Nexus and the US Electric Utility Sector. *Energy Law Journal* 30(1): 11–51.

Umar, B. (2013) Maharashtra: Between Thirst and Darkness. *Telhelka Magazine* 10(6).

United Nations Environmental Programme (UNEP) (2004). *Impacts of Summer 2003 Heat Wave in Europe*. www.grid.unep.ch/products/3_Reports/ew_heat_wave.en.pdf.

Union of Concerned Scientists (UCS) (2011). *Freshwater Use by US Power Plants Electricity's Thirst for a Precious Resource*. A Report of the Energy and Water in a Warming World Initiative.

Urban, F. and Mitchell, T. (2011). *Climate Change, Disasters, and Electricity Generation*. Institute of Development Studies. http://community.eldis.org/.59d5ba58/LatestClimate%20change%20disasters%20and%20electricity%20generation.pdf.

US Census Bureau (2010). Census 2010 Interactive Population Map. http://2010.census.gov/2010census/popmap/index.php.

US Department of Energy (2006). *Benefits of Demand Response in Electricity Earkets and Recommendations for Achieving Them*. Washington, DC: Department of Energy.

US Department of Energy (2008). *Residential Buildings Energy Efficiency Draft Summary Report 8*. Behavioral Research Workshop. Washington, DC: Office of Policy and International Affairs and Oakridge National Laboratory.

US General Accounting Office (2003). *Freshwater Supply: States' Views of How Federal Agencies Could Help Them Meet the Challenges of Expected Shortages*. Washington, DC: US General Accounting Office.

Wang, J., Rothausen, S. G. S. A., Conway, D., Zhang, L., Xiong, W., Holman, I. P. and Li, Y. (2012). China's Water–Energy Nexus: Greenhouse-Gas Emissions from Groundwater Use for Agriculture. *Environmental Research Letters* 7(1): 1–10.

Wellinghoff, J. and Morenoff, D. L. (2007). Recognizing the Importance of Demand Response: The Second Half of the Wholesale Electric Market Equation. *Energy Law Journal* 28(2): 389–96.

Xinhua News (2009). *China's 646 Counties Suffer Heavy Water and Soil Losses*. 25 January. http://english.people.com.cn/90001/90776/90882/6581089.html.

11

Cross-sectoral governance of the climate, energy and water sectors: A 'Rubik's cube' analysis of cross-sectoral co-ordination

GREG OLIVER AND KAREN HUSSEY

Introduction

As previous chapters in this volume attest, there is a compelling case for cross-sectoral management of climate, energy and water. The interdependencies are tangible, the trade-offs are becoming evident to the wider community and the framing of decisions that need to be taken are more focussed. This chapter approaches cross-sectoral co-ordination substantively as a matter of managing governance and institutional dynamics. Each of the sectors – climate (adaptation and mitigation), energy and water (with an emphasis here on urban water systems) – are considered within separate institution and governance frameworks. Each has their own institutional 'logic', comprising history, management culture, policy settings, statutory requirements, markets and operational practices (after Streeck and Thelen 2005). Many chapters in this volume are devoted to the technical aspects of the cross-sectoral nexus, but it is not until the institutional attributes of each of these sectors are defined that a systematic approach to alignment and co-ordination can be developed. Compared to the relatively mature technical analysis of climate, energy and water issues, the governance and institutional dimensions of cross-sectoral integration and co-ordination are relatively poorly examined and under-reported in the academic and policy literature (Kenway 2011). In this chapter, several key questions on cross-sectoral co-ordination are addressed:

- What are the governance and institutional components of the cross-sectoral 'nexus' that need to be managed?
- What are the inherent institutional similarities and differences between the climate, water and energy sectors? How do these enable or hinder effective management of this nexus?
- What would maximise the effectiveness of cross-sectoral governance?

In answering these questions, the chapter makes two significant contributions to the literature. First, we define and explore each of these three sectors as separate 'institutions' and consider cross-sectoral co-ordination as a task of institutional linkage and alignment; second, we frame this co-ordination task as one of co-ordinated governance or 'meta-governance', a task that requires purposeful and deliberative framing and enacting of co-ordination across institutionally different sectors, separate jurisdictions or levels of authority, and modes and mechanisms of governance. Defined this way, meta-governance requires the effective implementation of co-ordinating functions to connect sectors in policy, legislation, regulatory practices and program implementation (after Bell and Hindmoor 2009). This chapter, and indeed this book, suggests the need for deliberate, intentional and incidental forms of co-ordinating governance to optimise the outcomes at the cross-sectoral nexus between climate, energy and water.

We commence by briefly describing institutions and governance before examining the barriers to closer cross-sectoral alignment and co-ordination. The various dimensions of governance and institutions are then discussed, allowing a more nuanced institutional analysis of cross-sectoral co-ordination. Finally, a case study of cross-sectoral climate-energy-water management in California is used to illustrate institutional and governance mechanisms associated with cross-sectoral co-ordination (see also Chapter 16 for initiatives underway in the Western United States).

Institutions

Cross-sectoral co-ordination requires governance elements to operate and interact in multiple dimensions. But what is it that actually comprises these governance and institutional elements? What do we mean by governance, and what are institutions? Governance is associated with mechanisms and arrangements to achieve order and outcomes in society (Rhodes 1996; Pierre and Peters 2000). We adopt a practical definition that "Governance is about the use of institutions, structures of authority and collaboration to allocate resources and co-ordinate or control activity in society and the economy" (Bell 2002). Precisely what form that co-ordination or control takes can vary enormously, and may comprise any of the following governance instruments, operations or management tools:

- policies
- plans, strategies and frameworks
- management programs
- legislation and regulations
- codes and standards
- reporting requirements
- operating procedures
- markets and/or transferable rights
- regulation (asset, environmental and economic)
- governance oversight/monitoring arrangements
- community, customer and stakeholder engagement
- community, customer education and social-based marketing
- ministerial and departmental arrangements
- organisations and organisational structures
- partnerships/alliances/memorandums of understanding
- methods of contracting services/outsourcing

As we explore later in this chapter, these different types of governance instruments may be categorised into different *modes* of governance. The different use of and emphasis given to each of these mechanisms, within each of the three sectors, reflects their different institutional 'logic'. As we explore in the next section, co-ordination between these different institutional frameworks, each with their own unique emphasis on different governance mechanisms, is at the core of the inherent difficulty in cross-sectoral co-ordination and alignment.

Institutional dynamics

All management programs associated with climate (adaptation or mitigation), energy and water are derived institutionally and are expressions of particular governance modes and mechanisms. So, cross-sectoral co-ordination is, by definition, about institutions, how they function and the associated governance mechanisms. But what are institutions? Echoing Douglas North's early seminal work that defined institutions as the "rules of the game in a society or more formally, are the humanly devised constraints that shape human interaction" (North 1990), Connor and Dovers (2004, p. 11) conclude that institutions comprise:

[A]ll socially devised rules of governance such as articles of constitutions, statute and common law, regulations and by-laws, policies, legal rulings, contracts, codes of conduct and honour, and the myriad of social and cultural traditions and norms that constrain the way individuals and groups act in social contexts. This vast array of constraints and guides to appropriate behaviours forms an institutional matrix within which all social actors interact.

Connor and Dovers' description of an 'institutional matrix' highlights the layered complexity inherent within institutions, referring to this as an *institutional system* (2004, p. 18–20). Each of the three sectors (climate, energy and water) represents an institutional system in its own right; each has its rules and norms, formal and informal practices, cultures and requirements. Connor and Dovers refer to these separate institutional systems as having "embedded logic" (2004, p. 17). Streeck and Thelen (2005, p. 19–33) develop this further suggesting that interaction between different institutions (such as climate, energy and water) is represented by the interaction of these different, pre-existing institutional 'logics' that co-exist simultaneously, but which may emanate from very different contexts, paradigms and time periods. Importantly, *institutions* need to be clearly differentiated from *organisations* – a separation of the 'rules of the game' from the players (or actors). While in common language these two terms are often used interchangeably, the difference is much more than semantics. While generally, institutions are characterised by less change and greater longevity, organisations (working within institutional contexts) are subject to more rapid and frequent change according to the administrative arrangements of the day. Institutional elements are the variables, the rules and norms that provide the context for the emergence, form and actions of organisations (Connor and Dovers 2004), though an alternative definition might simply state that institutions are 'the rules of the game'.

Barriers to cross-sectoral governance and institutional alignment

In institutional terms, 'barriers' to cross-sectoral co-ordination arise from misalignment of existing sector-specific governance mechanisms. Overcoming or addressing such barriers by aligning or modifying existing institutions is often problematic. Some governance components and elements will align better than others; some pathways to cross-sectoral co-ordination will be more apparent and readily accessible than others. Those aspects that are incompatible or more difficult to align will likely remain entrenched as institutional barriers.

In an analysis regarding environmental policy integration, Ross and Dovers (2008, p. 246) characterise three principal inherent challenges with cross-sectoral co-ordination, identifying (i) the issue of overcoming existing institutional inertia associated with single-focus individual institutional systems; (ii) the levels of greater complexity associated with framing policy issues more widely; and (iii) the need for effective authority and steering of such cross-sectoral co-ordination. Each of these highlights the characteristics of effective meta-governance required for cross-sectoral co-ordination, a subject we will return to later. Other broader institutional barriers are highlighted by Scott et al. (2011), particularly those associated with governance 'mismatch' between the systemic interdependencies between water, energy and climate (GHG emission management) and the *scale* or level at which their respective governance and institutions operate. This institutional disconnect operates at various geographic scales and between global drivers for greater cross-sectoral co-ordination, local resource management, and individual consumer and end-user behaviours.

With respect to the climate-energy-water nexus, key governance 'barriers' highlighted in the literature are identified in Table 11.1 below. While described as *barriers*, it is apparent that this discussion on barriers highlights possible cross-sectoral co-ordination opportunities and approaches required. Other chapters in this collection expand on many of these barriers.

A 'Rubik's cube' analysis of cross-sectoral co-ordination

Attempts to link related but separate policy areas and sectors have been made. In a deliberative approach to cross-sector integration, the concept of environmental policy integration (EPI) recognises the levels of complexity associated with integrated environmental management. Lafferty and Hovden identify three criteria for evaluating policy integration (2003, p. 8):

- *comprehensiveness:* across temporal and spatial scales, inclusive of all relevant actors and issues
- *aggregation:* regarding the evaluation of policy from an overall perspective
- *consistency:* the different components of integration align with each other (and across different levels of governance)

The basis of effective cross-sectoral, co-ordinated governance is clearly evident in this definition: that co-ordination is most usefully analysed in multiple dimensions. This

Table 11.1 Typical barriers to cross-sectoral alignment between climate, energy and water.

Barrier	Description
Insufficient data and information	There is a chronic lack of relevant information and analysis, specifically regarding water and energy consumption and GHG emissions for various end use sectors (Hussey and Pittock 2012; Kenway et al. 2011a).
Regulatory fragmentation and separation	There is a long tradition of separate regulation and different regulatory practices. Additionally, there have been different approaches and cultures of regulatory implementation, capacity development and resourcing. These separate policy and regulatory practices now hinder the development and implementation of cross-cutting programs (Hussey and Pittock 2012; and GEI 2012, p. 9).
Separate funding and management responsibilities	Fragmented management responsibilities between water and energy within city or metropolitan systems means joint and simultaneous optimisation is difficult (Kenway 2011a, p. 254; Hussey and Pittock 2012).
Separate policy development and strategic decision practices	Separate systems means it is difficult to achieve an assessment of the costs and benefits of various options 'through the lens of a single resource'. There is separate and isolated policy development and strategic decision making (Hussey and Pittock 2012).
Lack of financial incentives (and possibly disincentives)	Existing 'business' or 'jurisdictional' boundaries between water and energy have the effect of limiting the ability of organisations (and governments) to take responsibility for outcomes and impacts outside defined core responsibilities. In effect, one sector's 'externalities' are the other's core business – and vice versa. Cross-sector policy development and integration require matters to be addressed, such as: Who pays for what and how? Who gets credit (eg for energy savings or GHG emissions reductions)? What is the role of customers? How are reduced revenues in one or both sectors accommodated?
Cultural inertia and path dependency	Each with long histories and separate traditions, the cultural differences between the water and energy sectors create a distinctive path dependency for current decisions in each sector (Hussey et al. 2013; Hussey and Pittock 2012).
Lack of integrative research	There is insufficient cross cutting and integrative research on technical issues, systems issues and integrative governance issues (Hussey and Pittock 2012).

chapter extends this concept by adding additional dimensions of institutional dynamics drawn from contemporary literature. Not dissimilar to the faces of a Rubik's cube, components of institutions can be thought of as arranged in multiple *dimensions*. Each of the three sectors (climate, energy and water) may therefore be described *institutionally*, as the combination and sum of elements and processes from each of these dimensions. Specifically, the five dimensions of institutional dynamics used to investigate climate-energy-water governance co-ordination and integration are the following:

1. **Spatial** dimensions: geographic scale at which governance systems are developed and managed
2. **Temporal** dimensions: time scale at which key management decisions are principally taken
3. **Scale** dimensions: dominant scales of governance at which management primarily operates
4. **Functional** dimensions: different functional areas operating within institutional frameworks
5. **Governance** modes: dominant modes (and mechanisms) of governance

Co-ordinated outcomes and management at the climate, energy and water nexus can be represented *institutionally*, by the alignment of various governance and institutional aspects within and between each of these dimensions. What follows is a brief cross-sectoral institutional comparison utilising these dimensions. Within each of these dimensions, key considerations and opportunities for greater degrees of alignment and co-ordination between climate, energy and water institutions are suggested. Within the 'water' sector, our focus here is on urban water systems which include potable water produced for and used within large and small urban communities, as well as wastewater treatment, reuse and any disposal. However, we do note that a similar analysis could readily be applied to rural water systems for agriculture. 'Energy' refers to power generation for urban communities, predominantly large-scale, centralised schemes for electricity and gas generation. 'Climate' in this context refers to the development and implementation of policy, regulatory and direct action processes and initiatives aimed at addressing greenhouse gas (GHG) emission reduction and climate change adaptation.

The following sections illustrate and characterise the dominant institutional features of these sectors with respect to various institutional dimensions. For the purposes of this chapter, this is more illustrative than intended as a definitive, detailed analysis.

Spatial dimension of climate, energy and water institutions

A critical consideration for cross-sectoral co-ordination is the geographic scale(s) at which climate, energy and water institutions operate, because a mismatch in scale can and does impede greater co-ordination and, ultimately, integration. The spatial institutional dimensions of climate, energy and urban water are illustrated in Figure 11.1, based on the geographic scale at which government authority is exercised, and at which decisions are

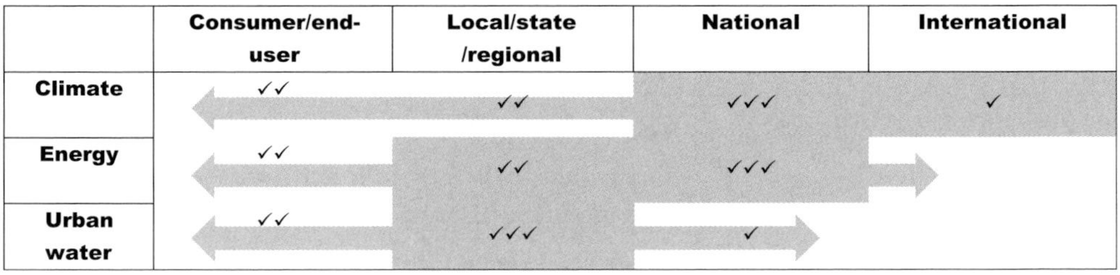

Figure 11.1 Spatial institutional dimensions of climate, energy and urban water.

taken that determine policies, plans and program implementation within individual sectors. In respect to government operations, institutional operations for climate adaptation and mitigation are generally more prevalent internationally and nationally, while urban water and energy institutions operate more regionally and nationally, respectively.

Opportunities for Cross-Sectoral Co-ordination

At the international/global scale:

- Little opportunity for co-ordination except in cases of trans-national river basin management

At the national scale:

- Energy generation, transmission and consumption policies and regulatory practices to be inclusive of water conservation and efficiency

At the local authority/regional/state scale:

- Co-ordinated water resource and energy generation and consumption planning and programs

At the consumer end-use scale:

- Consumer and community decisions to simultaneously optimise efficiencies for water and energy contribution to climate mitigation
- Pricing that reflects the true cost (ie including externalities)
- Climate adaptation through energy and water-efficient design (water- and energy-sensitive developments)

Temporal dimensions of climate, energy and water institutions

In addition to geographic scale, a further consideration for cross-sectoral co-ordination is the temporal scales at which institutional functions operate, such as policy development, program implementation, technological development and performance reporting. The differences in temporal scale illustrated below indicate the scale at which key policy, planning, regulatory and operational decisions are made that may impede or enhance cross-sector co-ordination. Urban water, energy and climate systems operate at different temporal scales. Decisions regarding climate adaptation and mitigation measures most

	Hourly/daily	Weekly/monthly	Annually	Decadally
Climate	✓		✓	✓✓✓
Energy	✓✓	✓	✓	✓✓
Urban water	✓	✓✓	✓	✓✓✓

Figure 11.2 Temporal institutional dimensions of climate, energy and urban water.

commonly operate with longer-term planning and implementation horizons and processes. Real-time consumer choices and behaviour drive the operation of the energy system and forward planning by utilities for capacity development. Institutional operations for water are generally more prevalent at weekly/monthly timescales, and like energy, forward planning for system capacity development undertaken at decadal scales.

Opportunities for Cross-Sectoral Co-ordination

At the decadal/generational scale:

- Co-ordinated sectoral resource management planning and policy development incorporates allowance for weather cycles and climate variability
- Linked research and development associated with phases of technological development
- Cross-sector strategic objectives established by governments, with clear performance targets and measures

At the annual scale:

- Annual operational planning by energy and water utilities undertaken in a co-ordinated manner, with transparency regarding meeting cross-sector objectives and performance targets

At the weekly/monthly scale:

- Integrated and comparative reporting on water, energy and GHG emissions aggregated at city, regional and/or national scale

At the daily/hourly scale:

- Integrated time of use data (eg smart meters to record and report peak power, peak water and GHG emissions)

Scale of operations for climate, energy and water institutions

Key issues for co-ordination

Institutional functions, influences and operations interact simultaneously at multiple levels or scales. In this chapter, we distinguish three levels. This approach enables the locus of actions and decisions to be defined, and also illustrates the wider context in which individuals and organisations operate. The literature on institutional dynamics and

	Micro-level	Mid-level	Macro-level
Climate	✓	✓✓	✓✓
Energy	✓	✓✓	✓
Urban water	✓✓	✓✓	✓✓

Figure 11.3 Institutional scale dimensions of climate, energy and urban water.

transitions is replete with examples of co-ordination and integration at one level not necessarily translating to other levels (Brown et al. 2009). On the one hand, there are often difficulties up-scaling institutional changes from the micro-level to wider-scale implementation. Also, decision makers often lack the authority to effectively transfer higher or macro-level co-ordination mechanisms to mid- and micro-levels for practical implementation. The differences of institutional scale illustrated below represent the scale at which key policy, planning, regulatory and operational decisions are made that impact cross-sector co-ordination.

Opportunities for Cross-Sectoral Co-ordination

Macro-institutional scale:

- Efficacy of international agreements on national GHG emission targets
- International safety regulation regarding non-fossil fuels (eg nuclear power)
- Operations of carbon markets (eg the pricing of carbon and the flow-on implications for incentivising clean energy)

Mid-level institutional scale:

- River basin co-ordinated or integrated water-energy-climate adaptation resource management strategies and plans
- Development of market-based instruments that account for cross-sector outcomes (eg 'pollution' tax that includes carbon and nutrients)
- Statutory reporting requirements for water and energy efficiency and GHG emissions
- Power and water transmission and distribution networks account for local and decentralised source supply (photovoltaic power generation, stormwater harvesting and storage)

Micro-level institutional scale:

- Implementation of new technology in manufacturing for simultaneous energy efficiency and water efficiency, with concomitant reductions in GHG emissions
- Corporate reporting processes and methods that link GHG emissions, energy use and water consumption
- Smart-meters at point of use simultaneously reporting energy and water consumption, and GHG emissions to facilitate improved consumer decisions (though this may be driven by the incentive to reduce costs, rather than a desire to reduce consumption per se).

Functional areas within climate, energy and water institutions

Cross-sectoral co-ordination, viewed through the lens of the various functions operating within climate, energy and water institutions, offers a useful insight into the different layers within which cross-sectoral co-ordination can and needs to occur. Table 11.2 highlights the typical operations that occur within these functions.

While it is critical to identify the different types or groups of functions that comprise institutions, a key consideration for better co-ordination is whether some of these functions offer greater opportunity to enhance integration and co-ordination than others. In the next section, we 'unpack' some of the five functional areas identified earlier and identify specific opportunities for cross-sectoral co-ordination.

While the energy and water sectors each have a long institutional history, institutions dealing with climate are relatively new. This different level of maturity has important implications for cross-sectoral co-ordination. In particular, the depth and breadth of institutional operations within the organisational, procedural, methodological and regulatory domains regarding climate adaptation and mitigation is generally under-developed, representing both challenges and opportunities for cross-sectoral co-ordination. On the one hand, the lack of institutional development in the climate sector, and the accompanying uncertainty, presents difficulties in connecting across sectors. Conversely, however, there may be greater opportunities to integrate or develop co-ordinating measures into and alongside existing energy and water institutional mechanisms. Referring back to the earlier discussion on separate institutional 'logics' for each of the three sectors, and the necessary intersection between these required for co-ordinated governance, it may be that adapting existing institutions for integrated, cross-sectoral outcomes is therefore a more likely scenario in the majority of (but not all) circumstances than developing a host of new and separate institutional operations.

Table 11.2 Typical institutional operations requiring co-ordination between climate, energy and water.

Functional areas (operating within institutions)	Represented by the following typical institutional processes or practices (adapted from Hussey and Schram 2011)
***Substantive* functions**	Policy and program outcomes and strategic objectives
***Regulatory* functions**	Regulatory policy, practices and instruments
***Methodological* functions**	Evaluation and assessment methods and practices Data and information collation methods and tools
***Procedural* functions**	Organisational processes, procedures and methods
***Organisational* functions (roles and responsibilities)**	Organisational role definition Capabilities within and between organisations

	Substantive	Regulatory	Methodological	Procedural	Organisational
Climate	✓✓	✓✓	✓	✓	✓
Energy	✓				✓✓
Water	✓	✓✓	✓✓	✓✓	✓✓

Figure 11.4 Institutional operations requiring co-ordination between climate, energy and urban water.

Opportunities for Cross-Sectoral Co-ordination

Examples of the types of cross-sectoral alignment and co-ordination possible in each of the five functional areas are given below:

Substantive functions

- Formal cross-sector recognition within sector-specific policy/programs objectives
- An overarching meta-policy or planning framework that establishes the framework for sector-specific policies and programs

Regulatory functions

- Formal cross-sector recognition within sector-specific policy/programs objectives
- An overarching meta-policy or planning framework that establishes the framework for sector-specific policies and programs

Methodological functions

- Establishing common metrics to enable comparative assessment between sectors (eg tonnes of GHG emissions/unit of water or energy consumption; dollar cost of abatement per unit of consumption)

Procedural functions

- Common, integrated or aligned approaches to stakeholder information and engagement activities

Organisational functions

- Establishing cross-sectoral organisational objectives (eg energy utilities becoming agents of water efficiency and conservation. Conversely, water utilities becoming agents for energy efficiency as well as net energy producers via biogas production or micro-hydro generation.)

Governance mechanisms within climate, energy and water institutions

In section 4, we 'unpacked' the different scales at which the climate, energy and water institutions operate, distinguishing between the geographic, temporal and operational scales, and in so doing we shed light on the 'institutional mismatch' that exists between the three sectors and which in turn impedes co-ordination. In addition, in section 5, we disaggregated the different institutional functions operating within each sector, with a view to identifying where opportunities for better co-ordination exist, using existing functional arrangements. While the analysis hitherto has analysed the *institutional* arrangements in the three sectors, what remains to be done is an analysis of the dominant modes and methods of *governance* used in each of the three sectors; and, importantly, understanding how those modes of governance enable or impede better cross-sector co-ordination.

As a direct result of their different institutional histories, climate, energy and water are subject to different suites of governance modes. A functional definition of multi-modal governance by Bell and Hindmoor (2009, p. 16–19) identifies five modes: hierarchy, markets, associations, community engagement and persuasion. This distinction will be used here to help identify and categorise the wide range of governance methods and styles deployed by different levels of government.

Hierarchy – Top-down government authority utilising traditional government management programs
Markets – Creation and use of market mechanisms to manage resource distribution, allocation and pricing
Associations – Devolution of governance responsibility to a special interest or stakeholder group
Community engagement – Community engagement to add legitimacy to government decision making
Persuasion – Information, education and marketing programs to influence consumer behaviour and choices

Figure 11.5 highlights the different *primary* and other governance modes most commonly utilised in each of the three sectors.

The differences in the primary governance characteristics of each of these three sectors may be summarised as follows:

Climate – Predominantly traditional government hierarchy (plans, abatement programs, regulations, tax, carbon price control), with a secondary emphasis on market mechanisms with emerging trading schemes.
Energy – Predominantly markets with considerable private enterprise investment and customer choice for energy supply source and retail services. However, there is a secondary primary mode of government regulation regarding price determination and geographic segmentation of bulk generation supply areas.
Urban water – Predominantly traditional government hierarchy (plans, regulations, price control), with a secondary but potentially an emerging emphasis on market mechanisms.

	Hierarchy	Markets	Associations	Community engagement	Persuasion
Climate	✓✓	✓		✓	✓
Energy	✓	✓✓✓	✓		✓
Urban water	✓✓✓	✓✓	✓	✓	✓

Figure 11.5 Primary governance modes in climate, energy and urban water.

Meta-governance of climate, energy and water co-ordination

The preceding discussion highlights the complexity and diversification of governance and institutional dynamics within and between the climate, energy and water sectors. Before providing an illustrative case study of this cross-sectoral co-ordination, we will briefly examine how the necessary depth and breadth of co-ordination occurs.

With shifting governance trends over the past thirty years, there remains some conjecture within the contemporary governance literature (and current government policy and program management practices) concerning the locus and extent of governance authority. Often, this is accompanied by a notion of governance 'plurality': a blurring of boundaries between the state and other actors (as represented by scholars such as Rhodes 1996, p. 660). Such framing of governance is characterised by a shifting focus away from state (government)-centred programs, with a devolution and diversification of governance featuring non-state actors and non-government-based modes of governance (Schout and Jordan 2005). This shifting emphasis has been dominated by a focus on economic reform and markets (market modes of governance) on the one hand, and so-called devolution to and 'empowerment' of civil society on the other. Here, the emphasis and role of the state is sometimes viewed as shrinking – being 'hollowed out'. An alternate position in the governance literature, and the one adopted in this chapter, approaches this in a more nuanced manner. It emphasises the adapting, yet continuing, central role of the state in oversight and co-ordination in increasingly complex multi-modal and multi-stakeholder governance settings (Marinetto 2003 p. 598–600; Davis 2005 p. 331; Bell and Hindmoor 2009 p. 69–70).

With this in mind, reflecting on the diversity of these five *dimensions* of climate, energy and water institutions discussed earlier, effective cross-sectoral meta-governance needs to operate:

- across institutional arrangements, operating at different spatial scales and between jurisdictions;
- between institutions with different temporal patterns of operation and decision-making;
- within and between the different scales or levels of institutional contexts, constraints and influences;
- across the five different functional areas operating within institutions; and
- between different governance modes and mechanisms within individual sectors and between sectors.

Viewed this way, co-ordinated cross-sectoral governance presents as a rather complex mosaic and may be considered from at least three perspectives, as shown in the examples given in the tables below:

1. Alignment between different governance mechanisms within each sector (Table 11.3);
2. Alignment of similar governance mechanisms across the different sectors (Table 11.4); and/or
3. Complementarity between different governance mechanisms across the different sectors (Table 11.4).

Table 11.3 Examples of opportunities for alignment between different governance mechanisms within each sector.

Climate	Government setting household GHG reduction targets *absent* an overarching national objective for emissions and/or industry sector emission targets
Energy	Government setting targets for renewable energy generation while maintaining regulatory barriers to entry into national grid and market for renewable energy suppliers
Water	Governments or water authorities setting urban water efficiency targets while retaining predominantly non-consumptive-based water pricing tariffs

Table 11.4 Examples of opportunities for cross-sectoral alignment and complementarity.

Sector	B: *Alignment* of similar governance mechanisms across different sectors	C: *Complementarity* between different governance mechanisms across sectors
Climate-energy co-ordination	Linked information to end-users regarding energy consumption, energy sources (conventional or renewable) and associated GHG emissions	GHG emission targets and price incentives for high energy consumption Market incentives for electric vehicles and capacity of grid for charging
Climate-water co-ordination	Linked information to end-users from smart water meters, measuring real-time use of hot and cold water and calculating GHG emissions, as well as the associated cost of use.	Regional climate adaptation plans establishing areas for additional or reduced irrigation and the expansion of water markets
Energy-water co-ordination	End-use targets for water efficiency and energy efficiency	Future regional water demand forecasts and available energy capacity to deliver additional transmission and treatment requirements

Some of the specific integrative governance mechanisms reported in the literature include the following:

- enhanced operation and broader cross-sectoral participation in markets (see Chapter 13; Pittock et al. 2013)
- target-setting (sector-specific and cross-sectoral) (Hussey and Schram 2011)
- enhanced cross-sectoral regulation (see Chapter 12)
- enhanced cross-sectoral information and policy 'dashboards' (Pittock et al. 2013)
- regional 'resource coupling' and trade-off scenarios (Scott et al. 2011)
- long-term, sustainable development strategies, with specific timelines and milestones (Hussey and Schram 2011)
- strategic environmental assessment (Chapter 12; Connor and Dovers 2004)
- robust (sustainability) indicators and reporting (Dovers and Hezri 2010)

However, rather than just developing a list of possible integrative measures, how do we know what cross-sectoral co-ordinating measures actually work? What indicators show that effective meta-governance is being implemented? What are the most likely practices for successful cross-sectoral co-ordination and alignment?

Functions of effective meta-governance

As illustrated in this chapter and elsewhere in this book, contemporary *governance plurality* in sectors such as climate, energy and water necessarily involves a wide suite of governance modes and actors. Accepting this, cross-sectoral co-ordination or alignment (meta-governance) represents the *management* of this variety of governance modes and mechanisms; or in other words, the 'government of governance' (Bell and Hindmoor 2009). The state's practical role in ensuring effective cross-sectoral co-ordination can be defined as the performance of six core functions: *steering, effectiveness, resourcing, democracy, accountability* and *legitimacy* (Bell and Hindmoor 2009; and Jessop 2003). The functions shown in Table 11.5 suggest how effective cross-sectoral co-ordination might be implemented.

The following case study highlights the complexity of cross-sectoral co-ordination and gives examples demonstrating the efficacy of integrative governance mechanisms.

Table 11.5 Functions of effective cross-sectoral meta-governance.

Functions of Meta-governance	Description*
Steering	Establishing long-term objectives and goals (strategic planning) Creation of regulations for markets and market mechanisms
Effectiveness	Approval of goals and objectives Establishing performance standards/measures Monitoring outputs and outcomes and adapting goals and strategies to reflect experience and new information Mitigation of inadequacies/shortcomings
Resourcing	Ensuring that adequate capacity and resources are available (from all sources) to achieve the outcomes and meet performance requirements
Democracy	Ensuring compliance with democratic practices and norms Ensuring democratic legitimacy Guarding community at large against unrepresentative sectional interests
Accountability	Clear lines of responsibility and transparency Responsiveness and control Safeguarding accountability arrangements
Legitimacy	Frequently a cumulative outcome from the exercise of the above functions, eg: Government decisions derived from community mandate (democratic legitimacy) Programs are delivered as promised/expected (effectiveness and resourcing legitimacy)

(*Adapted from Bell and Hindmoor 2009)

Case study: Co-ordinating carbon emissions, energy and water in California

The policy to reduce carbon emissions in the water cycle in California represents a case of regulatory practices and quasi-market settings to establish targets that deliver benefits within and across sectors. The example highlights the difference in emphasis of the dominant governance modes and mechanisms used in the cross-sectoral integration.

With a forecast population increase of nearly 60 per cent to 60 million by 2050 (CEC 2005), California faces significant management challenges with energy and water supply. High levels of embedded energy in the supply of treated water, together with the high energy associated with end use, positions the Californian water sector as a key constraint to the state's energy security and GHG inventory. The main relationships between climate, energy and water in California are as follows:

- Seventy per cent of the California's electricity consumption is produced within the state (with considerable water demand associated with thermal power production, gas production and oil processing).
- Energy demand for water transport is high (two-thirds of rain falls in the north and the majority of the population resides in the south).
- California is the twelfth largest emitter of GHG emissions internationally.
- Total *water-related* energy consumption is estimated as 20 per cent of all electricity consumption and 30 per cent of (non-power plant) natural gas consumption (CEC 2005).
- Of this, the California Energy Commission estimated that nearly 80 per cent was associated with urban use.
- The break-up of this energy usage through the water cycle is estimated at 34 per cent in supply and conveyance, 62 per cent in end use and 4 per cent in retail water and wastewater systems (CEC 2005).
- Energy utilised by utilities (in supply, conveyance and wastewater systems) accounts for nearly 8 per cent of statewide annual electricity consumption (CPUC 2010).
- Energy used in the end-use of water represents approximately 13 per cent of statewide energy consumption.

Institutional and governance differences

As highlighted in Table 11.6, there are significant institutional differences between the energy and water sectors in California. With over thirty agencies and entities involved, the energy and water sectors in California represent two distinctly different institutions and governance regimes.

The Californian government's approach to cross-sectoral co-ordination

Intentional co-ordination initiatives in regard to cross-sectoral climate-energy-water management in California have developed incrementally on three platforms, of which we briefly explore each in turn.

Table 11.6 Primary governance differences between the Californian water and energy sectors.

Sector element	Water sector	Energy sector
Market Players	80% supply by public agencies (over 1000 agencies) 20% supply by private Investor-Owned Utilities (IOUs)	75% supply from IOUs (three large private utilities) 25% supply from a number of small, local municipal utilities
Market Structure	Public resource to which parties have rights for beneficial uses as defined by regulation at both state and local government levels	Defined as a commodity sold by regulated monopolies (IOUs) (and some municipal utilities in special districts)
Determining Prices to Customers (Rates)	Public agencies must comply with Proposition 218 (1996) and obtain local approval; IOUs must obtain CPUC approval	IOUs subject to CPUC approval; Public agencies must comply with Proposition 218 (1996)
Funding Capital Improvements	Public utilities historically dependent on federal, state and local government funding	Predominately funded through rates and market investment
Regulatory Oversight	Concentrated at the local level and highly disaggregated	Concentrated at the state level through the CPUC

Statutory requirements for energy security

The 2003 *Energy Action Plan* and the Governor's 'Loading Order' formally established energy efficiency as California's top priority resource (GEI 2012, p. 13). The CPUC's Public Utilities Code requires utilities to first meet their "unmet resource needs through all available energy efficiency and demand reduction resources that are cost effective, reliable, and feasible" (CPUC 2008, p. 1). Specifically, the CEC's energy policy recommendations from 2005 included (CEC 2005):

1. pursue water conservation and efficiency as an effective means of saving energy;
2. reduce peak demand in the water sector to reduce peak demands in the energy sector;
3. generate energy resources co-located and onsite at water-related sites to reduce grid dependency of the water facilities; and
4. target research and development to bring new water-energy beneficial technologies to market.

Requirements for water efficiency measures

Efficiency in the water sector, while not developed as formally as a statutory governance mechanism (as in the energy sector), has nonetheless developed through the work of the Californian Department of Water Resources (DWR). Specifically:

- the 2009 *California Water Plan* (and 2013 update in public review) developed by the Department of Water Resources (DWRs) requires that water authorities include water efficiency measures into their program of works (for which state grant funding is available);
- a 20 per cent water savings target is included in the DWR '2020 Strategy'.

Establishing a GHG emission target

In 2006, the Californian *Global Warming Solutions Act 2006* (*Assembly Bill* [*AB*] *32*) introduced a 25 per cent reduction target in statewide GHG emissions (on 2005 levels) by 2020. As a co-ordinating governance instrument, AB 32 made significant gains and established:

- an executive-level, interagency Climate Action Team (CAT) reporting to the Governor;
- nine sector-specific working groups reporting to the CAT;
- a cross-sector working group (the Water-Energy Team [WET] of the CAT [the 'WET-CAT']), which became the primary vehicle for the investigations into the climate-energy-water nexus, particularly water-related energy consumption; and
- the preparation of a *Scoping Plan* in conjunction with the CAT enunciating how to achieve reductions in GHG emissions in California.

The key co-ordinating functions of AB 32 and the associated WET-CAT and *Scoping Plan* processes provided specific focus for cross-sectoral planning and management. In the first instance, a statewide 33 per cent renewable electricity target was adopted and three significant strategic plans were developed, including the *Water-Energy Strategic Plan* (WET-CAT 2011), the Californian *Bioenergy Action Plan* (2011 and 2012) and the mandatory *Integrated Regional Water management Plans* were required to address the water-energy-climate nexus. In addition, the state's private energy utilities were directed to include water-energy nexus considerations in their 2013–14 Energy Efficiency Portfolios, and energy savings from water-energy measures had to include embedded energy from all private energy utilities. Moreover, higher standards for energy and water efficiency were adopted in the state's building efficiency standards, regulatory approvals for cross-sector investment programs were introduced (ie energy utilities investing in water efficiency), and regulatory approvals were revised so as to decouple regulated investment returns for private energy utilities from just relying on the amount of electricity sold to also including investment in energy efficiency.

Efficacy of cross-sectoral governance

The primary issues to resolve regarding effective cross-sectoral co-ordination climate, energy and water in California remain as:

- different scale of organisations and associated lack of alignment between spatial operating areas;

- different authority for price setting and program investment;
- significantly fragmented authority for decision making in the water sector; and
- different performance measures and metrics between the sectors.

Analysing this process through the *institutional* lens developed earlier in this chapter, the cross-sectoral co-ordination can be evaluated within each of the five institutional dimensions and against the six functions of effective meta-governance. We turn our attention to this analysis, now, by briefly summarising the status of existing cross-sectoral co-ordination in California within each of the five institutional dimensions.

Evaluation within five institutional dimensions

Spatial alignment

The spatial dimension of cross-sectoral co-ordination remains an area for further institutional development. The discrepancy in alignment between the designated operating areas of water and energy utilities represents a significant barrier, particularly in regard to consumer pricing and economic regulatory arrangements. Alignment of energy and water efficiency measures associated with the embedded energy and end-user energy use within the water supply system has been very difficult to achieve. With current utility service area boundaries, differential pricing within each of the energy and water utilities' supply areas would be required. Changing economic regulatory arrangements to allow for this level of co-ordination has to date been considered as socially and politically unacceptable.

Temporal alignment

At least to 2020, a common temporal frame has been established for cross-sectoral co-ordination and alignment. While the three sectors usually operate at significantly different temporal scales (for policy, planning and operations), the basis of a more aligned temporal dimension was established in 2006 with the setting of a statutory 25 per cent reduction target for GHG emissions by 2020. This disrupted the 'business as usual' approach within these sectors and provided a common temporal focus for policy, planning, regulation and operations in all sectors.

Scale alignment

The locus of the cross-sector initiatives has been at the level of unilateral action by the Californian State Government. Had a national approach to carbon emission reduction been implemented (for example, a carbon market), the Californian State Government may have opted to implement different governance co-ordination mechanisms (to those in AB 32).

Transfer and alignment of these initiatives between different institutional scales remains a challenge. At the macro-scale, there is an absence of an effective national policy framework on carbon emissions, and little cross-sectoral co-ordination between water and energy in the United States. Transfer (or up-scaling) of the Californian model nationally seems unlikely. Conversely, alignment with (or state-based implementation of) any future national programs or cross-sectoral framework (should this ever eventuate) could potentially present an issue. Similarly, translating and transferring the state-level targets and policy settings to lower-level institutional scales of operation within California is proving challenging (eg within sector-specific planning and regulatory frameworks for energy and water).

Functional area alignment

Significant cross-sectoral co-ordination and alignment initiatives across each of the functional areas operating within climate, energy and water institutions have been undertaken, as illustrated in Table 11.7.

Governance modes and mechanisms

The misalignment of dominant governance modes, specifically between water and energy, remains a barrier and an area for further institutional development. The significant fundamental governance framework and structural differences between the dominant modes (summarised in Table 11.5) have generated significant issues in achieving effective cross-sectoral alignment. The key implication is that the existing governance framework of the energy and water sectors cannot readily accommodate cross-sectoral outcomes. Policy and regulatory amendments have been required to achieve degrees of alignment and co-ordination. However, the introduction of third sector (climate) and a new governance mechanism (a GHG emission target and cross-agency co-ordination process) has been very instrumental in establishing common ground, linked or aligned governance practices (eg driving modifications to both regulated market investment policies, practices for private energy utilities and forms and practices of management by public water utilities).

Evaluation against the six functions of effective meta-governance

As discussed earlier in the chapter, it is not so much the *presence* of a co-ordinating governance process or mechanism that is significant but how effective it is in developing a closer nexus between the on-going management of climate (specifically GHG emissions), energy and water. By way of example, the *Global Warming Solutions Act 2006* (*Assembly Bill [AB] 32*) and the associated WET-CAT process is evaluated against the six functions of effective meta-governance in Table 11.8. This evaluation has significant instrumental value. While it highlights the significant contribution this initiative has made, it also helps identify and characterise some of the inherent shortcomings in meta-governance effectiveness. A more complete analysis of all the mechanisms and processes highlighted in this case study is warranted, but beyond the scope of this chapter.

A case of purposeful and authoritative cross-sectoral co-ordination

An understanding of the climate-energy-water nexus in California developed from concerns regarding energy security and a realisation of the complex physical resource management interconnections between energy production and consumption and water.

Table 11.7 Examples of co-ordination and alignment mechanisms in institutional functional areas.

Functional Areas (operating within institutions)	Examples of co-ordination and alignment mechanisms or processes
Substantive *Policy and program outcomes and strategic objectives*	AB 32, essentially a climate abatement statute, established the legislative authority and common policy objectives (emission reduction target) and strategic processes (WET-CAT) for cross-sector co-ordination to deliver specified climate abatement outcomes – including in the energy and water sectors.
Regulation *Regulatory policy, practices and regulatory instruments*	CPUC's 'decoupling' of return-on-investment (RoI) income for private energy utilities from dependence solely on volume of energy sold to provide a mechanism for investment in energy efficiency by end-users as an allowable cost with a RoI. The CPUC then went a step further by allowing trials for cross-sector investment by private energy utilities in water efficiency programs (implemented by publicly owned water utilities) in order to gain an energy efficiency dividend.
Methodological *Evaluation and assessment methods and practices.* *Data and information collation methods and tools.*	Following the initial assessment of the energy footprint throughout the entire water cycle by the CEC in 2003 (and subsequent studies by the CEC and CPUC), common metrics (regarding GHG emissions) are being established for reporting efficiency gains and emissions across both water and energy sectors.
Procedural *Organisational processes, procedures and methods*	Large public water and wastewater utilities reducing GHG emissions and boosting the state's renewal energy production through reducing reliance on the traditional power grid by developing capability in large scale energy production through biogas and within-pipe mini-hydro.
Organisational *Organisational role definition.* *Capabilities within and between organisations.*	Organisational capability for both water and energy utilities being enhanced and extended through participation in associative governance arrangements (networks) through such organisations as the Association of Californian Water Agencies and the Californian Water and Energy Coalition.

However, it was only with the introduction of authoritative statewide GHG emissions target and the associated WET-CAT process that more purposeful meta-governance co-ordination was injected into managing the nexus for a specified primary purpose within a defined timescale – 25 per cent GHG emission reductions by 2020.

Notwithstanding the significance of AB 32 and WET-CAT, as outlined, several key barriers remain across several institutional domains. In this case at least, the process of developing increased levels of cross-sectoral co-ordination between climate-energy and water is highly incremental. The historically disparate institutional 'logics' inherent within

Table 11.8 Meta-governance Effectiveness of Assembly Bill 32 and WET-CAT.

Meta-governance function	Description of function	Comment on meta-governance strength Californian *Global Warming Solutions Act 2006 (Assembly Bill [AB] 32)* and the associated WET-CAT process
Steering	Establishing clear objectives and goals (strategic planning) Creation of clear and adequate regulations for markets and market mechanisms	The broader state government objectives of this initiative are very clear and precise – the 25% GHG emission reduction target by 2020. Additionally, the cross-sectoral co-ordination processes through the Governor's Climate Action Team (CAT) are similarly well defined. However, the same level of 'steering' clarity does not extend to all relevant within-sector processes. As this is where the specific implementation occurs, the lack of clarity and direction at this lower level has limited progress in cross-sectoral climate-energy-water co-ordination.
Effectiveness	Formal approval of goals and objectives and outcomes Establishing performance standards/ measures Monitoring outputs and outcomes Mitigation of inadequacies/ shortcomings	Measurement of energy consumption and calculation of GHG emissions as a common platform across the entire energy and water systems has proved very useful. This 'common currency' has fostered and helped legitimise cross-sectoral (energy-water) programs (eg cross-sector investment in efficiency measures).
Resourcing	Ensuring that adequate capacity and resources are available (from all sources) to achieve the outcomes and meet performance requirements	Security of sufficient on-going state government resourcing to manage this program appears doubtful and subject to policy changes and annual budget decisions by the Californian State Legislature.
Democracy	Ensuring compliance with democratic practices/norms Ensuring democratic legitimacy Guarding community at large against un-representative sectional interests	The issue of managing carbon/GHG emissions is well supported in the community and politically. However, there is a disconnect as possible flow-on impacts in the form of increases to utility pricing (energy and water) is met with community and regulator resistance.

Table 11.8 (cont.)

Meta-governance function	Description of function	Comment on meta-governance strength Californian *Global Warming Solutions Act 2006 (Assembly Bill [AB] 32)* and the associated WET-CAT process
Accountability	Clear lines of responsibility and transparency Responsiveness and control Safeguarding accountability arrangements	Relatively clear and transparent at higher levels. Annual reporting by all sectors through the CAT process allows reasonably transparent tracking of progress by the community and elected officials. However, lines and process of accountability of lower institutional levels are less clear.
Legitimacy	Frequently a cumulative outcome from the exercise of the above functions, eg: Government decisions derived from community mandate Programs are delivered as promised/ expected	This program appears to generally have wide support and possibly sufficient, if limited, resourcing to achieve the wider objectives. The extent of its success in the specific water-energy nexus with respect to carbon emissions is yet to be fully realised.
* **Overall Conclusion on the meta-governance capacity of this mechanism** *(Weak/Moderate/Strong)*		**Moderate**

***synthesised qualitative descriptors of meta-governance strength**

Weak	While operable, the instrument or process is largely ineffective in facilitating or driving integrated or co-ordinated resource management in regard to urban water and energy and/or climate, particularly GHG emissions.
Moderate	The instrument or process has some utility value beyond its immediate purpose and intent in supporting incremental steps towards greater alignment between urban water and/or climate, particularly GHG emissions
Strong	The instrument intentionally or incidentally engenders alignment between urban water and/or climate, particularly GHG emissions. New management regimes arise as a direct result of the instrument and its effectiveness in aligning these separate sectors.

the energy and water sectors are increasingly sharing common and linked elements. Specifically, the experience in California represents a case of how water and energy management embodied new objectives and adapted existing governance mechanisms to new demands and objectives (energy security and GHG emission reductions). While AB 32 and WET-CAT represented new institutions at the higher level, cross-sectoral co-ordination has primarily been achieved through the layering of amendments, revisions or additions to existing energy and water institutional processes.

However, it is clear in this case that this layering and co-ordination was achieved through direction and co-ordination (meta-governance) by key actors in the Californian State Government and its agencies. Intentional and purposeful actions were required to co-ordinate across more than thirty government agencies, energy and water utilities and quasi-government organisations. Using the authority of the office of the Governor and the Californian state legislature, specific co-ordination instruments were deployed within the energy and water sectors. State Government authority has been used to establish statutory targets, implement mandatory cross-agency co-ordination processes and adapt regulatory market rules.

Conclusion

Beyond the technical considerations, co-ordinated cross-sectoral management of climate, energy and water is presented in this chapter as a matter of institutional co-ordination and alignment, described as meta-governance. Too often governance and institutional analyses of integration or cross-sectoral management are viewed as a recipe: develop and use the right ingredients (governance methods and tools), and the necessary cross-sectoral co-ordination will follow. In this chapter, we have examined how this view is only partly correct. The 'ingredients' approach leaves the more complex notions of institutional dynamics and questions of how complex institutional frameworks form and adapt under-developed. We have suggested that such an understanding lies at the very heart of designing, developing and implementing new (or modified) institutional functions that achieve effective cross-sectoral co-ordination between climate, energy and water.

Thus, effective management 'at the nexus' requires a more nuanced analysis of what it is that is actually being co-ordinated, how this is being done and by whom. The brief analysis in this chapter of the five *dimensions* in which institutions operate points to a more complex and interrelated suite of cross-sectoral institutional relationships. These have also been described as intersecting historic institutional 'logics', each with their own culture, historic patterns, formal rules, and informal protocols and practices. In this regard, co-ordination here likened the cross-sectoral co-ordination between institutional dimensions to getting the multiple faces of the Rubik's cube to align simultaneously.

Despite, or as suggested in this chapter, because of, the plurality and increasing complexity and disaggregation of governance processes, particularly in regard to the increasing emphasis on markets and market based instruments, the overall responsibility (and accountability) for the design of this cross-sectoral co-ordination remains state-centric. However, in recognition of the need for a systems approach to co-ordination, meta-governance functions need to be inclusive of a variety of mechanisms and non-government actors across a diversity of areas including markets, information, policy, planning and regulation.

From our analysis and others' in this volume and elsewhere, four summary considerations on effective cross-sectoral co-ordination can be distilled. First, it is clear that often it is the enhanced, consolidated or incrementally expanded use of *existing* mechanisms rather than the development of completely *new* mechanisms that is required to facilitate more effective cross-sectoral co-ordination and alignment. Second, with the complexity of existing governance arrangements, there is seldom a single solution; rather, a combination and diversity of mechanisms operating at different scales and in different institutional domains is required. Third, a combination of government-led and market-led mechanisms and transition processes is likely to be needed, and understanding the links between those mechanisms is crucial. Fourth, the mere *presence* of co-ordinating mechanisms is insufficient: not only does the efficacy of meta-governance require the right mix or range of measures, but it requires sufficient capacity, authority (government), community legitimacy, resourcing, transparency and accountability as to be effective. Finally, even then, achieving the desired cross-sectoral co-ordination outcomes is highly relational to the appropriate *strength* (authority and capacity) of meta-governance measures.

References

Bell, S., ed. (2002). *Economic Governance and Institutional Dynamics*. Melbourne: Oxford University Press.

Bell, S. and Hindmoor, A. (2009). *Rethinking Governance. The Centrality of the State in Modern Society*. New York: Cambridge University Press.

Brown, R. R., Keath, N. and Wong, T. H. F. (2009). Urban Water Management in Cities: Historical, Current and Future Regimes. *Water Science and Technology* 59(5): 847–55.

California Energy Commission (CEC) (2005). *California's Water-Energy Relationship*. Final Staff Report. Prepared in Support of the Integrated Energy Policy Report Proceeding (04-IEPR-01E). California: California Energy Commission.

California Energy Commission (CEC) (2005). *Integrated Energy Policy Report*, November 2005, CEC – 100-2005-007.

Californian Public Utilities Commission (CPUC) (2008). *Long Term Energy Efficiency Strategic Plan: Achieving Maximum Energy Savings in California for 2009 and Beyond*. California: Californian Public Utilities Commission.

Californian Public Utilities Commission (CPUC) (2010). *Embedded Energy in Water Studies, Study 1: Statewide and Regional Water-Energy Relationship*. California: Californian Public Utilities Commission.

Connor, R. and Dovers, S. (2004). *Institutional Change for Sustainable Development*. Cheltenham, UK: Edward Elgar Publishing.

Davis, J. (2005). Local Governance and the Dialectics of Hierarchy, Market and Network. *Policy Studies* 26(3/4): 311–35.

GEI Consultants (2012). *California's Water-Energy Nexus: Pathways to Implementation*. California: GEI Consultants.

Dovers, S. R. and Hezri, A. A. (2010). Institutions and Policy Processes: The Means to the End of Adaptation. *Wiley Interdisciplinary Reviews: Climate Change* 1: 212–31.

Hussey, K. and Pittock, J. (2012). The Energy-Water Nexus: Managing the Links between Energy and Water for a Sustainable Future. *Ecology & Society* 17(1): 31.

Hussey, K. and Schram, A. (2011). Policy Integration and the Energy-Water Nexus: Accounting for, and Managing, the Links. In *Securing Sustainable Energy Futures in Europe and Australia*, ed. Winand, P. and Pearman, G. Brussels: P.I.E PeterLang Publishers.

Jessop, B. (2003). *Governance and Metagovernance: On Reflexivity, Requisite Variety, and Requisite Irony*. Lancaster University, www.comp.lancs.ac.uk/sociology/papers/Jessop-Governance-and-Metagovernance.pdf.

Kenway, S. J., Priestley, A., Cook, S., Seo, S., Inman, M., Gregory, A. and Hall, M. (2008). *Energy Use in the Provision and Consumption of Urban Water in Australia and New Zealand*. Canberra: CSIRO Publishing.

Kenway, S. J., Lant P. A. and Priestley, A. (2011). Quantifying Water-Energy Links and Related Carbon Emissions in Cities. *Journal of Water and Climate Change* 2(4): 247–59.

Kenway, S. J., Lant, P., Priestley, A. and Daniels, P. (2011). The Connection between Water and Energy in Cities: A Review. *Water Science and Technology* 63(9): 1983–90.

Lafferty, W. and Hovden, E. (2003). Environmental Policy Integration: Towards an Analytical Framework. *Environmental Politics* 12(3): 1–22.

Marinetto, M. (2003). Governing Beyond the Centre: A Critique of the Anglo-Governance School. *Political Studies* 51: 592–608.

North, D. (1990). *Institutions, Institutional Change and Economic Performance*. Cambridge University Press.

Pierre, J. and Peters, B. G. (2000). *Governance, Politics and the State*. New York: St Martin's Press.

Pittock, J., Hussey, K. and McGlennon, S. (2013). Australian Climate, Energy and Water Policies: Conflicts and Synergies. *Australian Geographer* 44(1): 3–22.

Rhodes, R. (1996). The New Governance: Governing without Government. *Political Studies* 44: 652–67.

Ross, A. and Dovers, S. (2008). Making the Harder Yards: Environmental Policy Integration in Australia. *Australian Journal of Public Administration* 67(3): 245–60.

Schout, A. and Jordan, A. (2005). Coordinated European Governance: Self-Organizing or Centrally Steered? *Public Administration* 83(1): 201–20.

Scott, C., Pierce, S., Pasqualetti, M., Jones, A., Montz, B. and Hoover, J. (2011). Policy and Institutional Dimensions of the Water-Energy Nexus. *Energy Policy* 39: 6622–30.

Wolfgang, S. and Thelen, K. (2005). Institutional Change in Advanced Political Economies. In *Beyond Continuity*, ed. W. Streeck, S. Wolfgang and K. Thelen. Oxford University Press.

12

Regulation of the nexus

JACQUELINE PEEL

Introduction

The climate-energy-water nexus is a relatively new focus of interdisciplinary research and policy attention (Proust et al 2007; WBCSD 2010; PMSEIC 2010; Water in the West 2013). Research and practice regarding *regulation* of the nexus is at an even more nascent stage. Nevertheless, regulation – establishing rules, laws, and institutional and decision-making frameworks – is emerging as a critical dimension of managing the trade-offs that arise at the intersection of climate, energy and water policies. On the one hand, well-designed regulation that takes account of all elements of the nexus can facilitate more integrated decision making and management, in some cases leading to outcomes that optimise benefits for all three sectors (Pittock et al. 2013). On the other hand, regulatory systems in different sectors that operate independently of one another – a situation described as "regulatory fragmentation" (Craig 2008) – may present barriers to management of the nexus and to the implementation of innovative technological options.

This chapter examines regulation of the climate-energy-water nexus. It focusses on the barriers to integrated climate-energy-water decision making posed by regulatory fragmentation, as well as case examples of ways that existing regulatory frameworks for climate, energy and water might be adapted to better manage the nexus. Case studies are drawn from specific jurisdictional contexts – Australia, the United States and the European Union – but are analysed as examples of regulatory practice that might offer useful lessons for regulation of the nexus in other regions and countries. The final section of the chapter considers some of the new challenges facing regulatory efforts at the nexus. These include current or imminent developments: for example, the climate change and water impacts of efforts to harvest 'unconventional' energy resources, including through the use of techniques of hydraulic fracturing (fracking) (Wiseman 2009; NWC 2010). Other challenges lie on the horizon: for example, the water and energy implications of proposed technologies to deal with carbon pollution and climate change, such as carbon capture and storage (Pittock et al 2013), or geoengineering (Royal Society 2009). The chapter concludes that with targeted reforms designed to enhance integration and flexibility, existing regulatory frameworks and tools offer significant potential to address and manage the trade-offs that arise at the climate-energy-water nexus.

Regulatory fragmentation in the climate, energy and water sectors

Regulation is a broad term understood in different ways by different actors and institutions. In a narrow, legal sense, regulation might be confined to the rules developed by government departments or independent agencies to implement a legislative mandate (Morgan and Yeung 2007). More commonly, regulation is conceived broadly as "the intentional activity of attempting to control, order or influence the behaviour of others" (Black 2001, p. 142). This activity may be undertaken by governments directly, through the promulgation of standards backed up by sanctions, or indirectly, through creating incentives for behavioural change by private actors (Gunningham, Grabosky and Sinclair 1998). Regulation is also frequently involved in the establishment of institutions and associated decision-making frameworks for managing a particular problem. For example, markets for energy, water and carbon are invariably accompanied by a detailed regulatory regime designed to constitute elements of the market (eg water rights or carbon pollution permits) and to police the market's operation (Godden and Peel 2010).

A defining feature of regulatory systems that have been adopted to manage water and energy production is their sectoral focus and largely self-contained nature (Pittock, Hussey and McGlennon 2013). For example, many developed countries have evolved, over time, quite sophisticated systems for the regulation of water treatment, distribution, use and conservation on the one hand, and for energy production, transmission and distribution on the other. Constitutive instruments for each regulatory regime generally comprise separate pieces of legislation and associated subordinate rules that make little reference to issues arising in other sectors. More recently, several jurisdictions have added an additional layer to this regulatory complex in the form of dedicated climate change legislation, which seeks to introduce an economy-wide carbon price or otherwise to regulate emissions of greenhouse gases (Zahar, Peel and Godden 2013). Efforts to adapt to inevitable changes to the climate, including the implications of climate change for water and energy use, are generally regulated separately and apart from mitigation efforts (Bonyhady, MacIntosh and McDonald 2010; Gerrard and Kuh 2012). The result is a fragmented array of regulatory requirements that, at best, neglects opportunities for enhancing interactions at the climate-energy-water nexus and, at worst, leads to antagonistic or perverse outcomes (Pittock 2012), sometimes referred to as 'maladaptation' (Barnett and O'Neill 2010). Further complications are introduced where issues relevant to the climate-energy-water nexus, such as greenhouse gas emissions reduction requirements, are regulated at multiple governance levels. This may include subnational regulations (issued by local and state or provincial governments), national or federal regulations, regional or transnational regulations, and international-level regulations (Osofsky 2011; Peel, Godden and Keenan 2012).

As a federal nation, which is an active participant in transnational and international governance arrangements, Australia provides an example of how regulatory fragmentation manifests at the climate-energy-water nexus. The following table (see Table 12.1) illustrates the variety of regulatory instruments impacting upon Australian climate, energy and water decision making. Victoria is used as an example of state regulation, but in practice each of the six states and two territories have separate regulatory schemes.

Table 12.1 Regulatory fragmentation at the climate-energy-water nexus in Australia.

International regulation	**Transnational regulation**
United Nations Framework Convention on Climate Change Kyoto Protocol Ramsar International Wetlands Convention Convention on Biological Diversity	European Union Emissions Trading Scheme
Federal regulation	**State regulation (eg in the state of Victoria)**
National Water Initiative Water Act 2007 National Electricity Law and National Electricity Rules Environment Protection and Biodiversity Conservation Act 1999 Clean Energy Act 2011 Energy Efficiency Opportunities Act 2006 Renewable Energy (Electricity) Act 2000 Carbon Credits (Carbon Farming Initiative) Act 2011 National Climate Change Adaptation Framework 2007	Water Act 1989 Catchment and Land Protection Authority Act 1994 National Electricity (Victoria) Act 2005 Electricity Industry Act 2000 Environment Protection Act 1970 Planning and Environment Act 1987 Environmental Effects Act 1978 Climate Change Act 2010 Victorian Energy Efficiency Target Act 2007 Victorian Renewable Energy Amendment Act 2009 Victorian Coastal Strategy 2008 Victorian Climate Change Adaptation Plan 2013

Other jurisdictions exhibit higher levels of regulatory fragmentation. Indeed, Australia is somewhat unusual in the degree of integration that has been achieved within the separate regulatory frameworks applicable in the climate, energy and water sectors due to the introduction of national markets for carbon, electricity and water respectively (Pittock et al. 2013). As a ‘quasi-federal’ governance system with a high level of policy integration among its constituent member states (Schmidt 2005), the European Union has also achieved significant regulatory harmonisation in the climate, energy and water sectors. Even so, integration between EU legislation comprising the Climate and Energy Package (2009) and water-related legislation, including the Water Framework Directive (2000), Urban Wastewater Treatment Directive (1991) and the Drinking Water Directive (1998), is more limited (Henriksen, Hussey and Holm 2011; WSSTP 2011). Jurisdictions lacking integrated regulatory systems, such as the United States, have a veritable patchwork of laws applicable across the sectors that vary considerably from issue to issue, between state and federal jurisdictions, and from state to state. Describing regulation of the energy-water nexus, prominent US water and climate law academic, Robin Kundis Craig, comments:

In the United States, the very concept of developing energy and water policies together is relatively new and not fully implemented at any level of government – national, state or local.... [E]ven without the energy complication, water management is a classic example of regulatory fragmentation, often leading to uncoordinated and detrimental results (Craig 2010, pp. 226–7).

In this case, the level of regulatory fragmentation within sectors may be such as to pose barriers to effective management of the sector itself, let alone to integration with other sectors at the nexus. For example, Buzz Thompson summarises the situation of US water regulation as one of "significant fragmentation", which is both "substantive – with separate agencies holding responsibility over different but often closely related substantive issues – and geographic – with a single watershed or water basin often crisscrossed by multiple geopolitical boundaries" (Thompson 2012, pp. 202–3). In a recent article, Hari Osofsky and Hannah Wiseman describe a similar situation in US energy law. According to the authors, US energy law is 'deeply fragmented' with each source of energy having a distinct legal regime, and divergent federal, state and local approaches to regulation (Osofsky and Wiseman 2013).

Adapting regulatory frameworks to manage the nexus

Given a context of regulatory fragmentation, achieving better integration of relevant regulatory frameworks becomes a key task in improving regulation of the nexus (ATSE 2012; Olsson 2012). Integrated regulation in the context of interdependent ecosystems is by no means a new concept. Integrated forms of environmental planning and natural resource management have long been advocated in environmental law and policy (Cairns Jr and Crawford 1991; Guruswamy 1991; RAC 1993; Farrier 2002; Jones 2002; Campbell and Sayer 2003; Hanna and Slocombe 2007). In some cases, they have been implemented with a reasonable level of success, most notably in integrated catchment (or water resources) management (Mitchell and Hollick 1993; Nelson 2005; Hassing et al 2009; Thompson 2012). More broadly, the overarching concept of sustainable development that lies at the core of international environmental law and many domestic environmental regulatory systems is an inherently integrative notion (Dovers 2005). Existing models of integrated environmental management and sustainability practices thus offer useful tools that can be applied to the task of integration of regulation at the climate-energy-water nexus.

The three examples discussed below illustrate different potential approaches (and their limitations) for integrating regulation of the nexus. They range from weak integration (ie a requirement to consider – but not optimise outcomes for – other elements of the nexus) to strong integration (regulations that seek to maximise synergistic, mutually beneficial outcomes for all elements of the nexus). Figure 12.1 illustrates this notion of a spectrum of regulatory options for integration at the nexus.

Integrating consideration of the nexus in decision making

A weak form of regulatory integration can be achieved through the inclusion, in the legislation or other regulatory instruments applicable to one element of the nexus, of requirements for decision-making to take account of other elements of the nexus. Australian regulation of the rural water sector offers an example of this approach (see Box 12.1).

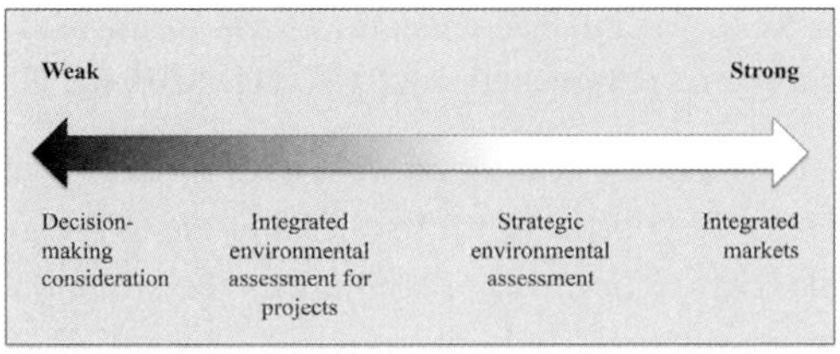

Figure 12.1 Spectrum of regulatory approaches for integration at the nexus.

Box 12.1 Integrating climate change considerations into rural water planning in Australia

Australia has undergone a series of water law reforms prompted by water scarcity (eg the major 1997–2009 drought), coupled with concerns over how water availability might be further impacted by climate change. These efforts included the National Water Reform Framework concluded in 1994, and the 2004 National Water Initiative that established the foundations of Australia's national water market (Foerster 2009; Pilz 2011). The Murray–Darling Basin, Australia's major inland water system that traverses the territory of five of the country's six states, has been the epicentre of these reforms, culminating with the enactment of the national *Water Act* in 2007 (Connell 2011). This legislation provides for the establishment of environmentally sustainable limits on the quantity of water that can be extracted from the Basin's water resources. Such limits are set out in a Basin Plan devised, and regularly reviewed, by the Murray–Darling Basin Authority (Water Act 2007, s 19). This agency is also responsible for making recommendations to the federal Water Minister regarding whether to accredit state water resources plans under the Basin Plan (Water Act 2007, s 63). The legislation specifies certain mandatory content for the Plan. This includes a requirement to identify risks to the condition or availability of Basin water resources arising from the effects of climate change (Water Act 2007, s 22[1]). The statutory obligation to take account of climate change in planning for sustainable water use in the Basin represents "a certain level of responsiveness" of the regulatory framework for rural water allocation to climate change concerns (Foerster 2011, p. 4013). The first Basin Plan, approved in November 2012, includes as part of its environmental objectives the goal of ensuring that water-dependent ecosystems are resilient to climate change (Basin Plan 2012, paras. 5.03, 8.04, 8.07). In relation to state water resources plans seeking accreditation, the Basin Plan also requires a description of how water will be managed should climatic extremes occur (Basin Plan 2012, para. 10.51).

Outside the rural sector, integration of climate and energy considerations in Australian water regulatory frameworks is at a more nascent stage. Dedicated bulk water entitlements for major fossil fuel power stations – preserved even under conditions of drought – point to a major gap in the integration of climate, energy and water planning. The Australian Government's 2012 Energy White Paper implicitly acknowledges this, stating:

In the future, there is likely to be a greater need to allow generators and other water users in the energy sector to manage risks and minimise costs flexibly. Giving generators the option to purchase tradable water rights in open markets can increase flexibility in using water for the production of energy in response to changing water availability (Australian Government 2012, p. 202).

The importance of integrating water and energy decision making is also only just beginning to surface in Australian urban water policy documents. The 2004 National Water Initiative called for the development of national guidelines on "water sensitive urban design," including the use of recycled water and stormwater (Commonwealth of Australia 2004, para. 92) – technologies that may have significant energy costs. During the 1997–2009 drought, policies designed to make cities drought-proof saw the development of costly, energy-intensive water supply projects such as desalination plants, albeit with their additional energy needs supplied by renewable energy (NWC 2012). Overall, the energy implications of policies regarding urban water supply and use (eg for air-conditioning cooling towers) are not at the forefront of the regulatory agenda in this area.

One of the interesting features of the Australian experience has been the extent to which concern over climate change has been a driver for the introduction of more integrated policy and decision making (as opposed to the United States, for example, where energy security has been the more prominent issue).[1] New climate change legislation in Australia has thus provided an opportunity to improve the integration between climate-related policy and decision making in other sectors through interjecting climate change considerations into existing regulations for those sectors. An example is the *Climate Change Act 2010* that applies in the Australian state of Victoria. While this legislation has been 'streamlined' by the present Victorian government, removing a number of its substantive requirements, it retains a procedural requirement for decision making under other specified statutes to "have regard to" the potential impacts of climate change relevant to the decision and the potential contribution to the state's greenhouse gas emissions (Climate Change Act 2010, s 15). Specified legislation coming within the scope of the provision includes water-relevant legislation such as the *Catchment and Land Management Act 1994* (important for integrated catchment management in Victoria), the *Coastal Management Act 1994* and the *Water Act 1989* (although the requirement to take climate change into account is limited to the consideration of draft Sustainable Water Strategies – a form of strategic planning under the Act).

Integrated and strategic environmental assessment

Well-established sustainability mechanisms, such as project-level environmental impact assessment (EIA) and strategic environmental assessment (SEA) for policies, plans and programs, offer significant potential for integrating decision making across the climate, energy and water sectors (Hussey and Pittock 2012). Both EIA and SEA offer a structured

[1] See, however, the interagency report of the Bureau of Reclamation and US Department of Interior on Managing Water in the West (2013), which responds to requirements under the SECURE Water Act 2009 (codified at 42 U.S.C. § 10363) for various federal agencies to gather information on the potential impacts of climate change on western water resources and to identify and study potential adaptation strategies. These powers are discussed in Benson (2013) who notes that the SECURE Water Act is "at best unclear ... regarding authority to *implement* these strategies" (id, 1058). See also The President's Climate Action Plan (2013), which mentions issues of water use and quality as part of actions to improve 'climate resilience'.

approach for obtaining and evaluating information in order to identify and ameliorate potential environmental impacts in advance of development taking place (Abaza, Bisset and Sadler 2004). Whereas EIA is applied most commonly in respect of physical developments and discrete projects, SEA targets 'higher' level policies, plans and programs that may have broad-scale, cumulative environmental impacts (Dovers and Marsden 2002). The two tools are closely related, particularly with the trend, over time, for EIA to integrate consideration of the indirect and cumulative impacts of development on the environment (eg CEQ 1997). For regulation of the climate-energy-water nexus, EIA and SEA offer the prospect of moderate integration by providing a decision-making framework within which climate, energy and water impacts and trade-offs can be evaluated.

An advantage of looking to EIA and SEA as approaches for improving integrated regulation of the nexus is that these mechanisms are already entrenched in environmental regulatory frameworks. EIA is the older tool. It originated in the US National Environmental Policy Act of 1969 and has been progressively adopted in a very large number of national legal systems (Thomas and Elliott 2009; Sands and Peel 2012). The technique also applies internationally. In 2010, the International Court of Justice declared that EIA "may now be considered a requirement under general international law" where proposed industrial activity threatens significant transboundary impacts, especially for shared resources such as boundary rivers (ICJ 2010, para. 203).

SEA is a more recent addition to the toolkit of environmental law. It was first adopted in the national laws of developed countries, but since 2009 there has been a sharp increase in the use of SEA by developing countries (OECD 2012). Transnational laws, such as the 2001 European Union SEA directive, and international laws, such as the 2003 SEA Protocol to the Espoo Convention on Transboundary Environmental Impact Assessment, have also adopted the technique.

SEA has been seen as an important mechanism in addressing "the policy and institutional challenge of sustainability" (Dovers 2002). Current forms of SEA tend to target particular sectors or regions in a kind of scaled-up version of EIA, or what Dovers (2002) labels 'shallow' SEA. For example, the EU SEA directive requires only the assessment of plans and programs, excluding policies unless member states choose to include policy assessment in implementing legislation (Marsden and de Mulder 2004). As commentators have noted, scope exists to apply SEA to River Basin Management Plans prepared under the EU Water Framework Directive, which may promote greater integration of water and land-use planning (Carter and Howe 2006). However, stronger integration of regulation at the climate-energy-water nexus will likely require the use of deep forms of SEA. As Dovers (2002) explains, 'deep SEA' is characterised by a focus on the actions, policies and strategic planning processes of government, rather than private actors, and considers not just direct environmental impacts but also the economic and social structures that underpin unsustainable patterns of development. Implementation of this kind of SEA for interlinked climate-energy-water policies would require attention to issues of institutional capacity, such as whether suitable organisations exist or can be re-tasked for the purpose, and whether there are suitable people and knowledge available to undertake integrated assessments of this kind.

Harnessing markets

Integrated decision-making processes, including EIA and SEA, offer informational and procedural tools for improving regulation of the nexus. Such tools expand the scope of the decision-making process but do not mandate synergistic climate-energy-water policy outcomes. Likewise, markets do not *require* such outcomes but can provide powerful economic incentives for the development of win–win solutions by rendering the benefits visible (and profitable) to a proponent. The establishment of carbon markets internationally, under the Kyoto Protocol to the Framework Climate Change Convention, as well as by countries and regional organisations like the European Union, has opened up opportunities for harnessing the power of markets to incentivise beneficial climate-energy policy options. The carbon markets, by and large, do not address water outcomes, but do have consequences for water both positive and negative, as the following example illustrates (see Box 11.2).

Box 12.2 International carbon credits for energy and water savings projects

At the international level, the Kyoto Protocol's Clean Development Mechanism (CDM) allows project proponents to earn carbon credits for emissions-reducing projects undertaken in developing countries (Zahar et al. 2013). Over the past few years, some companies have sought CDM accreditation for projects involving the installation of simple water purification systems in homes in developing countries such as Indonesia and Kenya (Rosenberg 2010). The projects seek certified emissions reduction (CER) credits under the CDM on the basis that use of the technology avoids the need for water to be boiled, in turn reducing the burning of firewood and the associated greenhouse gas emissions (Stanford Global Health Insight 2012). CER credits can then be sold on the international carbon market to, for example, an Australian company with liabilities under the national carbon pricing mechanism (Caripis et al. 2011). Credits from such projects can also be generated for the purposes of the voluntary carbon market (Gold Standard Foundation 2012), which supplies offsets to individuals and businesses who wish to offset emissions on a voluntary basis, rather than for legal compliance purposes (eg passengers offsetting their flight emissions).

Building on improved knowledge and research about the energy savings and emissions reductions that can be achieved through water conservation efforts (Kenway 2008; EEA 2012; Sanders and Webber 2012), other regulatory proposals are beginning to emerge for harnessing markets to improve outcomes at the climate-energy water nexus (see Box 12.3). These proposals call for the issue of credits under mandatory carbon market and cap-and-trade schemes for activities that reduce the energy-intensity of water use through improving water use efficiency or enhancing water conservation (eg for discussion of a pilot project of this kind being conducted in Japan, see Shimizu et al. 2012).

At this stage there are no indications that a similar strategy is under investigation in other major carbon market schemes such as those in Australia and the European Union.

Box 12.3 Carbon credits for water savings under the Californian cap-and-trade scheme

California has shown particular interest in the possibilities for reducing energy use and greenhouse gas emissions through reducing water use (Spivy-Weber 2011). In California, water-related energy use consumes 19 per cent of the state's electricity, 30 per cent of its natural gas and 88 billion gallons of diesel fuel every year, with demand growing (CEC 2005). Consequently, in its Scoping Plan, discussing proposed actions for implementing the emissions reduction targets under California's *Global Warming Solutions Act 2006*, the California Air Resources Board (CARB) included a chapter on possible actions that could be taken in the water sector (CARB 2009). Mentioned in this chapter is the possible introduction of "a mechanism to make allowances available in a cap-and-trade program" in order to "provide additional incentives for local governments, water suppliers and third party providers to bundle water and energy efficiency improvements" (CARB 2009, p. 66). CARB's current regulations for the California cap-and-trade program, which commenced in January 2013, do not elaborate such a mechanism, but this matter could be addressed in future rule makings of the agency. If so, it could incentivise actions by water utilities to decrease water use, with associated energy savings and emissions reduction benefits.

Most likely this is because the water sector in these jurisdictions is not such a significant user of energy as in California. Nonetheless, as emissions reductions in other sectors are realized and regulators start the search for other emissions reduction possibilities, this model is one that might be considered. Water providers could be included in a cap-and-trade scheme through encompassing them within the scheme's emissions cap or by accrediting their water and energy savings activities as offsets that can be used to discharge liabilities under the scheme. In Australia, several large water utilities are already covered by the national carbon pricing mechanism (Liable Entities Public Information Database 2012–13) and thus have a financial incentive to decrease their direct and indirect emissions in order to reduce their carbon price payments. Other water utilities, notably the Water Corporation in Western Australia and Yarra Valley Water in Victoria, use full-cost accounting of economic, environmental and social issues (triple-bottom-line analysis) as an input into investment decisions (ATSE 2012). These strategies provide a useful starting point for incorporating a variety of externalities (including impacts on other sectors) into decision making and investment frameworks.

New challenges for regulation of the nexus

Rapid technological development presents constant challenges for environmental regulation seeking to ensure that new technologies do not present unacceptable environmental or social impacts. For regulation of the climate-energy-water nexus, some immediate, as well as longer-term, challenges loom, as new technologies are developed to enhance energy production and water supply, and to mitigate or adapt to climate change. Ideally a process of *ex ante* assessment, encompassing potential climate-energy-water impacts, would be followed prior to the deployment of new technologies (which deep forms of SEA, discussed in the previous section, might allow). In some instances, scientific and public concern over the possible impacts of a new technology *has* promoted regulatory development in advance of widespread use of the technology (eg the development of international and EU rules on genetically modified organisms). More often, though, technological development outpaces regulatory development, leaving novel problems to be dealt with under existing or evolving regulatory frameworks. Several such examples are discussed below. They focus upon challenges created by new energy or climate mitigation technologies for water resources (water for energy); however, myriad issues may also arise in the 'energy for water' context. These include debates over whether to accredit biogas harvested by wastewater facilities as a form of renewable energy that will contribute to climate change mitigation goals, or how to account for and accredit energy embedded in water under energy efficiency programs where water savings are made (GEI Consultants 2012).

Current challenges: Unconventional energy resources and fracking

New technologies of hydraulic fracturing (or fracking) have underpinned a 'boom' in the production of 'unconventional' energy sources in a number of countries, most notably in the United States (EIA 2011; IEA 2012). Fracking – a process whereby a high-pressure mixture of water, sand and chemicals is injected into a reservoir to release gas or oil – is seen as the key to tapping large quantities of non-conventional gas, particularly from shale but also in some cases from coal (coal seam gas [CSG] or coalbed methane; Cooke 2011). [2] Yet fracking and unconventional resource extraction projects have also attracted controversy and opposition on the basis of their impacts on air, land and water resources (Williams, Stubbs and Milligan 2012). Concerns have been raised about the large quantities of water required for fracking, especially in areas already subject to water stress, as well as the effects on water quality (Williams and Pittock 2012; NWC 2010; Miller 2011; MacDonnell and Guerra 2011). Other research has highlighted the significant greenhouse gas footprint of CSG and shale gas extraction given the large volume of methane emissions produced during the process (Howarth et al. 2011; ABC News 2012).

Regulatory systems in Australia and the United States have struggled to keep pace with the rapid expansion of CSG and fracking projects in those nations (ABA Section of Environment, Energy and Resources 2013). In both countries, state governments are the primary regulator of oil and gas extraction, although the federal government will be involved in decision making where extraction is proposed in areas subject to federal jurisdiction (eg offshore areas in Australia or Bureau of Land Management lands in the United States), or where federal environmental laws apply. Overall, the regulatory

[2] In the case of CSG, fracking is not needed but may be helpful in some cases. For CSG mining in the Surat Basin in Queensland, significant amounts of water are also produced in the mining process. This wastewater is often highly saline and must be appropriately disposed of by operators under a licence issued by regulatory authorities (Day 2009; NWC 2011).

landscape for unconventional energy resources varies in both countries. For shale gas fracking in the United States, for example, fairly comprehensive regulations are found in some states, like Pennsylvania and New York, but there is an absence of regulatory controls in other states (Wiseman 2009; Wiseman and Gradijan 2012); for example, California released a preliminary set of draft regulations on the subject in December 2012. In addition, the complicated regime of water rights in the United States – under the control of states – may present considerable problems in terms of sourcing the quantities of water required for large-scale fracking operations, setting up the potential for conflict with other (primarily agricultural) water users (Healy 2012).

National-level regulation for extraction of unconventional resources is contemplated in both Australia and the United States. Australia has progressed much further down this path, with the recent passage of an amendment to the national EIA legislation, the *Environment Protection and Biodiversity Conservation Act 1999*, requiring EIA for CSG and large coal mining projects that are likely to have a significant impact on water resources (DSEWPC 2013). In addition, the Council of Australian Governments' (COAG) Standing Council on Energy and Resources has released a 'National Harmonised Regulatory Framework' for CSG, which provides guidance to governments on 'leading practice' for CSG management (SCER 2013). In the United States, the federal Environmental Protection Agency has initiated rule makings to set discharge standards for wastewater from both coalbed methane and shale gas extraction, with proposals to be released for public comment in 2013 and 2014, respectively. Nonetheless, significant regulatory gaps remain, including the so-called Halliburton loophole, which exempts oil and gas fracking from the requirements of the Safe Drinking Water Act (*Energy Policy Act 2005*, §322).

Potential future development of hydrocarbon-rich shale resources also presents a significant regulatory challenge for both countries. In California, the Monterey Shale is estimated to contain 15.5 billion barrels of recoverable oil (64 per cent of total US shale oil resources) (EIA 2011) and is being touted as the site of the next US oil boom (Hargreaves 2013). Likewise, intensive extraction operations for shale oil and gas are on the horizon in Australia, with the discovery of significant deposits located in some of most arid areas of the country (EIA 2011; Vivoda 2013).

Future challenges: CCS and geoengineering

Longer-term challenges for regulation of the nexus arise from technologies under development albeit not yet in full-scale operation. One example is carbon capture and storage/sequestration (CCS) (see Box 12.4), which has been promoted as "one of the core technologies that will be needed to handle the challenge of climate change" (Havercroft, Macrory and Stewart 2011).

In the case of geoengineering – involving "the deliberate large-scale manipulation of the planetary environment to counteract anthropogenic climate change" (Royal Society 2009, p. 1) – both the technical and regulatory discussion is at a more preliminary stage than in the case of CCS. Geoengineering technologies fall into two broad categories: solar

Box 12.4 CCS: A potential new challenge for regulation of the nexus

CCS allows the distillation of the carbon dioxide produced by a coal-fired power plant or other processes, its compression into a liquid-like state and injection into microscopic pore spaces in stable geological foundations or other underground storage areas (eg depleted oil and gas fields) (House Standing Committee 2007). While this technology holds out the promise of prolonging the life of high emission sources such as coal, it is likely to be both water and energy-intensive (Averyt 2011). For example, coal-fired power plants incorporating CCS could be anywhere from one-quarter to one-third more water-intensive (NWC 2012). Around the world, a number of CCS pilot projects have been implemented, but to date there is no large-scale commercial project in operation. Best estimates suggest the technology is still at least a decade away from full deployment (Australian Government 2012) and even then, associated energy and water costs may render it economically unviable. The difficulties encountered in up-scaling CCS technology have allowed, unusually, the development of a supporting regulatory framework in advance of the technology's deployment. This includes international regulation under treaties applicable to ocean dumping of wastes and marine pollution (Sands and Peel 2012), as well as the introduction of new CCS laws in several jurisdictions, including Australia and the European Union. These laws focus on the health and environmental risks associated with long-term underground storage of carbon dioxide, as well as questions of how liability for future accidents or harms is to be allocated (Zahar et al. 2013). However, no specific consideration has been given to the water or energy intensity of the capture process and how to factor such impacts into decisions on CCS projects.

radiation management and carbon dioxide removal (GAO 2010). At this time, geoengineering technologies are considered immature and significant concerns have been raised over the potential risks associated with their deployment, especially in the case of solar radiation management options (Royal Society 2009; GAO 2010). Some of the technologies under consideration could also have implications for the climate-energy-water nexus, due to effects on water quality, or through increasing demand for energy or water resources (see earlier chapters in this volume). In regulatory terms, the large-scale nature of geoengineering interventions is most likely to implicate laws governing transboundary or international environmental protection (Redgwell 2011). Already, concerns over the potential environmental effects of some geoengineering techniques, such as ocean iron fertilisation, have spurred precautionary restrictions under relevant international treaties (Sands and Peel 2012).

Future-proofing regulation of the nexus

As the examples discussed earlier illustrate, ensuring that regulation of the nexus is able to manage future challenges posed by new technologies requires flexibility within regulatory systems so as to be able to account for and respond to emerging risks to climate, energy and water systems. The question of how to make policy and regulatory systems more flexible, resilient and nimble in face of novel, complex and evolving threats is the subject of a considerable literature in fields such as climate change policy and regulation (Oikonomou and Jepma 2008; Craig 2010; Doremus 2010; Ruhl 2010; Ruhl and Saltzmann 2010), sustainable water management (ATSE 2012; Williams et al. 2012), new environmental governance (Holley, Gunningham and Shearing 2011), systems analysis (PMSEIC 2011; Newell, Marsh and Sharma 2011) and institutions for sustainability and policy integration (Dovers and Connors 2006; Dovers 2005). While this body of literature covers a vast territory and proffers many diverse proposals, a few common elements emerge that hold particular relevance for designing climate-energy-water regulatory systems that will be resilient in the face of change. These common elements – which might be cast as principles of best practice regulation for integrated policy issues such as the climate-energy-water nexus – include the following:

- Utilisation of a diverse range of policy instruments (ie not relying on a single mechanism such as a market-based approach);
- Overlapping, flexible distribution of decision-making authority between national, state and local agencies with a preference for decentralised approaches and solutions;
- Creation of an information-rich environment incorporating continuous monitoring of environmental conditions and robust scenarios planning;
- Multi-stakeholder processes with broad participation and multiple perspectives; and
- Shifting from conventional 'front-end' decision making towards a more adaptive 'back-end' approach.

To a large extent, these proposals draw on existing regulatory tools, albeit organised in different ways. In the case of the first four principles, the overall aim is to diversify sources of information and policy responses in an attempt to match the complexity and dynamic nature of the regulatory problems under consideration. They call for greater policy integration; dynamic or cooperative forms of power-sharing; widespread use of information-gathering tools such as reporting, monitoring and strategic planning; and participative and collaborative decision-making processes. The fifth principle suggests the need for the reorientation of conventional decision-making processes away from front-end exercises, such as EIA and to some extent SEA, which seek to predict and manage all impacts in advance of development, towards a back-end approach that focusses on adaptation and adjustment over time.

Adaptive management of this kind is often described as an approach of 'learning while doing' or 'policy experimentalism' (Walters and Holling 1990; Doremus 2007). An activity with uncertain impacts is allowed to proceed, but with systematic monitoring of results and feedback processes in place that allow on-going decision adjustments (Doremus et al 2011). While adaptive management – based on concepts developed by ecologists C. S. Hollings and Carl Walters in the 1970s – has enjoyed considerable popularity in environmental law and natural resource management, it is increasingly recognised that it is "not a panacea for the problems that plague natural resource management. It is appropriate in

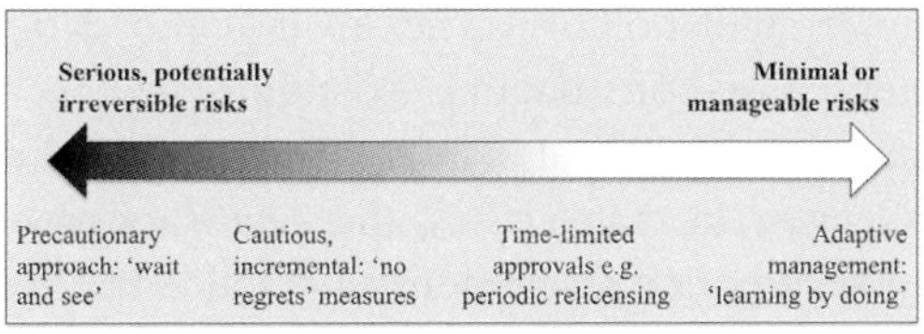

Figure 12.2 Approaches for dealing with emerging risks in regulation of the nexus.

some contexts, but not in others" (Doremus et al 2011, p. 3). As in the case of integration discussed earlier, there is arguably a spectrum of approaches to enhancing the flexibility of regulation to deal with emerging risks. Adaptive management is at one end of the spectrum, arguably being most appropriate in circumstances where risks or predicted impacts are perceived as minimal or readily manageable and reversible. As Doremus et al. (2011) emphasise, it is also critical that the regulatory framework under which decisions are made allows opportunities for adjustment. Laws that call for a single decision, not open to later reconsideration, will not provide a suitable institutional environment for adaptive management (Peel 2005).

At the other end of the spectrum are situations where risks and impacts are perceived as serious and where the activities under consideration have potentially irreversible consequences. Here, a precautionary approach of 'wait and see' may be the most desirable option, with an activity subject to a moratorium until more information regarding impacts is available (Peel 2005). Responses under international environmental treaty instruments to geoengineering proposals, such as iron ocean fertilisation, are an example of this approach, as are recent bans on fracking for shale gas enacted in France and Bulgaria (BBC News 2012). Lying in between precaution and adaptive management are other 'adaptive' options. These include (1) a cautious, incremental approach based on the initial adoption of 'no regrets' measures – eg measures to improve water or energy efficiency (Craig 2010); and (2) time-limited approvals – eg requirements for periodic relicensing of hydroelectric power activities under the US Federal Power Act (16 U.S.C. §799) that "create a policy-reform window for periodic review of the licensed activity" (Pittock and Hartmann 2011).[3] This spectrum of approaches is summarised in Figure 12.2.

Conclusion

The challenge of ensuring effective, integrated regulation of the climate-energy-water nexus is a significant, but not insurmountable, one. The current fragmentation of regulation across different environmental and resource-management sectors frequently poses barriers to the implementation of synergistic climate, energy and water policies. The chapter has discussed examples of promising integrative techniques that can be used to reduce such

[3] The Federal Power Act (16 U.S.C. § 791–828[c]) authorizes the Federal Energy Regulatory Commission to licence hydroelectric facilities for a maximum period of fifty years, after which time the licence must be renegotiated. This power could be used to incorporate climate change considerations in decision making, though this has not been the practice of the Commission to date (see Viers 2011).

fragmentation and optimise regulation of the nexus. In approaching this task, the author's preference is for a strategy based on adapting existing regulatory frameworks to explore ways that they can better work together and incorporate tools that enhance their flexibility to respond to new challenges. In many cases, the necessary architecture for integration exists, although moving from policy discussions to practice will require a more concerted and systematic focus on the nexus between climate, energy and water issues. The alternative approach – designing a new, all-encompassing integrated regulatory scheme for nexus decision making – would require significant political commitment and may run the risk of creating a system too complex to administer in practice.

Of course, an adequate regulatory framework is only ever one piece of the broader picture of what is necessary for delivering effective, integrated outcomes at the climate-energy-water nexus. Attention will also need to be given to questions of institutional capacity (which institutions, which people, what knowledge) and resourcing, as well as the crucial issue of appropriate strategies for ensuring compliance and the enforcement of regulatory requirements (Gunningham 2011). Nonetheless, existing regulatory frameworks and environmental and sustainability tools offer a solid foundation for designing regulation to deal with the current and future challenges facing policy making at the climate-energy-water nexus. Flexible, better-integrated regulatory systems for climate, energy and water can ensure that law and regulation contribute constructively to effective implementation of policies in these closely linked areas.

References

Abaza, H., Bisset, R. and Sadler, B. (2004). Environmental Impact Assessment and Strategic Environmental Assessment: Towards an Integrated Approach. Geneva.

ABC News (2012). The Coal Seam Gas Rush. *ABC Radio*, April 2012, www.abc.net.au/news/specials/coal-seam-gas-by-the-numbers/.

American Bar Association (ABA) Section on Environment, Energy and Resources (2013). *International Environmental and Resources Law Committee Newsletter* 15(4): 1–24.

Australian Academy of Technical Sciences and Engineering (ATSE) (2012). *Sustainable Water Management: Securing Australia's Future in a Green Economy*. Melbourne: ATSE.

Australian Government (2012). *Energy White Paper 2012: Australia's Energy Transformation*. Canberra: Australian Government.

Averyt, K. (2011). The Coal Conundrum. In *The Water-Energy Nexus in the American West*, eds. Kenney, D. S. and Wilkinson, R. Cheltenham: Edward Elgar.

Barnett, J. and O'Neill, S. (2010). Maladaptation. *Global Environmental Change* 20: 211–13.

BBC News (2012). Bulgaria Bans Shale Gas Drilling with 'Fracking' Methods. *BBC News* 19 January 19.

Benson, R. D. (2013). Federal Water Law and the "Double Whammy": How the Bureau of Reclamation Can Help the West Adapt to Drought and Climate Change. *Ecology Law Quarterly* 39: 1047–81.

Black, J. (2001). Decentring Regulation: Understanding the Role of Regulation and Self-Regulation in a 'Post-Regulatory' World. *Current Legal Problems* 54: 103–46.

Bonyhady, T., MacIntosh, A. and McDonald, J. (eds.) (2010). *Adaptation to Climate Change: Law and Policy*. Sydney: Federation Press.

Bureau of Reclamation and US Department of Interior. (2013). *Managing Water in the West*. Washington, DC.

Cairns Jnr, J .R. and Crawford, T. V. (eds.) 1991. *Integrated Environmental Management*, Chelsea, MI: Lewis Publishers Inc.

California Air Resources Board (CARB) (2009). *Climate Change Scoping Plan: A Frame-Work for Change*. Sacramento, CA: CARB.

California Energy Commission (CEC) (2005). *California's Water-Energy Relationship*. Sacramento, CA: CEC.

Campbell, B. M. and Sayer, J. A. (eds.) (2003). *Integrated Natural Resource Management: Linking Productivity, the Environment and Development*, Wallingford: CABI Publishing.

Caripis, L., Peel, J., Godden, L. and Keenan, R. J. (2011). Australia's Carbon Pricing Mechanism. *Climate Law* 2: 1–21.

Carter, J. and Howe, J. (2006). The Water Framework Directive and the Strategic Environmental Assessment Directive: Exploring the Linkages. *Environmental Impact Assessment Review* 26: 287–300.

Connell, D. (2011). Water Reform and the Federal System in the Murray–Darling Basin. *Water Resource Management* 25: 3993–4002.

Cooke, D. (2011). Explainer: Coal Seam Gas, Shale Gas and Fracking in Australia. *The Conversation*, 16 August, https://theconversation.edu.au/explainer-coal-seam-gas-shale-gas-and-fracking-in-australia-2585.

Commonwealth of Australia (2012). *Water Act 2007* Basin Plan, November.

Council of Australian Governments (COAG) (2004). *Intergovernmental Agreement on a National Water Initiative*.

Council on Environmental Quality (CEQ) (1997). Considering Cumulative Effects under the National Environmental Policy Act. Washington, DC: Executive Office of the President.

Craig, R. K. (2008). Climate Change, Regulatory Fragmentation, and Water Triage. *University of Colorado Law Review* 79: 825–927.

Craig, R. K. (2010). "Stationarity Is Dead" – Long Live Transformation: Five Principles for Climate Change Adaptation Law. *Harv. Envtl. L. Rev*. 34: 9.

Day, R. W. (2009). Coal Seam Gas Booms in Eastern Australia. *Australia ASEG Preview* June: 26–32.

Department of Sustainability, Environment, Water, Population and Communities (DSEWPC). (2013). 2013 EPBC Act Amendment – Water Trigger, www.environment.gov.au/epbc/about/water-trigger.html.

Doremus, H. (2007). Precaution, Science and Learning While Doing in Natural Resource Management. *Washington Law Review* 82: 547–79.

Doremus, H. (2010). Adapting to Climate Change with Law That Bends without Breaking. *San Diego J. of Climate & Energy L*., 2: 45–85.

Doremus, H., Andreen, W. L., Camacho, A., Farber, D. A., Glicksman, R., Goble, D., Karkkainen, B., Rohlf, D., Tarlock, A. D., Zellmer, S., Jones, S. and Huang, Y. (2011). Making Good Use of Adaptive Management. In Center for Progressive Reform.

Dovers, S. (2002). Too Deep a SEA? Strategic Environmental Assessment in the Era of Sustainability. In *Strategic Environmental Assessment in Australasia*, eds. Marsden, S. and Dovers, S. Sydney: Federation Press.

Dovers, S. (2005). *Environment and Sustainability Policy: Creation, Implementation, Evaluation*. Sydney: Federation Press.

Dovers, S. and Connor, R. (2006). Institutional and Policy Change for Sustainability. In *Environmental Law for Sustainability*, eds. Richardson, B. J. and Wood, S. Portland: Hart Publishing.

Energy Information Administration (EIA) (2011). *World Shale Gas Resources: An Initial Assessment of Fourteen Regions Outside the United States*. Washington, DC: US Energy Information Administration.

European Environment Agency (EEA) (2012). *Towards Efficient Use of Water Resources in Europe*. Copenhagen: EEA.

Executive Office of the President. (2013). *The President's Climate Action Plan*. Washington, DC: The White House.

Farrier, D. (2002). Fragmented Law in Fragmented Landscapes: The Slow Evolution of Integrated Natural Resource Management Legislation in NSW. *Environmental And Planning Law Journal* 19: 89.

Foerster, A. (2009). Progress on Environmental Flows in Southeastern Australia in Light of Climate Change. *Environmental Law Reporter* 39: 10426–34.

Foerster, A. (2011). Developing Purposeful and Adaptive Institutions for Effective Environmental Water Governance. *Water Resource Management* 25: 4005–18.

Government Accountability Office (GAO). (2010). Climate Change: A Coordinated Strategy Could Focus Federal Geoengineering Research and Inform Governance Efforts. *Report to the Chairman, Committee on Science and Technology, House of Representatives*. Washington, DC: United States Government Accountability Office.

GEI Consultants (2012). *California's Water-Energy Nexus: Pathways to Implementation*. WETCAT.

Gerrard, M. and Kuh, K. F. (eds.) (2012). *The Law of Adaptation to Climate Change: United States and International Aspects*. Chicago: American Bar Association.

Godden, L. and Peel, J. (2010). *Environmental Law: Scientific, Policy and Regulatory Dimensions*, Melbourne: Oxford University Press.

Gold Standard Foundation (2012). World's First Carbon-Financed Sustainable Water Programme Helps Millions in Africa, www.cdmgoldstandard.org/worlds-first-carbon-financed-sustainable-water-programme-helps-millions-in-africa/.

Gunningham, N. (2011). Enforcing Environmental Regulation. *Journal of Environmental Law* 23: 169–201.

Gunningham, N., Grabosky, P. and Sinclair, D. (1998). *Smart Regulation: Designing Environmental Policy*. Oxford: Clarendon Press.

Guruswamy, L. (1991). The Case for Integrated Pollution Control. *Law and Contemporary Problems* 54: 41.

Hanna, K. S. and Slocombe, D. S. (eds.) (2007). *Integrated Resource and Environmental Management: Concepts and Practice*. Toronto: Oxford University Press.

Hargreaves, S. (2013). California Could Be Next Oil Boom State. *CNN Money*, 14 January, http://money.cnn.com/2013/01/14/news/economy/california-oil-boom/index.html.

Hassing, J., Ipsen, N., Clausen, T., Larsen, H. and Lindgaard-Jorgensen, P. (2009). *Integrated Water Resources Management in Action*. Geneva: UNEP.

Havercroft, I., Macrory, R. and Stewart, R. B. (eds.) (2011). *Carbon Capture and Storage: Emerging Legal and Regulatory Issues*. Oxford: Hart Publishing.

Healy, J. (2012). For Farms in the West, Oil Wells are Thirsty Rivals. *The New York Times*, 5 September, www.nytimes.com/2012/09/06/us/struggle-for-water-in-colorado-with-rise-in-fracking.html?pagewanted=all&_r=0.

Henriksen, C. B., Hussey, K. and Holm, P. E. (2011). Exploiting Soil-Management Strategies for Climate Mitigation in the European Union: Maximizing "Win-Win" Solutions across Policy Regimes. *Ecology and Society* 16: 22.

Holley, C., Gunningham, N. and Shearing, C. (eds.) (2011). *The New Environmental Governance*. Milton Park: Routledge Earthscan.

House Standing Committee on Science and Innovation (2007). *Between a Rock and a Hard Place: The Science of Geosequestration* Canberra: Parliament of Australia.

Howarth, R. W., Santoro, R. and Ingraffea, A. (2011). Methane and the Greenhouse-Gas Footprint of Natural Gas from Shale Formations. *Climatic Change* 106: 679–90.

Hussey, K. and Pittock, J. (2012). The Energy-Water Nexus: Managing the Links between Energy and Water for a Sustainable Future. *Ecology and Society* 17: 31.

International Court of Justice (ICJ) (2010). *Case Concerning Pulp Mills on the River Uruguay (Argentina v Uruguay)*, Judgment 20 April 2010, *I.C.J. Reports*, 14.

International Energy Agency (IEA) 2012. *World Energy Outlook 2012*. Paris: OECD/IEA.

Jones, D. (2002). The Kyoto Protocol, Carbon Sinks and Integrated Environmental Regulation: An Australian Perspective. *Environmental & Planning Law Journal* 19: 109.

Kenway, S. J., Priestley, A., Cook, S., Seo, S., Inman, M., Gregory, A. and Hall, M. (2008). *Energy Use in the Provision and Consumption of Urban Water in Australia and New Zealand*. Canberra: CSIRO.

Liable Entities Public Information Database (2012–13). *LEPID for the financial year 2012–2013*. Canberra: Clean Energy Regulator, www.cleanenergyregulator.gov.au/Carbon-Pricing-Mechanism/Liable-Entities-Public-Information-Database/LEPID-for-2012-13-Financial-year/Pages/default.aspx.

MacDonnell L. J. and Guerra, K. L. (2011). Managing produced water from coalbed methane production. In *The Water-Energy Nexus in the American West*, eds. Kenney, D. S. and Wilkinson, R. Cheltenham: Edward Elgar.

Marsden, S. and de Mulder, J. (2005). Strategic Environmental Assessment and Sustainability in Europe – How Bright Is the Future? *Review of European Community and International Environmental Law* 14: 50–62.

Marsden, S. and Dovers, S. (eds.) (2002). *Strategic Environmental Assessment in Australasia*. Sydney: Federation Press.

Miller, B. (2011). Oil Shale and Water. In *The Water-Energy Nexus in the American West*, eds. Kenney, D. S. and Wilkinson, R. Cheltenham: Edward Elgar.

Mitchell, B. and Hollick, M. (1993). Integrated Catchment Management in Western Australia: Transition from Concept to Implementation. *Environmental Management* 17: 735–43.

Morgan, B. and Yeung, K. (2007). *An Introduction to Law and Regulation: Text and Materials*. Cambridge University Press.

National Water Commission (NWC) (2010). *The Coal Seam Gas and Water Challenge: National Water Commission Position*. Canberra: NWC.

National Water Commission (NWC) (2011). *Onshore Co-Produced Water: Extent and Management*. Waterlines Report Series No. 54 September 2011. Canberra: NWC.

National Water Commission (NWC) (2012). *Water Policy and Climate Change in Australia*. Canberra: Australian Government.

Nelson, R. (2005). Legislation for ICM: Advancing Water Resources Sustainability? *Environmental and Planning Law Journal* 22: 96–129.

Newell, B., Marsh, D. M. and Sharma, D. (2011). Enhancing the Resilience of the Australian National Electricity Market: Taking a Systems Approach in Policy Development. *Ecology and Society* 16: 15.

OECD (2012). *Strategic Environmental Assessment in Development Practice: A Review of Recent Experience*. OECD.

Oikonomou, V. and Jepma, C. J. (2008). A Framework on Interactions of Climate and Energy Policy Instruments. *Mitig Adapt Strat Glob Change* 13: 131–56.

Olsson, G. (2012). *Water and Energy: Threats and Opportunities*. London: IWA Publishing.

Osofsky, H. (2011). Diagonal Federalism and Climate Change: Implications for the Obama Administration. *Alabama Law Review* 62: 237–304.

Osofsky, H. M. and Wiseman, H. J. (2013). Dynamic Energy Federalism. *Maryland Law Review* 72(3): 773–843.

Peel, J. (2005). *The Precautionary Principle in Practice: Environmental Decision-making and Scientific Uncertainty*. Sydney: Federation Press.

Peel, J., Godden, L. and Keenan, R. J. (2012). Climate Change Law in an Era of Multi-Level Governance. *Transnational Environmental Law* 1: 245–80.

Pilz, R. D. (2011). Lessons in Water Policy Innovation from the World's Driest Inhabited Continent: Using Water Allocation Plans and Water Markets to Manage Water Scarcity. *Water Law Review* 14: 97–130.

Pittock, J. (2012). National Climate Change Policies and Sustainable Water Management: Conflicts and Synergies. *Ecology and Society* 16: 25.

Pittock, J. and Hartmann, J. (2011). Taking a Second Look: Climate Change, Periodic Relicensing and Improved Management of Dams. *Marine and Freshwater Research* 62: 312–20.

Pittock, J., Hussey, K. and McGlennon, S. (2013). Australian Climate, Energy and Water Policies: conflicts and synergies. *Australian Geographer* 44: 3–22.

Prime Minister's Science, Engineering and Innovation Council (PMSEIC) (2010). *Challenges at Energy-Water-Carbon Intersections*. Canberra: Australian Government.

Proust, K., Dovers, S., Foran, B., Newell, B., Steffen, W. and Troy, P. (2007). *Climate, Energy and Water: Accounting for the Links*. Canberra: Fenner School of Environment and Society.

Redgwell, C. (2011). Geoengineering the Climate: Technological Solutions to Mitigation – Failure or Continuing Carbon Addiction. *Carbon & Climate Law Review* 2: 178–89.

Resource Assessment Commission (RAC) (1993). *Coastal Zone Inquiry: Integrated Resource Management in Australia*. Canberra: Resource Assessment Commission.

Rosenberg, T. (2010). Clean Water at No Cost? Just Add Carbon Credits. *The New York Times*, 15 November.

Royal Society (2009). *Geoengineering the Climate: Science, Governance and Uncertainty*. London.

Ruhl, J. B. (2010). Climate Change Adaptation and the Structural Transformation of Environmental Law. *Environmental Law* 40: 343.

Ruhl, J. B. and Saltzmann, J. (2010). Climate Change, Dead Zones and Massive Problems in the Administrative State: A Guide for Whittling Away. *California Law Review* 98: 59.

Sanders, K. T. and Webber, M. E. (2012). Evaluating the Energy Consumed for Water Use in the United States. *Environmental Research Letters* 7: 034034.

Sands, P. and Peel, J. (2012). *Principles of International Environmental Law*. Cambridge University Press.

Schmidt, V. A. (2005). Democracy in Europe: The Impact of European Integration. *Perspectives on Politics* 3: 761–79.

Shimizu, Y., Toyosada, K., Yoshitaka, M. and Sakaue, K. (2012). Creation of Carbon Credits by Water Saving. *Water* 4: 533–44.

Spivy-Weber, F. (2011). Water-Energy Integration in California. In *The Water-Energy Nexus in the American West*, eds. Kenney, D. S. and Wilkinson, R. Cheltenham: Edward Elgar.

Standing Council on Energy and Resources (SCER) (2013). The National Harmonised Regulatory Framework: Coal Seam Gas. Canberra.

Stanford Global Health Insight (2012). Lifestraw Carbon for Water: Sustainable Funding for a Public Health Intervention, October, www.google.com/search?client=safari&rls=en&q=Lifestraw+Carbon+for+Water:+Sustainable+Funding+for+a+Public+Health+Intervention&ie=UTF-8&oe=UTF-8.

Thomas, I. and Elliott, M. (2009). *Environmental Impact Assessment in Australia: Theory and Practice*. Sydney: Federation Press.

Thompson, B. H. (2012). A Federal Act to Promote Integrated Water Management: Is the CZMA a Useful Model? *Environmental Law* 42: 201.

Viers, J. H. (2011). Hydropower Relicensing and Climate Change. *Journal of the American Water Resources Association* 1–7.

Vivoda, V. (2013). Can Shale Oil Fuel Australia? *The Conversation* 25 January.

Walters, C. J. and Holling, C. S. (1990). Large-Scale Management Experiments and Learning by Doing. *Ecology* 71: 2060–8.

Water in the West (2013). *Water and Energy Nexus: A Literature Review*. Stanford: Water in the West, a joint program of Stanford Woods Institute for the Environment and Bill Land Center for the American West.

Williams, J. and Pittock, J. (2012). *Unconventional Gas Production and Water Resources.* Lessons from the United States on Better Governance – A Workshop for Australian Government Officials. Canberra: The Australian National University and United States Studies Centre.

Williams, J., Stubbs, T. and Milligan, A. (2012). *An Analysis of Coal Seam Gas Production and Natural Resource Management in Australia.* A report prepared for the Australian Council of Environmental Deans and Directors. Canberra: John Williams Scientific Services Pty Ltd.

Wiseman, H. (2009). Untested Waters: The Rise of Hydraulic Fracturing in Oil and Gas Production and the Need to Revisit Regulation. *Fordham Environmental Law Review* 20: 115–95.

Wiseman, H. and Gradijan, F. (2012). Regulation of Shale Gas Development, Including Hydraulic Fracturing. University of Texas Law School.

World Business Council for Sustainable Development (WBCSD). (2010). *Water, Energy and Climate Change: A Contribution from the Business Community*. Geneva: WBCSD.

The European Water Platform (WSSTP) (2011). *Water and Energy: Strategic Vision and Research Needs*. Brussels: European Commission.

Zahar, A., Peel, J. and Godden, L. (2013). *Australian Climate Law in Global Context*, Melbourne: Cambridge University Press.

13

Climate, energy and water: the potential roles and limitations of markets

NEIL BYRON

Introduction

We are all familiar with markets – the concepts of supply and demand. Spatially or virtually, markets are institutions where rights to use property (or to not use it) are exchanged between buyer and seller, for a financial consideration. During the twentieth century, people throughout most of the world learned to operate in real estate markets, labour markets, financial markets, capital markets, and machinery and auto markets.

In western market economies, decisions about

1. what goods and services to produce and sell, when, where and how, and
2. what goods and services to buy and consume/use

are basically left to the private decisions of billions of enterprises and consumers.

This is all based on the assumption that each of those market participants is capable of working out what is in her/his/its own best interests. Technically, the presumption is that each individual understands and is fully informed about its own requirements, and the attributes of all available goods and services and their ability to meet those specified requirements. Obviously, this set of presumptions is rarely completely true, but it seems to be 'close enough, much of the time' for markets to work 'reasonably well' in allocation of resources for production and consumption – what the eighteenth-century philosopher – economist Adam Smith called "the invisible hand of the marketplace".

Nevertheless, policy discussions and journals are now full of circumstances where the outcome from free exchange in markets is most unlikely to accord with what is in the long-term best interests of the whole society; the so-called *market failures*. For two centuries, economic theories assumed away all these 'unfortunate anomalies'. However, research on market failures is now a major feature of many economics textbooks. The most common of these are:

- the existence of *public goods* where everyone benefits from their provision whether or not they contributed to pay for them (eg defence and street lights). They are usually (but not always) provided by governments because of this 'non-excludability'; it would be unprofitable to provide goods and services that users do not need to pay for individually, so they are provided collectively;

- the existence of *externalities*, where one party is not responsible for the harm that their actions cause to others (negative externalities) or where the provider of services to others is not adequately recognised and compensated for the benefits their actions produce (positive externalities);
- the incomplete definition (even complete absence) of clear and enforceable *property rights* to be exchanged (for example, some natural resources are still 'open access', where whoever captures the resource first owns it, which creates incentives for excessive, rapid resource depletion and underinvestment in conservation and management); or
- the inability of buyers to understand what is in their own best interest (*cognitive dissonance*) – for example, many modern consumers are said to confuse their wants and their needs, or mistakenly think that harmful actions like drug taking will be beneficial for them.

All of these market failures abound in the relatively new field of environmental economics, and especially in discussions of climate change, energy and water.

Oliver and Hussey (Chapter 11) have described the suite of policy instruments available to deal with market failures: regulations and planning, market-based instruments and quasi-markets, and education and suasion. Their summary emphasised what effective policy solutions require, not only that all three types of policy instruments have a role to play but also that each of the policy measures must be aligned rather than working at cross-purposes. For example, to discourage people from smoking, governments often impose health warnings, reinforced by financial disincentives (excise taxes) and restrictions on access. The same applies to energy and water conservation; as the experience during brown-outs and droughts has repeatedly shown, it is very difficult to change people's behaviours through regulatory or educational measures if the price signal continues to shout out, "This stuff is *so* cheap, please use more!" Price signals do matter in people's choices and behaviours, so governments can select a combination of prices, regulations and education to influence aggregate demand for products in short supply ('demand management' as a complement to 'supply augmentation').

In Chapter 12, Peel has described how governments (almost without exception) have attempted to manage and resolve "market failures" in both the energy and water sectors, and now in regard to climate change, by regulations, plans and standard setting. As she notes, this has usually been with specific regulations within each sector, and usually at multiple levels (national, state-province and local). This has commonly led to "regulatory fragmentation" (Figure 12.1) one of the most common forms of *regulatory failure*.

This regulatory approach stands in stark contrast to the operation of markets. Many aspects of modern society and economy are complex, large-scale and multidisciplinary, but effective co-ordination is achieved through the operation of markets for capital, land, labour and technologies where buyers and sellers make choices based on their own self-interests. Centralised planning and regulation by governments has recorded very few successes – even the People's Republic of China is now a market economy (WTO 2001) including in regard to its management of water and energy resources (EIA 2013). Meanwhile, as explained elsewhere in this volume, the "home of the free market ideology", the

United States (especially California) continues to struggle with abortive plans and regulations that are inconsistent with, and which seem largely oblivious to, market forces in the crucial energy and water sectors.

The approach taken by most economists examining the problems of the energy and water sectors – separately or together – would be to 'get the prices right', as this gives all suppliers and users the right incentives to produce and consume rationally:

- rising prices normally encourage users and consumers to curb their demands for any product: to buy less and to use whatever they do buy as efficiently as possible;
- rising prices normally encourage producers to invest in producing more of the product concerned or to find new ways to create that product, with the prospect of higher profits;
- where there are externalities, make the perpetrator pay for the costs they impose on others, and conversely, ensure that those who provide benefits to the wider society are adequately and appropriately compensated for doing so;
- where public goods are to be provided, use a clear and explicit 'shadow' prices in decision making, and transparent mechanisms for tracking who is providing how much public goods, to which communities, and at what cost;
- where property rights are absent or incompletely defined, make the property rights explicit so that those who hold rights are free to enter into voluntary trades (which enhance the welfare of those on both sides of the trade).

As stated in Chapter 1, it is impossible to optimise the operation of either energy- or water-*supply* utilities in isolation from each other, or from the *demand* for energy and water, because of the numerous economic interdependencies. And this will be especially important for the future, as the supply of and demand for both will depend on climatic conditions. Not only does climate affect both the supply of and demand for water and energy, the energy conversion and water extraction processes have the potential to contribute to climate change (positively or negatively, depending on which policies are chosen).

Amazingly, for the past fifty years especially, Western societies have tried to manage the perceived market failures in electricity supply and use, and water supply and use, as if the two sectors were totally independent of each other and both totally independent of climate change effects, or had impacts anywhere else within the economy. And they have tried to use regulation to achieve this.

Indeed, the most striking feature of this debate is the extent to which regulatory controls, planning and standards setting dominate public policy with regards to energy, water and climate change. Almost all other aspects of the modern market economy are left to market forces, operating within a general minimalist regulatory regime (secure property rights, independent judiciary, etc.).

The emergence of energy markets

Although energy generation and distribution is now commonly left to markets, in many countries through the nineteenth and twentieth centuries these were 'utilities' or government services, rarely businesses to be run along commercial lines. These old institutional structures often *tried* to deliver high quality, highly reliable and low-priced energy to citizens and for industry and commerce – they typically emphasised their professional skills and community service motivations. Yet despite that, many failed to deliver the utility services efficiently, sometimes because of budget constraints; some had poor governance and accountability systems; some appeared to focus more on meeting their own needs than the community's.

Where the energy sector was privately owned and managed, it was typically subjected to antitrust regulation to avoid the abuse of market power – where monopolists could set excessively high prices to their customers and constrain the supply of energy. Some governments that operated their own electricity supply enterprises deliberately chose to set artificially low (below-cost) prices as a *hidden* subsidy to encourage industrial development, and so often operated their public utilities at a financial loss each year.

The policy objectives of such governments has been, at best, confused and conflicted, with tensions about why they were running public businesses or regulating the prices and behaviours of private businesses. The common pattern was not just to prevent energy costs becoming unreasonably expensive, but often to deliberately make it unreasonably cheap (whether to appeal to voter popularity or to promote industrial development, especially for large energy-intensive industries that were thought to create many new jobs).

Whether public sector utilities or regulated private utilities, the consequences of this common pattern were similar. Holding prices down to the lowest possible level led to situations where:

1. utilities often found it difficult to raise new capital to augment their production and distribution capacity; and
2. consumers (whether households or industries) had less incentive to economise on their energy inputs because the prices they paid did not reflect the true scarcity of energy or the environmental costs being imposed on society, as a consequence.

Numerous studies of energy conservation (especially since the 1973 'energy crisis') have observed that for many industries, the supply of abundant and cheap energy has been 'hard-wired' into their design and operations, especially in the United States, Canada and Australia. By contrast, quite different technologies for the manufacture and operation of cars, steel, paper, glass and aluminium were adopted in Europe, the United Kingdom and Japan, where there was no such presumption of abundant cheap energy. Adjustment to a sudden increase in energy prices can be extremely disruptive to many industries, because it is very difficult to quickly redesign huge industrial plants to reduce their energy consumption.

For many households, consumption patterns and behaviours are also relatively fixed and unresponsive to price changes in the short term. But over the longer term, behaviours can and do slowly change in response to higher real energy prices, especially where energy represents a significant proportion of the household budget (see Box 13.1).

Box 13.1 Do higher energy prices encourage conservation behaviour in consumers?

Economic theory suggests that consumers tend to reduce their purchases of any product that increases in price. However, there are at least four reasons why households may not respond to higher prices for energy (or water) by reducing consumption and using these resources more efficiently.

1. Many utility bills are in two parts: a fixed monthly charge for having the supply connection and a variable component based on actual usage over the billing period. Because utilities have very high fixed costs and small variable costs, they prefer to have most of their income 'locked in' by fixed charges. But this means that even if users drastically reduce their consumption, the total bill may not decrease by much.
2. Most utility bills are sent monthly or every three to six months, so there is a complete disconnect (in the user's mind) between when the consumption occurred (weeks or months ago) and when the bill is paid. This is in stark contrast to buying auto fuel at the service station, for example, where payment can be directly linked and identified with current consumption decisions and behaviour.
3. Utility prices are usually fixed for at least a year – the cost per unit of electricity, gas or water does not vary with time of day[1] or season – although the actual costs of production and delivery can vary greatly (as with almost all other goods and services) and the demand fluctuates greatly seasonally, even within a day. Thus, households and businesses may continue to use energy and water at their 'usual' rate even at times when it is extremely scarce and expensive to supply, while using much less than is readily and cheaply available at other times. The fact that most households and businesses are screened and insulated from the real varying costs of energy and water makes them much less responsive to supply shifts and is likely to exacerbate short-term shortages and surpluses.
4. For many households, the utility charges per unit of electricity or water are so small that, even if aware of conservation options that might reduce the utility bill, the household decides that it isn't worth the effort to save a few dollars per month. This also explains why many industrial and commercial energy users are not very interested in adopting energy-conservation measures – the potential savings just are not worth the effort or management time.

(*Source*: Australian Productivity Commission 2002)

Emergence of water markets

Given its success with microeconomic reforms since the 1980s (market liberalisation, privatisation or corporatisation of state-owned business enterprises, and competitive neutrality between public and private enterprises) in the energy sector, telecommunications

[1] It is not uncommon for electricity utilities to have an 'off-peak' rate, such as 10 PM to 6 AM, when the fixed price is lower because demand is very low at that time, and in many cases, surplus base load electricity would have had to be 'shorted to Earth' (ie wasted) as it could be neither sold nor stored for future use. Off-peak pricing does *not* occur for water or gas because they can be stored for sale later.

and transport sectors (road, rail, shipping and airlines/airports), Australia decided in the mid 1990s to extend such reforms to water supply for agriculture (but not yet to urban water supplies to industries and households) (see Productivity Commission 2002, 2007, 2011; NWC 2013). The active water markets now permit very significant trade, across state borders[2] and even between river systems, such that the marketplace now determines who uses how much water, where and for what purpose, rather than administrative allocations by officials in government agencies.

Water has often been seen as a public utility on the basis that water supply is a 'basic human need' and every household has a right to guaranteed low cost supply of water for drinking, cooking and washing. This may well be true, but in most Western countries, the majority of household water use is for other uses – taking a thirty minute shower, watering a half-acre of azaleas and roses during a drought or refilling the swimming pool are discretionary behaviours – there is no compelling social reason why these private choices should be subsidised by the public purse.

Economists have observed that water may be a gift of God but collecting, storing, purifying and distributing it is a human activity which costs billions of dollars, and so it should be supplied at a price that covers the costs. A frequent response to such a proposition is the *equity* issue: charging the full cost of supplying water to households in cities would adversely affect the poor. However, the Australian Productivity Commission found that, even for the poorest 20 per cent of the population in each state capital city, household expenditure on reticulated water was less than one dollar per week; much less than this quintile spend on discretionary items like cigarettes, alcohol and soft drinks, even bottled water (APC 2002).

For irrigated agriculture, water is just another critical input (like land, labour, capital, fuel, fertilizer and seeds). But unlike all the others, water has been routinely subsidised by the taxpayers, allocated by government officials or assigned by historic rights that are very difficult to amend or reallocate. When all other inputs and most production are traded in markets, why is water so different?

The third major category of users is industrial, particularly in electricity generation (whether from fossil fuels or nuclear), where vast quantities of potable water is used as a coolant. This is not because water is the best coolant in a technical sense (it isn't) but because, historically, it has been so cheap that any alternative arrangements were considered uneconomic.

Australia is regarded as a world leader in the use of market-based instruments for managing natural resources, particularly in water (but also a growing body of experience with markets for sequestered carbon, biodiversity, and water quality improvements) (Pannell et al 2006; Reeson 2009; Bjornlund, Wheeler and Cheesman 2011). Water markets have been operating for around twenty years now. Acceptance of water markets has increased over time with learning, changes in community attitudes, increased confidence

[2] The transition to full water markets has been slower than expected, partly due to reluctance by individual States to see economic activity move to other States (reflecting the politicians' desire to 'protect' their local communities by continued supply of cheap water). For example, the annual NWC market reports detail the restrictions on trade imposed by the NSW and Victorian governments.

in institutions, and changes in underlying economic benefits and costs to stakeholders (NWC 2013).

Woods et al. (2011) and Hardesty, Sivapalan and Humphries (2012) provide a valuable case study of how major Australian utilities (Sydney Water and Western Australia Water Corporation, respectively) use markets to integrate water and energy issues with other management imperatives.

Why has the energy-water interaction been so fraught?

One question plaguing those concerned about the energy-water nexus is, "Why does the energy sector use so much water, and vice versa?"

Energy and water are tightly entwined. It takes a great deal of energy to supply water, and a great deal of water to supply energy. With water stress spreading and intensifying around the globe, it's critical that policymakers do not promote water-intensive energy options. (S. Postel, cited by Lavelle and Grose 2013)

One inference from such statements is that governments should mandate certain energy technologies and prohibit others, to ensure that the world retains sufficient potable water supplies in future. Conversely, one could argue that governments should mandate water use to protect energy supplies. The arguments are not only circuitous, but fallacious. Concerns about particular technologies or industries 'using too much' or 'wasting' some particular input reflect a focus on *technical* use efficiency rather than *economic* efficiency. Industrial plants are designed to utilise inputs that are relatively inexpensive and abundant, and minimise consumption of those that are scarce and relatively expensive, in order to minimise production costs. But when relative prices change dramatically, any technology may appear wasteful of the input whose relative price has suddenly increased. This typically happens when plants were designed to use 'cheap' energy or water; but years later, when prices have changed, managers wonder how consumption of water and energy be can reduced cost effectively as they struggle to retrofit efficiency measures.

Central planning and regulatory controls have substantially failed in most (if not all) western countries in the energy sector and in the water sector (Hussey et al. 2013). We have argued here that this is mainly due to confused objectives, poor governance arrangements, and lack of transparency and accountability in each sector. But in combination, the effect has often been profoundly disappointing.

Other contributors to this volume have argued that the solution to the observed failures is to have better central planning, better administrative co-ordination within and across the two sectors and stronger regulations. There is a widespread but naïve belief in the 'perfectibility of regulation' – if only we had more complex (or more simple?), more national (or more local?) and more generic (or more targeted?) regulation. This is as common as the equally naïve belief in the perfectibility of free markets. The objective of governments is not to improve the standard of regulation per se, but rather to improve the outcomes for society (which may require more, less or different regulation). But regulations to encourage conservation and eco-efficiency in resource use is likely to be ineffective so long as the price signal is insisting "this stuff is so cheap, please use more!" The *real* price[3] of energy and water should determine what, where, how and whether to produce and consume, as occurs in most other aspects of the modern economy (except where politicians intervene and create distortions). Markets can, under the right conditions, generate such price signals. By contrast, even large teams of highly trained government officials have a generally poor track record of estimating the true value of water and energy to society at any point in time (and of course even if their estimate was correct at that moment, it would very soon be out of date and incorrect).

Extending markets to deal with environmental externalities

The adoption of market principles can offer significant benefits in the form of business flexibility, risk management and potentially an efficient means of achieving improved environmental outcomes. Lessons from experience with other markets are now helping to improve market design (of current and new markets) to enable better targeting and implementation of market-based instruments to regions and industries, and to promote greater understanding of the role and limitation of markets for environmental policy. There are still relatively few contexts in which markets have actually been created to address environmental externalities, but more are gradually emerging each year.

Markets for 'bads and disservices' as well as for 'goods and services'

While market economies deal very readily with inputs and outputs of goods and services, there are many cases where unwanted byproducts are disposed of by 'dumping them to the environment', which has negligible costs to the dumper but can impose very significant costs on the community, especially if those wastes are toxic or environmentally harmful (Sigman 1998; Porter 2002).

Over the past fifty years, economists have gained considerable experience in developing markets or quasi-markets for pollution. Some of the best examples are the emissions trading schemes or carbon prices for carbon dioxide emissions, the market in the United States for emissions of nitrous and sulphurous oxides (Nox and Sox). Some of the most successful examples involve turning 'noxious waste disposal' into a commercial resource – a feedstock for industry (such as for used lubricating oils, used car tyres, used aluminium cans and waste paper).

Box 13.2 is a simple example that illustrates the idea of multiple points of charging for the disposal of a waste product or the purchase of a desired product. This concept has potential application in bringing both the energy and water sector into better financial and environmental balance.

[3] Incorporating as many externalities as possible into the pricing.

Box 13.2. Input markets or product markets: A case study of green waste

The municipality was concerned about excessive dumping of green garden waste, either at a designated landfill site or even worse, by illegal dumping at local campgrounds and picnic sites. So it offered a service to all residents: they could deliver unwanted green waste to the depot where it would be shredded, composted and then offered for sale back to residents as compost or mulch, as well as being used on municipal parks and gardens. The concept was that the local government could offer a valuable service to residents on a commercially viable basis at no net cost.

Unfortunately, within a few months there were mountains of unsold mulch, mountains of green waste awaiting treatment and a considerable debt on municipal finances.

Rather than abandon the scheme completely, the municipality reversed the business model from one that accepted that disposal of 'bads' was free but purchase of 'goods' had to be paid for. Residents had to pay to drop off green waste, but could then help themselves to as much mulch as they wanted for free or a nominal charge.

The market soon cleared – both stockpiles disappeared and the financial accounts balanced. A price was found where the quantity of green waste to be disposed of was in balance with the demand for mulch. There are usually a number of pricing points along the value chain, but many are rarely used.

Another variant is in the emerging markets for ecosystems services, where landholders derive income by providing services like carbon sequestration, management of water catchment values and conservation of rare biodiversity. Rather than producing conventional goods and service like wheat, corn and domesticated livestock for private markets, these landholders are producing public goods and are compensated by the society at large who are the beneficiaries (Stoneham et al. 2003; Wunder 2005, Landell-Mills and Porras 2002; Pagiola et al. 2002)

Three main issues plague schemes for payments for environmental stewardship:

1. Reconciling payments with a basic notion of Duty of Care or a minimal standard or norm of "good practice" that should be met, without any payment. There may be perverse outcomes if people stop doing what they used to do voluntarily, unless they are paid.
2. The logistics of on-going monitoring and evaluation of stewardship to ensure that the contracted outcomes are actually being delivered.
3. Stewardship payments rarely capture the major impacts on resources, so even if beneficial, they tend to be operating at the margins.

How might this apply in the energy sector?

Many major users of water in the energy sector are charged for the water they use (if they have to pay at all) on the basis of the gross volume of water taken, rather than the net amount actually consumed and dissipated in their production processes. This may distort their decisions about how they use water and how much they take. In many cases, such as electric power stations (whether fossil fuel–fired or nuclear), it would be possible to operate a closed system for water, where they continuously refine and reuse the same stock of coolant, or by air-cooled or hybrid systems. But when large amounts of 'new' high-quality water is readily and consistently available at very low prices, it is not financially worthwhile for them to invest in the closed-loop technology. Similarly, coal mine and 'fracking' for coal seam and shale gas generates very large volumes of 'used' or wastewater. So long as the water consumed is available cheaply and costs almost nothing to dispose of, they will resist investing in treatment plants to treat and purify their wastewater so it can be reused by themselves or sold to others (even while regulators try to compel them to do so). The green waste example shows that it is possible to provide commercial incentives for industries to convert waste products to raw materials by investing in new waste management technologies, *provided* that the price of the energy or water inputs is not so low that waste management becomes uneconomic.

Similarly, considerable amounts of water used in industrial processes, or the return flows to streams or groundwater after irrigation, are contaminated with salts, sediments, heavy metals or agricultural chemicals (pesticides and fertilizers), which can damage human health, downstream users or the environment at large. In economic terms, this negative externality occurs because the polluters are not held liable for the costs their actions impose on others and so have little incentive to mitigate such pollution. The regulatory solution is to compel polluters to desist and require that they install and operate mitigation technologies (process-based regulation) or that their effluent achieve minimum standards (outcome-based regulation). While both approaches may work adequately, a transformational solution can be achieved by a combination of financial penalties for all effluent in excess of the minimum standards (at increasing rates proportional to the concentration or severity of the pollutant cocktail) and financial rewards (proportionate to the improvement in water quality in excessive of the minimum standards). Such financial incentives (carrot *and* stick) can justify the installation and operation of treatment and purification plants to deliver additional usable water of the required standards.

Energy and water markets in developing countries

Supplies of energy and water in developing countries have to be cheap enough to be affordable by populations on low incomes and by emerging industries with very limited financial resources and borrowing capacity. But they should not be so cheap that they are used wastefully, or that they reduce incentives to install and expand supply networks. Many developing countries' governments have very limited domestic investment funds for large-scale energy and water supply schemes, and most also lack private sector corporations with the resources and capability to build and operate or expand such systems (particularly when the prospective returns are very low because retail and industrial prices are set so low that they do not even permit efficient cost recovery).

The 'solution' in many cases has been bilateral development assistance (eg 'aid' grant) or more commonly large infrastructure loans from the multilateral development banks, to finance state-run utilities' expansion plans. Unfortunately, these typically suffer from not only the inherent flaws of public sector utilities in industrialised countries but also many other specific local issues – poor governance structures; ineffective metering and billing systems; susceptibility to massive and expensive disruptions in the event of natural disasters; and inability to minimise the social and environmental impacts created.

Limitations of markets and the risks of over-reliance on markets

Markets cannot operate successfully in a vacuum; they depend on the existence of (at least minimal) regulation that defines and confers property rights, that settles commercial disputes quickly and efficiently, that avoids abuse of market power by monopolies or collusion by oligopolies, that defines standards of weights and measures that facilitate private contracts, and so on. Markets are normally assumed to consist of a large number of buyers and sellers, each well aware of their own needs and preferences and none large enough to be able to affect the prices in the market. Yet it is well-known that this assumption is frequently violated.

Markets are not inherently ethical – they do not identify right and wrong outcomes – although governments often try to extinguish certain markets (eg illegal drugs, human slavery) and heavily regulate others that violate social norms (eg prostitution, alcohol, tobacco products). Energy and water markets will not spontaneously deliver the 'correct' outcomes, unless the 'rules of engagement' are well designed, clearly understood and enforced.

Because participants in markets 'vote with their dollars', it is inevitable that whoever has the most dollars influences the outcomes. Markets are typically blind to equity issues – some sellers and buyers start with an unusually large endowment of assets or annual incomes. Others competing in the same markets have few assets and much less income, and their preferences are muted in the final distribution of resources by the marketplace.

Conclusion

An economist's diagnosis of the energy-water imbroglio is that, in many localities, regions and countries, both industry structure for production and household consumption patterns in the use of energy and water have been significantly and systematically distorted for many years by:

- confused government objectives, poor governance structures and diminished accountability and transparency;
- administered prices for energy and water, which do not accurately reflect their true scarcity, their value to society and/or the environmental and economic costs of their production and consumption; and
- a consequent reduced incentive for users to conserve resources and use them efficiently, or for producers and suppliers to innovate or expand supply to meet industrial and consumer demands.

The corrective measures are relatively simple, although not entirely painless. They already apply in most other facets of the economy:

- Set up a robust system that continuously generates price signals, in real time, to producers and consumers about the true value to society of the energy and water that they produce or use.
- Monitor the outcomes to ensure that there are no unintended or inequitable outcomes.
- Incorporate environmental externalities into pricing wherever possible.
- Have a regulatory and supervisory capacity that intervenes *only* when there is a clear justification for doing so *and* conviction that the intervention will actually improve on the outcomes delivered by the market (because regulatory failure is even more common than market failures).

References

Australian Productivity Commission (2003). *Water Rights Arrangements in Australia and Overseas*. Productivity Commission Research Paper, October, Canberra.

Australian Productivity Commission (2005). *The Private Cost Effectiveness of Improving Energy Efficiency*. Productivity Commission Inquiry Report, Canberra.

Australian Productivity Commission (2006). *Rural Water Use and the Environment: The Role of Market Mechanisms*. Research Report, Canberra.

Australian Productivity Commission (2008). *Towards Urban Water Reform: A Discussion Paper*. PC Research Paper, Canberra.

Australian Productivity Commission (2011). *Australia's Urban Water Sector*. Productivity Commission Inquiry Report, Canberra.

Bjornlund, H., Wheeler, S., and Cheesman, J. (2011). Irrigators, Water Trading, the Environment and Debt: Buying water entitlements for the environment. In *Basin Futures Water reform in the Murray-Darling Basin*, ed. Connell, D. and Grafton, R. Q. ANU E Press.

Hardisty, P. E., Sivapalan, M. and Humphries, R. (2012). Determining a Sustainable and Economically Optimal Wastewater Treatment and Discharge Strategy. *Journal of Environmental Management*. 114: 285–92.

Hussey, K., Carter, N. and Reinhardt, W. (forthcoming). Energy Sector Transformation: Implications for Water Governance. *Australian Journal of Water Resources*. Accepted September 2013.

Landell-Mills, N. and Porras, I. T. (2002). *Silver Bullet of Fools' Gold? A Global Review of Forest Environmental Services and Their Impacts on the Poor*. London: IIED.

Lavelle, M. and Grose, T. K. (2013). Water Demand for Energy to Double by 2035. *National Geographic Daily News*, www.news.nationalgeographic.com/news/energy/2013/01/13/.

National Water Commission of Australia (NWC) (2011). *The National Water Initiative – Securing Australia's Water Future: 2011 Assessment*. Canberra: NWC.

Pagiola, S., Landell-Mills, N. and Bishop, J. (2002). Making Market-Based Mechanisms Work for Forests and People. In *Selling Forest Environmental Services: Market-Based Mechanisms for Conservation and Development*, ed. S. Pagiola, J. Bishop, and N. Landell-Mills. London: Earthscan, pp. 261–89.

Pannell, D., Marshall, G. R., Barr, N., Curtis, A., Vanclay, F. and Wilkinson, R. (2006). Understanding and Promoting Adoption of Conservation Practices by Rural Landholders. *Australian Journal of Experimental Agriculture* 46: 1407–24.

Porter, R. (2002). *The Economics of Waste*. Washington, DC: Resources for the Future.
Reeson, A. (2009). *Carbon Markets and Their Implications for Natural Resource Management in Australia*. Socio-Economics and the Environment in Discussion, CSIRO Working Paper Series, 2009–09, Canberra.
Sigman, H. (1998). Midnight Dumping: Public Policies and Illegal Disposal of Used Oil. *RAND Journal of Economics* 29(1): 157–78.
Stoneham, G., Chaudhri, V., Ha, A. and Strappazzon, L. (2003). Auctions for Conservation Contracts: An Empirical Examination of Victoria's Bush Tender Trial. *Australian Journal of Agricultural and Resource Economics* 47(4): 477–500.
US Energy Information Administration (2013). *China Energy Market Analysis*. www.eia.gov/countries/country-data.cfm?fips=CH.
Woods, P., Hynes, R., Jones, M., Walters, R., Sullivan, J. and Ferguson, M. (2011). Carbon Abatement at Sydney Water: Applying the cost of carbon abatement tool. Water Nov: 78–83.
World Trade Organisation (2001). *China Accession to WTO*. www.wto.org/english/thewto_e/countries_e/china_e.htm.
Wunder, S. (2005). *Paying for Environmental Services: Some Nuts and Bolts*, CIFOR Occasional Paper no 42, CIFOR. Bogor, Indonesia.

14

Strategies to mainstream climate change, energy, water and food security nexus knowledge and skills

MICHAEL SMITH AND GEORGES DYER

Many of the chapters in this book have highlighted insufficient progress on sustainable development since the publication of *Our Common Future* in 1987. The chapters have discussed reasons for this and proposed policy and other reforms to try to ensure greater progress is made over the next twenty-five years than has been made in the previous two and a half decades. This is very important because urgent progress needs to be made. The latest science warns that unless urgent action is taken, our civilisation will, over the next two decades, pass irreversible climatic, hydrological and ecological tipping points, including

- irreversible global warming,
- over extraction of groundwater supplies, and
- irreversible decline of ecological resilience in many parts of the world.

Achieving the required rate and scale of change at the global, national and local level involves the informed participation of many actors: investors, business, government, educators, as well as citizens and communities. This will be hard to achieve given the current level of knowledge and skills, evidenced as follows:

- **Investors and Climate Change:** A global survey of the institutional investors, who recognise the risks of climate change to their investments, found that only 26 per cent of these asset owners have made tangible changes to their investment strategy or investment decision making as a result of climate risk assessments (IGCC 2011).
- **Business and Climate Change:** Whilst businesses can profitably cut energy and water costs and greenhouse gas emissions through investing in energy and water efficiency, the majority of Australian businesses, when surveyed, have said they had not made such investments. Of the reasons given, 40 per cent said that they "Did not know that they should be investing in energy efficiency" and 23 per cent said they hadn't due to lack of skills and knowledge (Australia Industry Group 2012). Most small and medium-sized enterprises (SMEs) have not implemented energy efficiency opportunities in full.
- **Business and Water**: The 2011 Carbon Disclosure Project (CDP) Water Disclosure global report finds that nearly half (43 per cent) of the largest 190 publicly listed organisations in the world do not report board-level oversight of water policies, strategies or plans. This is surprising, given that respondents recognised the short-term nature of water-related risks and opportunities.

- **Business and Climate Change Adaptation:** The majority of businesses globally have not yet systematically assessed risks from climate impacts nor developed climate change adaptation plans. The World Economic Forum's Global Risk Report for 2013 ranked 'failure to recognise climate change adaptation' as one of the top risks for business.
- **Governments and Policy Makers**: As previous chapters have shown, otherwise smart and capable policy makers and government leaders are making climate change, energy, and water and food policy decisions without enough knowledge or awareness of how such policies could negatively or positively impact the other. According to the IEA, current energy policies will result in the volume of water consumed for energy production to almost double by 2035 (IEA 2012). This lack of understanding is partly why current national energy policies globally, overall, give no chance of avoiding passing the two degree Celsius guardrail, beyond which irreversible catastrophic global warming will take place.
- **Citizens**: Evidence suggests that current knowledge and skills on these issues is insufficient. For instance, in a 2013 survey of US attitudes to climate change found that only one in two agree that both climate change is occurring and that it is caused by human actions (Leiserowitz 2013). A total of fifty-seven per cent of the US public either disagreed or were unaware that scientists agree that the earth is very likely warming due to human activity (Pew 2012). Also, where knowledge and awareness of these issues is high, this does not always lead to the required level of behaviour change amongst citizens and households. (Kollmuss and Agyeman 2002).

Given the scale and speed of change needed, combined with the disturbingly low levels of understanding documented above, a multiple-track 'portfolio' approach is needed to:

- **Engage Decision Makers**: This will involve retraining decision makers so they are empowered with the skills and data they need to embed climate change, energy, water and food security nexus issues and solutions into their decision making and policy making as soon as possible. For this to occur rapidly, the following are needed:
  - International and national climate change, energy, water and food security nexus web platforms and information clearinghouses
  - Online communities of practice
  - Effective professional development programs for decision makers and policy makers
  - Better alignment of nexus issues with other major synergistic global initiatives such as the OECD's Green Growth agenda and the work of the Global Green Growth Institute.

- **Identify and Target Key Leverages For Rapid Change:** For instance, it is widely recognised that large institutional investors, who are responsible for investing trillions of superannuation over the next twenty to forty years, are particularly exposed to physical risks to their investments from the climate change, energy and water nexus due to the breadth of their portfolios (Gardiner 2011). These institutional investors also tend to own significant shares in the insurance industries. Mobilising institutional investors would rapidly elevate nexus issues and concerns as a priority amongst business leaders and governments, who tend to be very sensitive to institutional investors' concerns.

- **Reform the Traditional Educational System:** Rapid curricula renewal is required to update and embed nexus issues into the traditional educational system, especially in existing climate change, water management, sustainable energy and sustainable agriculture courses and majors. Other critical areas for embedding nexus considerations include economics, business and accounting schools, sustainable design-related disciplines such as engineering, teacher training, international relations and history majors. Ultimately, all students will need to have at least a baseline understanding of these nexus dynamics and a comprehensive sustainability perspective to bring to bear on every aspect of their personal and professional lives, regardless of their field. These nexus issues are not new. They have arisen ever since the first civilisation in Mesopotamia, which failed partly due to salinisation of farming land and loss of food security (Diamond 2005). Just as these nexus issues have been major factors in the rise and fall of ancient civilisations, there is formal recognition amongst national security agencies (US Director of National Intelligence 2013) and leading economic agencies today that better managing these issues will be critical to achieving long-term economic prosperity and geosecurity (US National Intelligence Council 2013).

The purpose of this chapter is to discuss these strategies in more detail and provide precedents of where such strategies are already starting to be implemented to help inspire further global action.

A portfolio approach to mainstreaming climate change, energy, water and food security nexus knowledge and skills

Strategy #1: Invest in climate, energy, water and food security nexus knowledge exchange web platforms

Fundamentally, for there to be better policy integration and measureable improvements in how nexus issues are managed, millions of decision makers need to better understand these issues and receive training in new skills. Becoming more effective at designing integrated solutions to nexus issues requires multiple literacies across not just the individual fields of climate change, energy, water and food security but also the nexus between them. Climate change, energy, water and food security are all significant fields of knowledge in their own right. Knowledge in each of these fields as well as their interconnectivity is expanding, and will continue to expand. Hence it is impossible for any formal university course or short professional development course to cover all the knowledge that this and the next generation of decision makers will need. Knowledge exchange web platforms and online communities of practice provide convenient mechanisms for busy decision makers to easily find and access the specific knowledge and information they need when they need it. Examples of such platforms include the following:

- The German Government, in partnership with the European Union and leading NGOs, has initiated the world's first Energy/Water/Food Security Nexus platform at www.water-energy-food.org/.

- IUCN and International Water Association (IWA) in 2013 launched the Nexus Dialogue on Water Infrastructure Solutions to build an online community of practice at www.waternexussolutions.org/1x8/home.html.

There are also online knowledge platforms in the United States focusing on aspects of these issues, such as the Climate Literacy and Energy Awareness Network (CLEAN, www.cleanet.org/), which focusses on teaching climate and energy science, and the Climate Adaptation Knowledge Exchange (CAKEX, www.cakex.org/), which focusses on managing natural and built systems in the face of rapid climate change. There are opportunities to explicitly integrate nexus issues into existing platforms such as these.

In addition, nexus web platforms and online communities at the subnational level are needed to make it easier for decision makers, educators and students to quickly find the best locally relevant data and information for their local environments.

Strategy #2: Invest in professional development for decision makers and government policy makers

Building a deeper understanding of the scale and speed of change required

Sociopolitically, it will be difficult to achieve the required scale and speed of policy and institutional changes without greater understanding by decision makers and citizens of why urgent action on nexus issues is needed. The non-linear aspects of how global climatic and ecological systems respond to environmental pressures and the implications of crossing irreversible tipping points is poorly understood (Sterman 2010 and Cronin, Gonzalez and Sterman 2009). Decision makers need to better understand the urgency of the challenge. Economists, who often have critical roles in key policy-making positions, are rarely trained in global climate change or ecological resilience theory. Therefore, they commonly assume that once pressures have pushed the global climate system (or groundwater-salinity balance, or an ecosystem) past a certain threshold, that reducing the pressure by a relatively small percentage is all that is needed to restore the system. The opposite is true. The OECD has warned that "without more ambitious policies, increasing pressures on the environment could cause irreversible damage within the next few decades" (OECD 2008). As the OECD warns, "Non-linear impacts, including the existence of ecological thresholds and irreversible changes, can have significant effects on the total costs of inaction" (OECD 2008). The costs of inaction are already mounting. With notable exceptions such as the Netherlands and Costa Rica, this understanding of the scale and speed of change needed to avoid crossing irreversible tipping points is not reflected in current policy settings in most countries globally.

Addressing nexus issues to unlock productivity growth

Historically, many economic policy makers have focused on policies to improve labour and capital productivity to improve national productivity growth, and efforts to improve resource productivity have been relatively neglected. This has been understandable, since, over the past century, global resource prices have overall fallen significantly. That appears

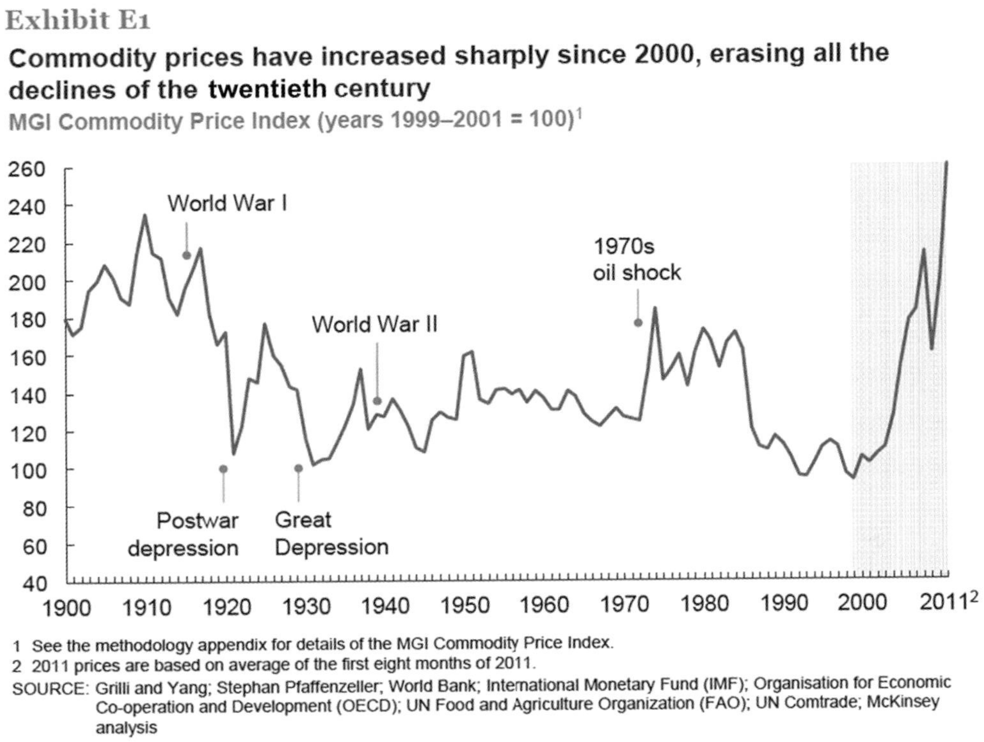

Figure 14.1 Commodity prices have increased sharply since 2000, erasing all the declines of the twentieth century (Dobbs et al. 2011).

to have changed in the past decade. Resource commodity price rises have wiped out the previous 100 years of commodity price reductions (Figure 14.1). This recent trend of rising fossil fuel–based energy, metals, water and food prices is evidence of the emerging climate change/energy/water/food/resource nexus challenge. It reflects the fact that

(i) With population growth and the demographic transition of millions of people out of poverty, especially in Asia, global energy, water and food resource consumption has grown and is likely to keep growing.
(ii) Demand for greater amounts of energy, water and resources in general is occurring at a time when it often is taking more energy, water inputs to extract a particular resource.
(iii) Due to (i) and (ii) increasing linkages between energy, water and other resources as outlined in previous chapters. Historically, the energy intensity of delivering water and the water intensity of delivering energy has never been higher.
(iv) Extreme weather shocks, such as long-term droughts, reduce water availability, increase food and electricity prices (Dobbs et al. 2011).

Addressing nexus issues through a focus on improving end-use energy, water and materials efficiency to reduce global demand has the potential to yield US$2.9 trillion in increased global productivity by 2030 (Dobbs et al. 2011; see Figure 14.2).

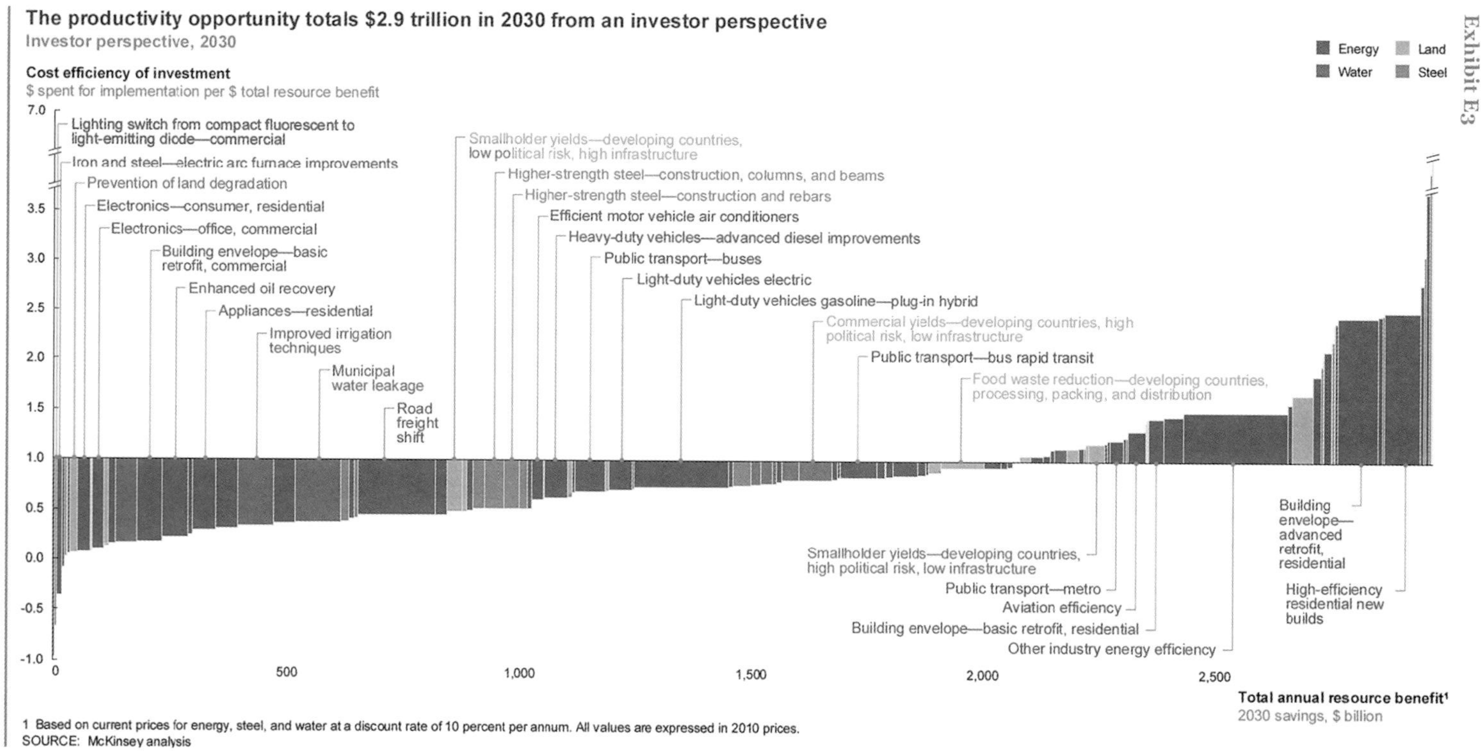

Figure 14.2 Productivity opportunities contribute US$2.9 trillion by 2030 – investor perspective(Dobbs et al. 2011).

Decision makers and policy makers would benefit from appropriate professional development in this area

- Given the historic shift in global resource prices over the past decade, and
- The role climate change, energy, water, metals and food production nexus is playing in it.

This professional development is needed to help ensure all decision makers understand how a focus on energy, water and resource productivity and other 'green' strategies can turn current drags on productivity into new sources of productivity growth (Smith 2013a). Doing so is also strategically important to ensure governments clearly understand the productivity benefits of proactively addressing the climate, energy, water and food production nexus. With this understanding, governments are more likely to commit to an integrated strategy to improve education and skills in this area (Smith 2013a).

Overcoming government silos, enhancing integrated policy development

A range of major reports on nexus issues by leading institutions (OECD 2010; World Economic Forum 2011; IEA 2012) all urge improved policy integration at the local, national, regional and global level. These reports evidence the need to improve skills in cost-benefit analysis as well as integrated policy development, co-ordination and harmonisation to account for trade-offs and synergies from the interconnectedness of climate change, water, energy and food production. A significant part of addressing nexus issues is identifying and eliminating contradictory policies (World Economic Forum 2011). Doing so will require enhanced communication and collaboration between traditionally distinct governmental departments and agencies. This book itself is a significant effort to address knowledge and skills gaps in this area.

Strategy #3: Align nexus issues with 'Green Growth'

Given the common motivations of both the OECD Green Growth initiative (OECD 2011) and the field of climate change, energy, water and food security nexus, there is an opportunity and a need to create a major Green Growth report on nexus issues. There are no major reports linking these two bodies of work to support policy makers' efforts to improve policy integration and design. Such a report would be a very effective investment by the OECD and or member governments to help both (a) integrate nexus issues into the OECD's Green Growth work (OECD 2011) and (b) provide clear guidance to OECD member nations on how to address nexus issues as part of their efforts to better integrate economic and environmental policy making. It would also be timely because the OECD now requires member nations to review their economic and environmental policies and improve policy integration through the lens of Green Growth every two years. Developing such a report would not be difficult given the OECD has published a report with recommendations for energy and water nexus policy reforms (OECD 2010).

Strategy #4: Inform and mobilise the investment community

Large institutional investors, who are responsible for investing billions of superannuation over the next twenty to forty years for their members' retirements, are particularly exposed to physical risks to their investments from the climate change, energy and water nexus. Large institutional investors invest in mining, oil and gas, construction and property, manufacturing and logistics, agriculture and forestry, water and energy sectors. Companies in these sectors often have long supply chains and capital-intense fixed assets, which are vulnerable to damages and operational losses from extreme weather events. Most of them also have significant energy and water requirements and are vulnerable to increased competition for water and drought (Gardiner 2011).

Institutional investors, who invest in construction and property markets, are at risk of having stranded building assets through:

- Re-zoning and changes to planning laws in response to risks from sea-level rises and storm surges, flooding and bushfires.
- Changes to climate change policy because between 40 and 50 per cent of the market value of all fossil fuel companies, including coal, is based on the reserves in the ground. If a global agreement on climate change is reached by 2020 to keep global warming below two degrees, then only 20 to 30 per cent of all remaining fossil fuels will be able to be burnt to ensure the two-degree guardrail is not crossed (Carbon Tracker Initiative 2013). If carbon capture and storage is not commercially competitive, this would lead to a significant devaluation of these companies and high systemic risk across financial markets.

To address these concerns, some institutional investors have formed regional peak bodies, and now a global peak body. The Global Investor Coalition on Climate Change (GICCC), which represents investor's worth over US$20 trillion, was formed in late 2012. They are aiming, amongst other goals, to:

- Raise awareness of the potential impacts, both positive and negative, of climate change to the investment industry, corporate, government and community sectors.
- Encourage best practice approaches to facilitate the inclusion of the climate change impacts in investment analysis by the investment industry
- Provide information to assist the investment industry to understand and incorporate climate change into the investment decision.

To meet these goals, GICCC's regional sister organisations, such as the Investor Group on Climate Change in Australia and New Zealand, have worked with the Australian National University to publish a series of 'Climate Change, (Energy, Water) Risks and Opportunity Guides' for investors on various business sectors (Smith 2013b, c, d and e). These guides are the first of their kind to give investors the information they need, not just on climate change, energy and water risk assessment but also whether or not companies are doing all they reasonably can to manage these risks and implement cost-effective adaptation and mitigation strategies.

In addition to this, over 530 major institutional investors, representing over US$57 trillion in assets are now signatories of the Carbon Disclosure Project and the Water Disclosure Project. Each year these two projects send questionnaires to thousands of major corporations globally. The resulting reports help investors assess a company's exposure to climate change, energy, water and supply chain input risks as well as their climate change adaptation and mitigation performance to manage those risks. The current questionnaires ask questions related to some, but not all aspects of the climate change, energy, water and food security nexus. There is an opportunity to improve both the Carbon and Water Disclosure Project's questionnaires by including questions to address currently ignored nexus issues and thereby prompt thousands of multinational businesses to proactively address these issues.

Strategy #5: Better inform business on risks and opportunities from nexus issues

Business will increasingly face physical risks to their operations and assets from more intense extreme weather events, changes to water availability, and higher energy and resource input costs due to the interlinked nature of the nexus issues. Despite this, few businesses have begun to comprehensively adapt to climate change as shown in the World Economic Forum's Global Risk Report for 2013, which ranked 'failure to recognise climate change adaptation' as one of the top risks for business.

There is an urgent need to better communicate and educate business on their sector's specific exposure to nexus risks and provide them with options for how to adapt and better to manage such risks (Smith, 2013b, c, d and e). There is also significant consensus that business can improve competitive advantage by addressing many of these risks proactively by:

- Improving energy and water efficiency to cut operational costs and reduce exposure to a carbon price by mitigating greenhouse gas emissions
- Reducing exposure to risks of extreme weather events through cost-effective climate change adaptation measures
- Innovating new 'greener', more eco-efficient products

Comprehensive educational resources have been developed for business leaders, operational business managers and engineers, as well as educators and students, which cover risks and opportunities for major business sectors such as mining and mineral processing, construction and property, manufacturing and food processing; guides for agriculture and the energy sector are forthcoming (Smith 2013f, g, h, i, j and k). A significant amount of climate change mitigation by 2020 can be achieved through energy efficiency. Likewise, water efficiency will play a significant role in reducing the demand–supply gaps in many nations. Energy and water efficiency and combined energy-water efficiency opportunities exist for all business sectors and offer significant ways to reduce operational business costs to improve competitive advantage. Despite this, surveys show that specific knowledge of these opportunities is poor amongst most business leaders and their staff. For instance, the month before Australia implemented a price on carbon, the country's largest member industry group conducted its largest survey ever. It found that over 50 per cent of members had not implemented energy management systems nor invested in energy efficiency partly due to lack of internal capacity or due to lack of reliable information (Figure 14.3).

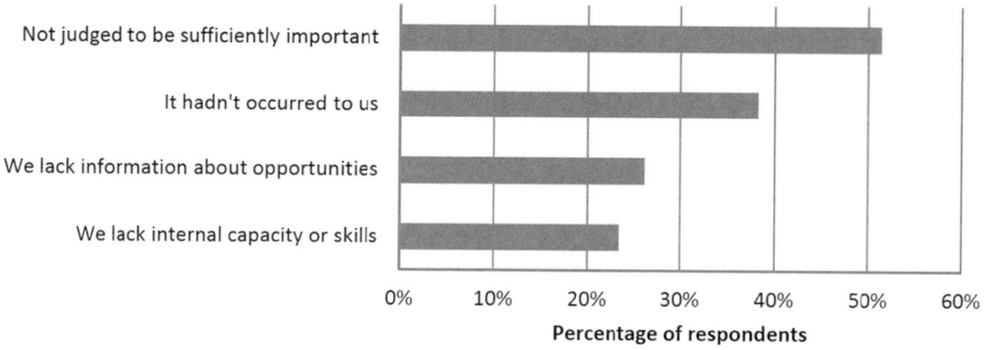

Figure 14.3 Reasons why business did not actively manage electricity use in response to carbon pricing arrangements, 2012 (AIG 2012).

To address this, governments around the world have invested in (a) energy and water efficiency web portal best practice guide information for multiple business sectors (COAG 2012; DSEWPC 2010) and (b) education and training packages for engineers and technical staff (Smith et al. 2007; Stasinopoulous et al. 2010). Other governments have chosen to actively engage with peak industry groups for key sectors and academic experts to work together to produce relevant information that is widely disseminated utilising the peak industry groups' strong networks with businesses in the sector.

The energy-water nexus is particularly important for businesses and professionals working in the energy and water sectors. Professionals need skills in how to better manage energy and water supply infrastructure in a changing climate through integration of these sectors. There are a range of new resources to help energy and water professionals achieve this (Cammerman 2009). Energy and water professionals can benefit from additional training on the potential for energy and water efficiency to understand likely demand-management trends. A wide range of educational resources exist to assist business, energy professionals and educators in improving skills in the area of energy efficiency and water efficiency (Smith 2007; Stasinopoulos 2010; COAG 2012; Smith 2013f, g, h, i, j and k). Whilst a transition to renewables will reduce water intensity in the energy sector, it is important also to reduce the water intensity of fossil fuel energy systems. Best practice educational guides also exist on how to reduce freshwater usage in coal mining (Smith 2010a), coal-fired power stations (Smith 2010b), and petroleum refining (Smith 2010c). These show significant potential for water savings. For instance, the first-ever water mass balance for water usage in the Bowen Basin coal mining fields of Australia found that freshwater usage could be reduced by 70 per cent and overall fresh and worked water usage could be reduced by 40 per cent (Smith 2010a).

Strategy #6: Embed nexus issues into existing climate change courses for current and future decision makers

One of the main opportunities to embed education and training on nexus issues is within existing climate change initiatives and courses. Over the past decade there has been a rapid rise in the number of such courses offered in the formal education sector as well as in the short professional development courses. Many universities now offer climate change mitigation

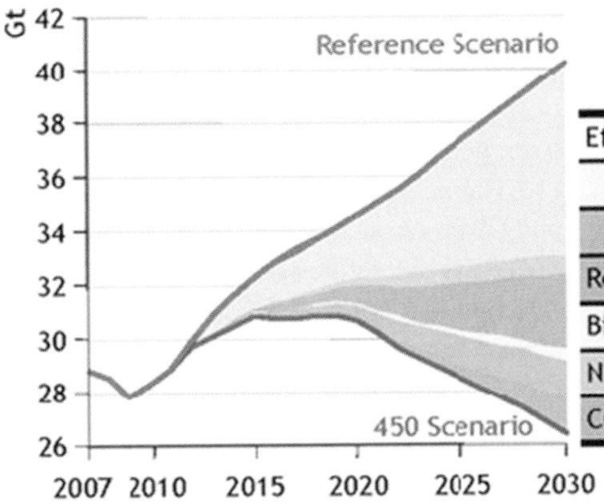

	Abatement (Mt CO_2)		Investment ($2008 billion)	
	2020	2030	2010-2020	2021-2030
Efficiency	2 517	7 880	1 999	5 586
End-use	*2 284*	*7 145*	*1 933*	*5 551*
Power plants	*233*	*735*	*66*	*35*
Renewables	680	2 741	527	2 260
Biofuels	57	429	27	378
Nuclear	493	1 380	125	491
CCS	102	1 410	56	646

Figure 14.4 International Energy Agency – carbon abatement strategies and stabilisation wedges (IEA 2009).

courses at both undergraduate and postgraduate master's level. In the United States, the non-profit Second Nature initiated the American College and University Presidents' Climate Commitment (ACUPCC). They invite US colleges and universities to commit to becoming net climate neutral from specified sources in campus operations, accelerating their research and educational efforts to equip society to mitigate and adapt to climate change. More than 670 colleges and universities have joined the ACUPCC network. Many of these are teaching undergraduate and postgraduate courses on climate change. Some universities, such as Princeton University and the University of Arizona, have also begun to develop and teach energy-water nexus courses. But these are the exception to the rule. And to date, there are very few climate change-energy-water-food security nexus courses available in either short course form for busy decision makers or, as course subjects, for the next generation at universities.

One of the goals of this publication is an attempt to create a rigorous resource from which educators will be able to develop new climate, energy, water and food security nexus courses for the formal educational sector and also as short courses for busy decision makers/policy makers. There is an opportunity and a significant need to create online peer reviewed educational resources to support the teaching of such courses. Whilst such courses are being created, there is also an opportunity and a need to create resources for educators who teach climate change courses that provide simple ways to embed nexus issues into their courses.

Strategy #6A: Embed energy/water/food security nexus into existing climate change mitigation courses.

Previous chapters have shown how most climate change mitigation frameworks rely, at least in part, on rapidly expanding biofuels, carbon capture and storage, nuclear power and natural gas, each of which could have significant negative impacts on water or food security (Figure 14.4). Most climate change frameworks have not recognised nor addressed these impacts, nor are they included in courses on climate change mitigation.

Table 14.1 Pacala and Socolow's fifteen stabilisation wedges.

(1) Double vehicle fuel efficiency (2) Reduced use of vehicles – by half (3) Efficient buildings and appliances – cut GHG emissions by ¼ (4) Efficient base-load coal plants – double conversion efficiency to 60% (5) Replace 1400 GW 50%-efficient coal plants with gas plants (four times the current production of gas-based power) (6–8) Carbon capture and storage (three wedges). (6) Capture CO_2 at baseload power plant. (7) Capture CO_2 at H_2 plants. (8) Capture CO_2 at coal-to-synfuels plant (9) Nuclear power – double current global output	(10) Wind – 2 million MW (50 times current) (11) PV – 2000 GW (700 times current) (12) Wind-generated hydrogen – to power hydrogen fuel cell cars. (13) Biomass for fuel – 100 times current Brazil and US cropland – 1/6th of world's crops. (14) Reduced deforestation, plus reforestation, afforestation and new plantations (15) Conservation tillage in agriculture

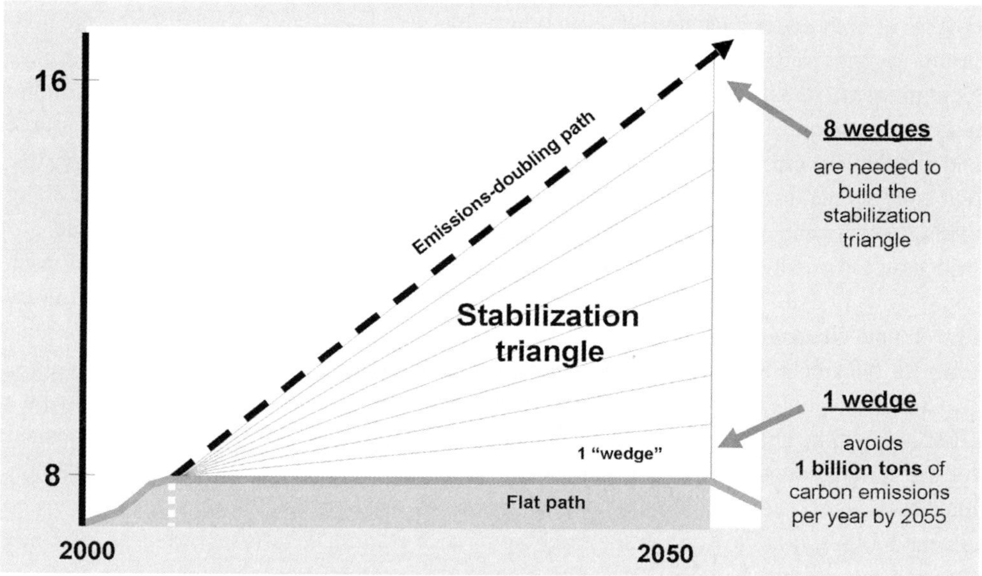

Figure 14.5 Each stabilisation 'wedge' triangle, measured in increments of 1 billion tonnes (or 1 gigatonne) of carbon per annum globally (Pascala and Socolow 2004).

Teaching example: Embedding nexus into climate change mitigation teaching and learning

The concept of 'stabilisation wedges' (Figure 14.5) from the highly cited *Science* journal paper by Pacala and Socolow (2004) is a popular tool for educators teaching climate change mitigation. The stabilisation wedges framework provides a simple way to convey the concept of a global, integrated climate change mitigation strategy. It was also highlighted prominently in Al Gore's presentation in the film *An Inconvenient Truth* – a common point of entry into the climate change issue for many people.

Stabilisation wedges are also a simple, elegant way to communicate to lay audiences the scale of the solutions required to stem the drivers of climate change. Unfortunately, the stabilisation wedges framework, like most published climate change mitigation frameworks, relies on water-intensive forms of energy (biofuels, carbon capture and storage, nuclear power) and does not explicitly address nexus risks and opportunities. However, by updating the stabilisation wedges approach with insights from the field of nexus studies, educators can better teach students how addressing the nexus challenge can help address climate change. The following example demonstrates one of many creative ways to embed nexus insights into traditional educational approaches to teaching climate change mitigation.

Pacala and Socolow (2004) postulated that humanity would need to choose seven out of fifteen potential climate change mitigation 'stabilisation wedges', where each wedge represented 1 gigatonne of carbon dioxide–equivalent reductions, to stabilise the earth's climate (Figure 14.5). Put together, they argued that these seven wedges demonstrated that it was possible, using existing technologies, to stabilise carbon dioxide emissions rapidly and keep them constant over the next fifty years.

More recent literature shows that at least fourteen to fifteen stabilisation wedges are needed, instead of seven, to stabilise the climate and avoid dangerous climate change (Romm 2008; Davis et al. 2013). This is largely because, almost a decade on, from the publication of the original Pascala and Socolow paper (Pascala and Socolow 2004),

- Greenhouse gas emissions are still tracking at the higher end of IPCC predictions.
- Negative feedback loops in the climate and biosphere are occurring at a faster rate than predicted by climate scientists a decade ago.

In addition, Pacala and Socolow's recommended wedges include several technologies that would have significant negative impacts on water consumption and food security, such as the following:

- Wedge #13: Adding 100 times the current Brazil or US ethanol production to use a sixth of the world's crop production;
- Wedge #5: Increasing four fold natural gas production globally, when more recent literature suggests such a transition from coal to natural gas will make negligible positive difference (Wigley 2011);
- Wedge #9: A doubling of global nuclear power production, which is a significant user of water;
- Wedge #14: Expansion of plantation forests, which also require significant amounts of water; and finally,
- Wedges #6, #7 and #8: Large-scale roll out of carbon capture and storage, which increases by 80 per cent water consumption at coal-fired electricity power stations.

Finally, a number of the stabilisation wedges rely on carbon capture and storage technologies that are not yet, and may never be, commercially viable.

This popular climate change mitigation framework provides educators with a 'teaching opportunity' to show how common it is for climate change experts to ignore the negative water and food security implications of their recommendations.

Given these considerations, does the overall conclusion of the Socolow and Pacala paper – that "humanity already possesses all the fundamental scientific, technical and industrial know how to solve the climate and carbon problem" – still hold true? More recent literature, which has identified new stabilisation wedges, shows that it does (Table 14.2). It is possible to both achieve the required levels of climate change mitigation and do so in ways that help water and food security – mainly through a focus on end-use energy, water efficiency and reducing food waste.

These can be combined with the following original Pacala and Socolow stabilisation wedges, which do not have significant negative impacts on water and food production

(17) Double vehicle fuel efficiency
(18) Reduced use of vehicles – by half
(19) Efficient base-load coal plants – double conversion efficiency to 60 per cent
(20) Wind – 2 million megawatts (50 times current)
(21) PV – 2,000 gigawatts (700 times current)
(22) Reduce deforestation
(23) Conservation tillage in agriculture

to overall achieve at least fourteen stabilisation wedges.

This updated framework for the original Pacala and Socolow stabilisation wedges framework provides busy educators with a new way to succinctly communicate many of the central findings of climate change, energy, water and food security nexus researchers. Importantly, these new stabilisation wedges largely involve energy and water efficiency improvements that provide good returns on investment. These additional stabilisation wedges listed in Table 14.2 happen to correspond with most of the activities evaluated by the McKinsey Global Institute, which calculated that together, they would add US$2.9 trillion to global productivity growth by 2030 (Dobbs et al. 2011; Figure 14.2).

Strategy #6B: Embed climate, energy, water and food security nexus into existing climate change adaptation courses

Energy, water and food security issues also sit naturally in climate change adaptation courses. Likewise, adaptation strategies need to be embedded into integrated water and energy resource management and agricultural courses and training. The climate change adaptation literature until recently has been dominated by biophysical analysis of the likely risks and potential negative impacts from unmitigated climate change, with much less emphasis on researching, trialling and assessing options to adapt to the risks of climate change in the energy, water and agriculture sectors. For instance, the IPCC AR4 WGII report published in 2007 had virtually no text addressing options for nations, business and communities of how to adapt to changes in water availability from climate change adaptation.

Table 14.2 New climate change mitigation strategies with water and food security co-benefits (total = 7–10 additional climate stabilisation wedges).

(1) End use energy efficiency #1 – A step change in energy efficiency of industrial processes. The energy intensity > 50% higher than the theoretical minimum determined by the laws of thermodynamics (IEA 2006).
(2) End use energy efficiency #2 – Upgrade and optimise industrial and commercial technologies, eg, electric motor driven systems such as pumps and fans and compressed air (IEA 2011).
(3) End use energy efficiency #3 - Reduce sub-optimal heat transfer (heat loss and heat gain) + waste heat and power recovery (including cogeneration) (Romm 2008).
(4) End Use Energy Efficiency #4 – Energy efficient street, traffic and neon signage lighting, such as LED, cuts energy usage by 85%.
(5) End use fuel efficiency #5 – Non-passenger vehicle freight transport (Trucks and Rail) (Mui 2007).
(6) End Use Fuel Efficiency #6 – Fuel Efficient air transport and alternatives to air transport (videoconferencing + very fast trains) (Mui 2007).
(7) End use energy efficiency #7 – Doubling of building efficiency, eg, Increasing building albedo reducing air-conditioning loads.
(8) End use efficiency #10 – Designing products to be more energy and water efficient.
(9) Energy/Water/Materials/Waste Nexus – Product stewardship, materials efficiency, recycling, methane capture from landfill (Bahor 2009).
(10) Decarbonising water supply and treatment through reducing water leakage water efficiency, demand management, methane to electricity from sewerage treatment plants. Investing in solar power generation structures on top of the water behind hydroelectric dams to generate electricity and reduce water losses due to evaporation from dam water.
(11) Significantly reduce 1.3 billion tonnes of annual global food waste (a 1/3 of global food production), currently responsible for 3.3 billion tonnes of CO equivalent emissions (FAO, 2013).
(12) Increased soil carbon storage in grazing systems through time controlled grazing principally in the$_2$ South – Africa, Australia and South America
(13) Decarbonising the Electricity Sector – Transitioning to 80–100% renewables by 2030–2050 through the use of additional renewable technologies including wind power at sea, solar thermal and geothermal in non-arid areas (Elliston B et al. 2012; IPCC 2012; IEA 2011, 2009).
(14) Smart grid technologies to enable plug-in hybrid or fully electric renewable transportation vehicles (cars, vans, motorbikes, trucks, buses) powered by renewable energy sources through vehicle to grid technologies (Romm, 2008; IEA 2009 and 2011; Went et al. 2008).
(15) Reducing Non-CO2 Greenhouse gas emissions (Reilly 2008).
(16) Behaviour Change (Romm 2008).

To date, colleges and universities as a whole have not sufficiently addressed the issue of climate preparedness. In the United States, there is a need to increase course offerings related to the societal implications of climate change impacts, and the strategies for preparing for and coping with those impacts. There are also opportunities for higher education institutions to serve as role models for implementing adaptation strategies, and to serve as hubs in their local communities for disseminating successful adaptation

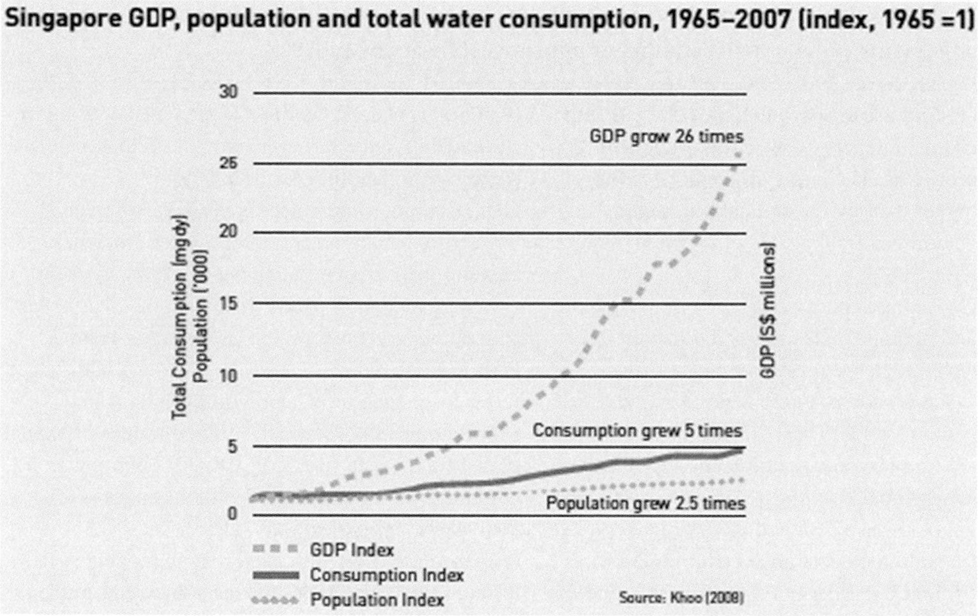

Figure 14.6 Singapore GDP, population and total water consumption growth (1965–2007) [1965 =1] (Khoo 2008).

strategies (Andrews and Dyer 2011). An explicit focus on critical nexus issues would make these emerging climate adaptation efforts more effective.

A range of projects have been undertaken in recent years to address this failing and move from climate change risk analysis into assessment of climate change adaptation options. For example, the online textbook, entitled *Water Transformed: Sustainable Water Solutions Through Climate Change Adaptation*, shows that it is possible to meet the supply–demand gap for water though achieving a step change in water efficiency, demand management, and greater recycling of stormwater and treated water and utilisation of managed aquifer recharge, storage and recovery (Smith et al. 2010d). More recent studies have backed up this claim (Water Resources Group 2012). Some nations like Singapore have applied these strategies to achieve remarkable levels of decoupling (Figure 14.6).

More of these types of educational resources and formal and informal offerings are needed to ensure that policy makers and decision makers understand the complex dynamics of nexus issues in the context of climate change adaptation. For example, some areas will need to move to higher levels of water treatment, water recycling and desalination in the face of increased drought. Some argue this will increase the energy and greenhouse gas intensity of supplying water too much and should be avoided. However, water treatment options such as managed aquifer storage and recovery are very energy efficient. Further, in both urban and agricultural settings, the energy required to treat and pump water is relatively small in most countries; the bulk of energy use related to water is in heating it. For instance, the amount of energy used to supply water to the urban water sector in Australia is only about 0.5 per cent of the total urban energy use (see Chapter 9). If all the additional water supplies were sourced from desalination over the next twenty years, the energy use will still be less than 1 per cent of the total urban energy use (Kenway et al. 2008). The Californian Energy Commission has found that the embodied energy in water in agriculture is significantly less than the urban water sector (California Energy Commission 2005). In China, where its agricultural sector has significant pumping requirements due to falling groundwater tables, agricultural pumping energy usage represents less than 1 per cent of total Chinese energy usage. And much can be done to reduce energy usage in irrigation for agriculture through the use of a holistic approach to (i) reducing leakage, (ii) irrigation scheduling, (iii) appropriate crop choice, (iv) mulch to reduce evaporation and (v) advanced deficit irrigation where only one side of a plant is watered at one time.

The one place where significant amounts of energy are used is in the heating of water. Traditional electric storage hot water systems consume a significant proportion of energy in residential homes around the world – representing 9 per cent in the European Union – 15 per cent in 2004, 11 per cent in the United States in 2005, 25 per cent in Australia and 27 per cent in China in 2000. Fifteen per cent savings in hot water heating across Australia would offset the entire energy use to supply water to the urban water sector (Kenway et al. 2008). As the IPCC AR4 explained:

> Energy requirements for domestic water heating can be reduced by at least 90 per cent, through a range of cost effective options, including: reducing consumption of hot water; improving thermal properties of hot water systems; recovering lost heat; and selecting low energy consumption hot water systems, such as solar thermal or heat pumps (Levine et al. 2007).

Like with domestic hot water systems and appliances, wherever else heated water is used – industry, commercial buildings and offices, restaurants, laundromats, swimming pools or sports centres – there are enabling technologies to reduce energy use and associated greenhouse gas emissions by (a) reducing the amount of hot water needed, (b) reducing heat losses through better insulation and (c) preheating water with thermal solar hot water systems. Energy reductions through improving the use of hot water can easily offset increased energy consumption from increased irrigation, water treatment and recycling or desalination needed to adapt to climate change to ensure water security.

This is but one critical example of the types of nexus issues that business, professionals and policy makers need to better understand to inspire greater efforts and ensure appropriate policies are implemented for achieving energy, water and greenhouse gas savings. New education and training packages are needed to address this to ensure these types of energy and greenhouse gas reduction opportunities are realised.

Strategy #7: Develop and implement national strategies for nexus education

While faculty integrate nexus issues into existing courses on climate change mitigation (Strategy 6A) and adaptation (Strategy 6B), steps can be taken to develop and implement strategies for ensuring that all students, in all disciplines, have a sufficient understanding of nexus issues – and holistic sustainability concepts – to factor into decision making in their personal and professional lives.

There is an opportunity and need to create national networks of senior administrators at colleges and universities to provide leadership on this process. Critical participants include chief executive officers, chief academic officers, and heads of departments and schools (eg presidents, provosts and deans in the US system). These networks can serve as communities of learning and action, leading to better understanding of these issues and sharing best practices on how to appropriately integrate them into student education.

Specific areas of focus could include the following:

- How to engage faculty and provide professional development offerings on a broad scale
- Developing resources for administrators and faculties to help them better understand these issues and the relevance to their disciplines
- Developing model learning outcomes for graduates who will be navigating a changing climate throughout the twenty-first century.

One such effort is in the early stages in the United States, where a group of provosts, presidents, deans and faculty members met in March of 2012 and developed the basis for a five-year national strategy to infuse sustainability into the educational experience of all students (ACUPCC 2012).

Conclusion

This chapter has demonstrated the need and opportunity to share knowledge on nexus risks and opportunities. The risks of insufficient understanding of the interconnections between climate, water, energy and food in the global economy, society and environment, and security are high (US National Intelligence Council 2013). It is vital, therefore, that a comprehensive education and retraining portfolio program begins including, at least, the strategies outlined earlier. This chapter has shown that, despite progress, the current United Nations Decade of Education in Sustainable Development has not yet achieved what is needed to ensure all peoples have the necessary knowledge and skills to play their part in achieving a transformation to sustainability. This chapter shows that there is an opportunity to re-invigorate education for sustainability through a new 'United Nations Decade of Education for Sustainable Development' (2015–25) with a strong focus on the climate change, energy, water and food production nexus. Such a strategy would build political will for the policy reforms outlined in this book. It will also ensure that investors, business and the community can better respond to economic incentives and regulatory changes to ensure that such policy reforms achieve their objectives.

References

Akbari, H., Menon, S. and Rosenfeld, A. (2009). Global Cooling: Increasing World-Wide Urban Albedos to Offset CO_2. *Climatic Change* 94: 275–86.

American College & University Presidents' Climate Commitment (ACUPCC) (2012). Academic Committee, www.presidentsclimatecommitment.org/academic-committee/.

Anderson, K. and Bows, A. (2011). Beyond 'Dangerous' Climate Change: Emission Scenarios for a New World. *Philosophical Transactions of the Royal Society A* 269: 20–44.

Andrews, J. and Dyer, G. (2011). *Higher Education's Role in Adapting to a Changing Climate*. American College & University Presidents' Climate Commitment, Boston, MA. www.presidentsclimatecommitment.org/files/documents/higher-ed-adaptation.pdf.
Australia Industry Group (2012). *Energy Shock*. Australia Industry Group.
Bahor, B. (2009). Integrated Waste Management as a Climate Change Stabilisation Wedge. *Waste Management & Research* 27: 839–49, www.seas.columbia.edu/earth/wtert/sofos/wmr_nov09_p839.pdf.
Bernstein, L., Roy, J., Delhotal, K. C., et al. (2007). Industry. In *Climate Change 2007: Mitigation*. Contribution of Working Group III to the Fourth Assessment Report of the Intergovernmental Panel on Climate Change, eds. B. Metz, O. R. Davidson, P. R. Bosch, R. Dave, L. A. Meyer, Cambridge University Press.
Bogner, J., Abdelrafie Ahmed, M., Diaz, C., et al. (2007). Waste Management. In *Climate Change 2007: Mitigation*. Contribution of Working Group III to the Fourth Assessment Report of the Intergovernmental Panel on Climate Change, eds. B. Metz, O. R. Davidson, P. R. Bosch, R. Dave, L. A. Meyer, Cambridge University Press.
Bye, A. (2011) Case Studies Demonstrating Value from Geometallurgy Initiatives. 1st International Geometallurgy Conference (GeoMet 2011), www.ausimm.com.au/publications/epublication.aspx?ID=12887.
Cammerman, N. (2009). *Integrated Water Resource Management and the Water, Energy, Climate Change Energy Nexus*. Masters Thesis. University of Queensland, www.watercentre.org/resources/publications/attachments/ncammerman/.
Canadell, J. G., LeQuere, C., Raupach, M. R., Field, C. B., Buitenhuis, E. T., Ciais, P., Conway, T. J., Gillett, N. P., Houghton, R. A. and Marland, G. (2007). Contributions to Accelerating Atmospheric CO_2 Growth from Economic Activity, Carbon Intensity, and Efficiency of Natural Sinks. *Proceedings of the National Academy of Sciences*.
California Energy Commission (2005). *California's Water-Energy Relationship*. Californian Energy Commission.
Carbon Tracker Initiative (2013). *Unburnable Carbon 2013: Wasted capital and stranded assets*. Carbon Tracker and the Grantham Research Institute, London School of Economics.
COAG Energy Efficiency Exchange. Web portal at www.eex.gov.au/.
Cook, J.; Nuccitelli, D.; Green, S. A.; Richardson, M.; Winkler, B.; Painting, R.; Way, R.; Jacobs, P. and Skuc, A. (2013). Quantifying the Consensus on Anthropogenic Global Warming in the Scientific Literature. *Environmental Research Letters* 8(2): 024024. doi:10.1088/1748-9326/8/2/024024.
Cox, P., Betts, R., Jones, C., Spall, S. and Totterdell, I. (2000). Acceleration of Global Warming Due to Carbon-Cycle Feedbacks in a Coupled Climate Model, *Nature* 408, 184–7.
Cronin, A., Gonzalez, C. and Sterman, J. (2009) Why Don't Well-Educated Adults Understand Accumulation? A Challenge to Researchers, Educators, and Citizens. *Organisational Behaviour and Human Division Processes* 108(1): 116–30.
Cuellar, A. D. and Webber, M. E. (2010). Wasted Food, Wasted Energy: The Embedded Energy in Food Waste in the United States. *Environmental Science Technology* 44, 6464–6469.
Dobbs, R., Oppenheim, J., Thompson, F., Brinkman, M. and Zornes, M. (2011) *Resource Revolution: Meeting the World's Energy, Materials, Food, and Water Needs*. Seoul: McKinsey Global Institute.
Davis, S., Cao, L., Caldeira, K. and Hoffert, M. I. (2013). Rethinking Wedges. *Environmental Research Letters* 8: 011001.
DeConto, R. M., Galeotti, S., Pagani, et al. (2012). Past Extreme Warming Events Linked to Massive Carbon Release from Thawing Permafrost. *Nature* 484:87–91, doi:10.1038/nature10929.

Defence Intelligence Agency, et al. (2012). *Global Water Security*. Intelligence Community Assessment Publication. 2012-02-02 at www.dni.gov/files/documents/Newsroom/Press%20Releases/ICA_Global%20Water%20Security.pdf.

Department of Sustainability, Environment, Water, Population and Communities (2014). *Tools for Planning Water Efficiency-Web Portal*. The Australian Government and Australian National University, 15 January.

Diamond, J. (2005). *Collapse: How Societies Choose to Fail or Succeed*. New York: Random House.

Dietz, T., Gardner, G. T., Gilligan, J., Stern, P. C. and Vandenbergh. M. P. (2009). Household Actions Can Provide a Behavioural Wedge to Rapidly Reduce US Carbon Emissions. *Proceedings of the National Academy of Sciences* 106(44): 18452–6.

Elliston B, Diesendorf M, MacGill I. (2012). Simulations of Scenarios with 100% Renewable Electricity in the Australian National Electricity Market. *Energy Policy* 45:606–13.

Food and Agriculture Organisation (2013). *Food Wastage Footprint: Impacts on Natural Resources*. Rome: Food and Agricultural Organisation.

Gardiner, D. (2011). *Physical Risks from Climate Change. A guide for companies and investors on disclosure and management of climate impacts*. Ceres: Oxfam America, Calvert Investments.

International Energy Agency (IEA) (2009). *Wind Energy Technology Roadmap*. Paris: IEA, www.iea.org/publications/freepublications/publication/Wind_2013_Roadmap.pdf.

International Energy Agency (IEA) (2010a). *Concentrated Solar Thermal Technology Roadmap*. Paris: IEA, www.iea.org/publications/freepublications/publication/csp_roadmap.pdf.

International Energy Agency (IEA) (2010b). *Solar Photovoltaic Technology Roadmap*. Paris: IEA, www.iea.org/publications/freepublications/publication/pv_roadmap.pdf.

International Energy Agency (IEA) (2011a) *Electric and Plug-In Hybrid Technology Roadmap*. Paris: IEA, www.iea.org/publications/freepublications/publication/EV_PHEV_Roadmap.pdf.

International Energy Agency (IEA) (2011b). *Electric and Plug-In Hybrid Electric Vehicle Technological Roadmap*. Paris: IEA, www.iea.org/publications/freepublications/publication/EV_PHEV_Roadmap.pdf.

International Energy Agency (IEA) (2011c). *World Energy Outlook*. Paris: IEA, www.iea.org/publications/worldenergyoutlook/publications/weo-2011/.

International Energy Agency (IEA) (2012). *World Energy Outlook*. Paris: IEA, www.iea.org/publications/worldenergyoutlook/publications/weo-2012/.

Intergovernmental Panel on Climate Change Mitigation (IPCC) (2011). *Special Report on Renewable Energy Sources and Climate Change Mitigation*. Prepared by Working Group III of the Intergovernmental Panel on Climate Change, eds. O. Edenhofer, R. Pichs-Madruga, Y. Sokona, K. Seyboth, P. Matschoss, S. Kadner, T. Zwickel, P. Eickemeier, G. Hansen, S. Schlömer, C. von Stechow. Cambridge University Press, 1075 pp., http://srren.ipcc-wg3.de/.

Institutional Investor Group on Climate Change (IIGCC), Investor Network on Climate Risk (INCR) and Investor Group on Climate Change (IGCC) Australia/New Zealand (2011). *Global Investor Survey on Climate Change*. IIGCC, IICC, INCR, IGCC, www.igcc.org.au/Resources/Documents/Global%20investor%20survey%20on%20climate%20change.pdf.

Kenway, S. J., Priestley, A., Cook, S., Seo, S., Inman, M., Gregory, A. and Hall, M. (2008). *Energy Use in the Provision and Consumption of Urban Water in Australia and New Zealand*. CSIRO: Water for a Healthy Country National Research Flagship.

Khoo, T. C., (2008) *Singapore Water, Yesterday, Today and Tomorrow*. New York: Public Utilities Board of Singapore Kingdom and New York.

Kollmuss, A., Agyeman, J., (2002). Mind the Gap: Why Do People Act Environmentally and What Are the Barriers to Pro-Environmental Behaviour? *Environmental Education Research* 8(3): 239–60.

Lavelle, M., Grose, T. (2013). Water Demand from Energy Production to Double by 2035 – The Great Energy Challenge. *National Geographic*, http://news.nationalgeographic.com.au/news/energy/2013/01/130130-water-demand-for-energy-to-double-by-2035/.

Leiserowitz, A., Maibach, E., Roser-Renouf, C., Feinberg, G., and Howe, P. (2013). *Climate Change in the American Mind: Americans' Global Warming Beliefs and Attitudes in April, 2013*. New Haven, CT: Yale University and George Mason University. Yale Project on Climate Change Communication, http://environment.yale.edu/climate-communication/files/Climate-Beliefs-April-2013.pdf.

Levine, M. D. Ürge-Vorsatz, K. Blok, L. et al. (2007). Residential and Commercial Buildings. In *Climate Change 2007: Mitigation*. Contribution of Working Group III to the Fourth Assessment Report of the Intergovernmental Panel on Climate Change eds. B. Metz, O. R. Davidson, P. R. Bosch, R. Dave, L. A. Meyer. Cambridge University Press.

Mathiesen, B. Lund, H. and Norgaard, P. (2008). Integrated Transport and Renewable Energy Systems. *Utilities Policy* 16: 107–16.

Organisation for Economic Co-operation and Development (OECD) (2010). *Coherence Between Water and Energy Policies*. Working Party on Global and Structural Policies. Paris: OECD. http://search.oecd.org/officialdocuments/displaydocumentpdf/?cote=ENV/EPOC/GSP(2010)21&docLanguage=En.

Organisation for Economic Co-operation and Development (OECD) (2011). *Towards Green Growth*. Paris: OECD.

Pacala, S. and Socolow, R. (2004). Stabilization Wedges: Solving the Climate Problem for the Next 50 Years with Current Technologies. *Science* 305: 968–70.

Pew Research Center (2012). *More Say There Is Solid Evidence of Global Warming*. Washington, DC: Pew Research Center for the People and the Press.

Reilly, J., Jacoby, H. and Prinn, R. (2004). *Multi-Gas Contributors to Global Climate Change: Climate Impacts and Mitigation Costs of Non-CO_2 Gases*. Boston: MIT, http://dspace.mit.edu/bitstream/handle/1721.1/5543/MITJPSPGC_Rpt114.pdf?sequence=1.

Romm, J. (2008). Cleaning Up on Carbon. Nature Reports: Climate Change. *Nature Online*, www.nature.com/climate/2008/0807/full/climate.2008.59.html.

Smith, M. (2010a). *Water Efficiency Opportunities – Best Practice Guide – Coal Mining*. DSEWPC and ANU FSES.

Smith, M. (2010b). *Water Efficiency Opportunities – Best Practice Guide – Coal Powered Electricity Generation*. DSEWPC and ANU FSES.

Smith, M. (2010c). *Water Efficiency Opportunities – Best Practice Guide – Petroleum Refining*. DSEWPC, ANU FSES.

Smith, M. (2010d). *Water Transformed: Sustainable Water Solutions for Climate Change Adaptation*. Griffith University and Australian National University, The Natural Edge Project.

Smith, M. (2013a). *Green Growth – Unlocking New Sources of Productivity through Education for Sustainability An Education and Skills for Sustainability Reform Agenda to Lift Australia's Productivity Discussion Paper*. United States Study Centre and Australian National University.

Smith, M. (2013b). *Assessing Climate Change Risks and Opportunities for Investors: Oil and Gas Sector*. Sydney: Investor Group on Climate Change.

Smith, M. (2013c). *Assessing Climate Change Risks and Opportunities for Investors: Property and Construction Sector*. Sydney: Investor Group on Climate Change.

Smith, M. (2013d). *Assessing Climate Change Risks and Opportunities for Investors: Mining and Minerals Processing Sector*. Sydney: Investor Group on Climate Change.

Smith, M. (2013e). *The Mining and Mineral Processing Sector – Climate Change Risks and Opportunities*. An Educational Guide to Assist Climate Change Risk, Adaptation and Mitigation Opportunity Assessment. Department of Industry and ANU.

Smith, M. (2013f). *The Oil and Gas Sector – Climate Change Risks and Opportunities. An Educational Guide to Assist Climate Change Risk, Adaptation and Mitigation Opportunity Assessment*. Department of Industry and ANU.

Smith, M. (2013g). *The Manufacturing Sector – Climate Change Risks and Opportunities. An Educational Guide to Assist Climate Change Risk, Adaptation and Mitigation Opportunity Assessment*. Department of Industry and ANU.

Smith, M. (2013h). *The Food and Beverage Processing Sector – Climate Change Risks and Opportunities*. An Educational Guide to Assist Climate Change Risk, Adaptation and Mitigation Opportunity Assessment. Department of Industry and ANU.

Smith, M. (2013i). *The Product Manufacturing Sector – Opportunities to Enable Climate Change Mitigation and Adaptation An Educational Guide*. Department of Industry and ANU.

Smith, M. (2013j). *Construction and Property – Climate Change Risks and Opportunities. An Educational Guide to Assist Climate Change Risk, Adaptation and Mitigation Opportunity Assessment*. Department of Industry and ANU.

Smith, M., Hargroves, K., Stasinopoulos, P., et al. (2007). *Energy Transformed: Sustainable Energy Solutions for Climate Change Mitigation*. ANU and Griffith University, CSIRO and TNEP.

Stasinopoulos, P., Smith, M., Hargroves, K. and Desha, C,. (2010). *Whole System Design: An Integrated Approach to Sustainable Engineering*. London: Routledge.

Sterman, J. D. (2010). Does Formal System Dynamics Training Improve People's Understanding of Accumulation? *System Dynamics Review* 26(4): 313–34.

UN Millennium Ecosystem Assessment (2005). Ecosystems and Human Well-Being. Washington, DC: Island Press.

US Director of National Intelligence (2013). *Statement for the Record, Worldwide Threat Assessment of the US Intelligence Community*. Senate Select Committee on Intelligence. US Director of National Intelligence.

US National Intelligence Council (2013). *Global Trends 2030: Mapping the Global Future*. US National Intelligence Council, www.dni.gov/index.php/about/organization/national-intelligence-council-global-trends/.

Water Resources Group (2011). *Charting our Water Future-Economic Frameworks to Inform Decision-Making*. Water Resources Group, www.mckinsey.com/client_service/sustainability/latest_thinking/charting_our_water_future/.

Went, A., Newman, P. and James, W. (2008). *Renewable Transport: How Renewable Energy and Electric Vehicles Using Vehicle to Grid Technology Can Make Carbon Free Urban Development*, Discussion Paper. Fremantle, Australia: Curtin University Sustainability Policy Institute.

Wigley, T. (2011). Coal to Gas: The Influence of Methane Leakage. *Climatic Change* 108: 601–8.

World Economic Forum Water Initiative (2011). *Water Security: The Water-Food-Energy-Climate Nexus*. Washington, DC: Island Press.

15

A nexus of nexuses: systemic governance for climate response

PHILIP WALLIS

Introduction

It has become a feature of the governance lexicon to refer to narrowly focused, single-issue policy domains as 'silos'. In its original context, the word refers to vertical grain storages that sit alongside each other, but are compartmentalised and largely function separately. From a governance perspective that embraces complexity and holism, the term 'silo' is often used pejoratively to denote approaches that can't deal with messy, real-world situations. One could extend the metaphor to the fact that most silos are filled from the top and emptied from the bottom: a reflection of the top-down nature of siloed policy domains. How is it that silos have become so entrenched in our current governance regimes? Silos could result from efforts to simplify management tasks, or from a tendency to classify and conceptually delineate our environment, or perhaps they emerge from a more fundamental nature–culture divide.

It is clear that the silo approach to governance can lead to serious unintended consequences. For example, in the well-studied case of the Kissimmee River, part of the Florida Everglades in the United States, the river and its surrounding floodplains were historically managed for the primary goal of flood control. The channel of the river was even refashioned into a canal for increased hydraulic efficiency, which severely damaged the ecological integrity of the river (Koebel 1995). A large-scale restoration program is currently being carried out to restore the floodplain and river ecosystem, which some have described as a successful transition to a more sustainable trajectory (Olsson et al. 2006). This is probably an extreme example, but it demonstrates the possible consequences of managing to one goal: in this case, flood protection. In Australia, there is a long history of separation, nowhere clearer than in water politics, where water itself has become a disembodied entity:

> *Australian water politics and decision-making processes are highly contested. Contemporary water politics is characterised by conceptual and bureaucratic separation of water – from land, and from social processes.* (Gibbs 2013, p. 469)

The above quote can be understood as saying that water policy has been separated from contexts of geography and society. Many would argue that the separation of water rights from land rights was a successful piece of public policy that led to water trading and

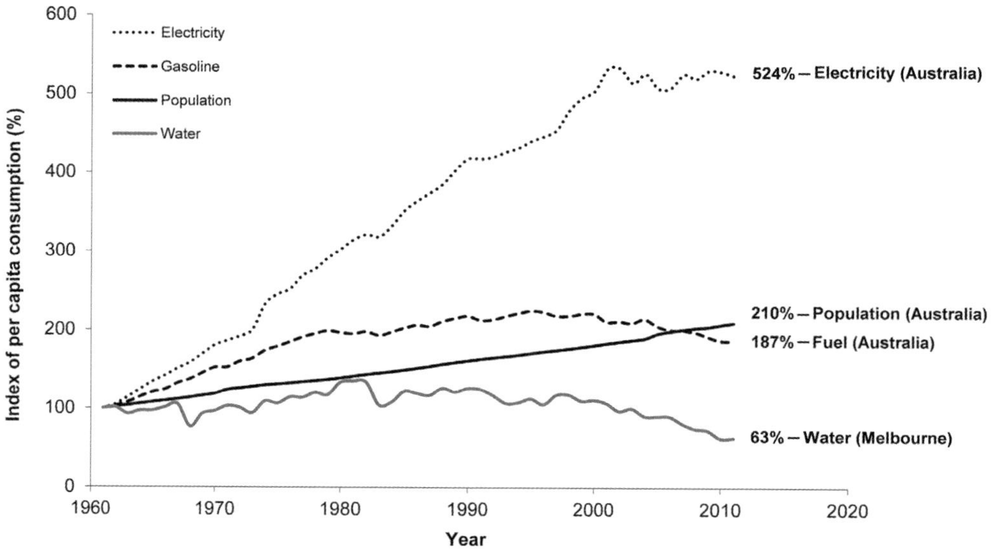

Figure 15.1 An index of per capita resource consumption of electricity (Australia), automotive fuel (Australia) and water (Melbourne), and an index of population (Australia) from 1961 to 2011. (Australian Bureau of Statistics, Bureau of Resources and Energy Economics, and Melbourne Water.)

assisted some irrigators in dealing with drought. The point I am trying to make is to highlight the tradition of separation in Australian policy. Figure 15.1 provides one demonstration of the consequences of separate governance regimes for energy and water. The graph depicts an index of Australian per-capita consumption of electricity and automotive fuel, and Melbourne consumption of water, plotted against an index of population. The different trends correlate with different histories and governance approaches.

The story behind the electricity curve is clearly one of unrestricted growth in use, until only recently (2000 onwards) when higher prices and a slowing economy dampened demand. Automotive fuel use also grew rapidly in the first two decades represented in the chart, and then effectively flattened out. One explanation is that the oil crises of the 1970s led to the production of more fuel-efficient cars, which was later offset to some degree by a preference for larger cars. The curve for water use in Melbourne – a city of, at present, more than 4 million residents – reflects the drivers of drought and long-term programs for water conservation, including restrictions on use and mandatory water-efficiency standards. Given the substantially different outcomes in electricity, transport fuel and water, it is tempting to say that the silos haven't worked – that electricity, fuel and water should be managed together somehow. However, the situation is undoubtedly more complex than this, with each example of resource use interconnected to a broader range of drivers and issues.

A new discourse has emerged around the concept of a 'nexus' of different policy domains, for instance climate, energy and water, and has prompted an interdisciplinary research effort into identifying ways of integrating these domains, or 'silos', and assessing the consequences of leaving them separate. The nexus approach looks for win–win (or win–win–win!) options, an exercise in optimisation in economic, social and environmental outcomes in different policy arenas. In practice, this approach is sometimes considered in discourses around 'whole-of-government' (eg the Council of Australian Governments – COAG), or 'joined up' approaches to public policy, 'horizontal' policy and management (Lindquist 2012) and even government departments that are clustered together (eg the Queensland Government Department of Energy and Water Supply).

In systemic terms, the nexus approach can also be conceptualised as a boundary-setting exercise, where the boundaries of analysis and action are broadened beyond traditional silos (eg separate electricity and water governance systems). Pittock et al. (2013) argue that a nexus approach can illuminate cross-sectoral issues, while maintaining the simplicity required to derive achievable actions. While this argument is likely to be an accurate assessment in governance contexts where silos are common, the nexus approach may constrain the limits of analysis and action in contexts with greater complexity and interconnectedness.

This complexity is most immediately apparent in efforts to respond to climate change. Responding to climate change, either through mitigation or adaptation, draws on many different, yet overlapping, governance domains. One could argue that mitigation (limiting the cause of climate change) is nested within adaptation (reducing vulnerability to effects of climate change). Another perspective is that mitigation policy is theoretically simple and centralised, 'only' politically hard, whereas adaptation is markedly different in being inherently decentralised and multi-sectoral. Hence, there are fundamental differences in policy form (Dovers and Hezri 2010). The governance of energy is central, as both a cause of climate change through combustion of fossil fuels, and also a key sector threatened by climate impacts. The governance of freshwater resources is also a key domain, being critical to equitable allocation among direct human consumption, for agriculture, in cooling for electricity generation, transport and livelihoods in the face of changes to the distribution of water resources (Vörösmarty et al. 2010). One can go through a list of governance domains that conceptually overlap, such as food security, catchment management, rural development and so on, and understand these as a nexus of nexuses.

In this chapter, I explore the ways that placing boundaries around particular nexuses of governance regimes yield different systemic outcomes. I present a governance framework that includes the nexus approach in a systemic, multi-level model. I use examples of policies of climate change response to illustrate the added value to the nexus approach that a systemic governance framework offers conceptually and practically.

A systemic governance framework

From a theoretical 'systems' perspective (Ison 2010), governance can be thought of as a process of steering an on-going, viable course in response to feedback from enacting activities. This 'viable course' is charted in relation to a purpose that is negotiated, collectively, in a co-evolutionary and unfolding context. Governance that is 'systemic' and 'adaptive' incorporates learning and change in response to unfolding circumstances,

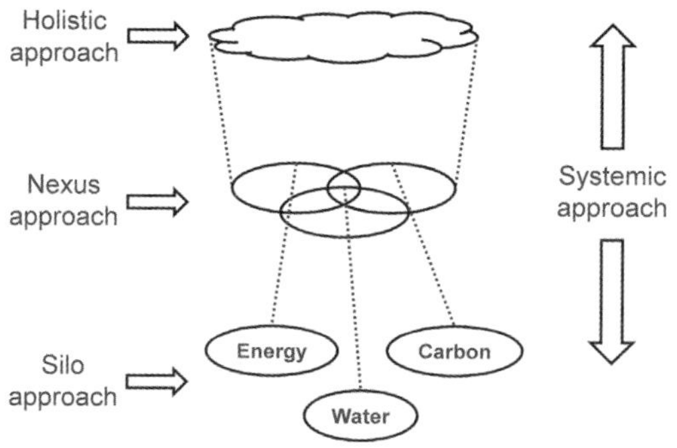

Figure 15.2 A systemic framework that positions the nexus approach above the traditional silo approach, and below a holistic approach. A systemic approach represents the ability to move between these levels.

and therefore demands understanding of, and managing, feedback processes. The concepts and language involved in the governance discourse opens up the assumptions and 'cognitive commitments' in the thinking about decision making (Hajer and Wagenaar 2003). In other words, governance entails a consideration of the worldviews, framing and assumptions of those governing.

Extending this systems thinking perspective, the nexus approach is typically situated at one systemic level: a supra-system of traditional siloed governance domains (Figure 15.2). While the level of abstraction rises with a nexus approach, the mode of understanding tends to remain grounded in a 'real systems' view of the world. In other words, the nexus is considered to be the sum of its parts, not complex, systematic (ie step by step) and objective. Above the nexus lies a holistic world of broad, confusing complexity and emergence, and below it, a world of simple, narrow focus.

The model portrayed in Figure 15.2 represents the key elements of a systemic governance framework. It is centred here on the nexus, which is understood as integrative of traditional siloed approaches to governance, or as a practical middle ground between silos and a full holistic view. The holistic approach is depicted with a 'cloudy' boundary to represent its diffuse and open nature. Systemic approaches are depicted as crossing all three levels.

This model shouldn't be confused with the landscape-regime-niche model that forms the multi-level perspective on transitions (discussed later), but does relate to the systems concept of different levels of abstraction. I explain my understanding of each element in more detail in Box 15.1.

In relation to the above understandings, I contend that the nexus approach ought to be situated within a systemic governance framework. A nexus approach alone could lead to some of the same unintended consequences that result from policy silos. If nexus governance is simply an exercise in optimisation, then it could possibly result in suboptimal outcomes in each separate domain or across a broader situation. Take, for instance, the example of the retrofit or substitution of water-cooled coal-fired thermal power plants with air-cooling technology. The application of this technology conserves water resources, but

Box 15.1 Key elements of a systemic governance framework

Siloed approaches: This represents the traditional way of governing, where one resource is controlled and managed by a single industry, often under specific government legislation. For example, the water industry in the state of Victoria, Australia, was managed for most of the twentieh century by the Melbourne Metropolitan Board of Works (MMBW) and the State Rivers and Water Supply Commission (SRWSC). These were organisations that were dominated by engineers and framed their activities around public health (MMBW) and water delivery (SRWSC). These could be considered good examples of powerful state water bureaucracies (Molle et al. 2009), to the point where infrastructure development becomes an end, and not a means to an end.

Nexus approaches: An approach that joins together governance of interconnected domains that are traditionally managed separately: for instance, climate-energy-water or agriculture-water-food security. A simple example of governing for both climate-energy-water outcomes is the showerhead exchange program in Victoria, which was designed primarily as an energy-saving (and carbon reducing) measure, but promoted through the Office of Water as a water-saving measure. The nexus approach can be thought of as an optimisation framework, searching for win-win-win outcomes (as per the showerhead example), or minimising trade-offs. Other chapters in this book provide more in-depth examples of governing climate, energy and water together.

Holistic approaches: Characterised by integration of existing governance functions, this approach sets the boundaries much broader than the area of immediate control. The main purpose of this approach is the transformation to an improved situation. It can also include a model of 'muddling through' (Lindblom 1959), engaging in learning by doing or adaptive management. For example, the state of Victoria has recently set up the *Office of Living Victoria*, which is mostly focused on water but aspires to a holistic approach. It seeks to integrate urban water supply with environmental renewal and urban amenity, and takes into account a broader set of costs and benefits in decision making, aiming for resilience and flexibility of the urban water system.

Systemic approaches: A systemic governance approach is, theoretically, one that can enable use of all the different levels mentioned above, and easily moves between them as needed. It can be characterised by open-ended inquiry into poorly defined areas of interest, learning approaches that are intended to generate an understanding of a situation and move towards its improvement. One example that appears regularly in the organisational governance and environmental management literature is the concept of 'social learning' (eg Collins and Ison 2009; Wallis and Ison 2013). This chapter explores social learning and other alternatives for systemic governance.

incurs an efficiency penalty in energy output. This might be considered an acceptable trade-off or optimisation; however, the technology also supports continued use of fossil fuels in power generation, which, from a holistic perspective on climate change mitigation, is not optimal at all.

In this section, I examine some of the alternative options for systemic governance approaches that are consistent with the multi-level framework in Figure 15.2. There are a plethora of theoretical frameworks for governing transformation. The two examples that follow are transition management and social learning.

Transition management

Transition management is a relatively recent field of scholarship emerging from fields of innovation systems and sociotechnical systems. Transition management involves the transformation of, and fundamental change to, societal systems (Geels and Schot 2007; Smith et al. 2005). The multi-level perspective is widely used to understand and describe transformations in societal regimes for managing particular resources or sectors. A brief summary of the multi-level perspective is given in Box 15.2, while more detail can be found in Smith, Voß and Grin (2010) and is explored in narratives of patterning in de Haan and Rotmans (2011).

Box 15.2 The multi-level perspective on transitions

The theoretical framework for transition management, as represented in the multi-level perspective, features several conceptual structures and processes that help make sense of changing sociotechnical systems. Transitions centre on *regimes*, which are the stable, established ways of governing, comprising institutional structures, cultures and practices (Rotmans and Loorbach 2006). The context in which a regime operates is referred to as a *landscape*, which drives change through evolving economic, biophysical and social processes. *Niches* are non-mainstream networks of actors, technologies, innovations and institutions that do not fit within the regime but can influence it. Four pathways of change have been described by Geels and Schot (2007): (1) *transformation*, which involves modification of the regime in response to a changed landscape; (2) *de-alignment and re-alignment*, where a destabilised regime is replaced by one or more competing niches; (3) *substitution*, where a niche replaces a regime; and (4) *reconfiguration*, where a regime adopts a niche.

Transition management has been used to describe the history and dynamics of natural resources management regimes internationally. In a study of the dynamics of the Dutch energy regime from 1960 to 2004, Verbong and Geels (1997) employed the multi-level transitions perspective to understand the complex drivers of change. Niche innovations in renewable energy technologies were seen to have some potential to transform the regime, but that there were no current windows of opportunity to effect this transformation. Recent examples of using a transitions framework for understanding water governance dynamics have been published in rural (Wallis and Ison 2011) and urban (Bos and Brown 2013) contexts in Australia. Both studies highlight the need to focus on social and institutional innovation, rather than the traditional technical innovation framing that is characteristic of the transitions field.

Few examples exist of using a transitions lens to understand the governance of multiple regimes (eg climate, energy and water). One study by Konrad et al. (2008) examined interrelated regimes for telecommunications, electricity, gas and water in Germany. They identified the benefits of governing multi-regime dynamics, for example reducing water consumed in washing clothes has beneficial outcomes for water, sanitation and energy regimes. Teschner et al. (2012) explored the transition of energy and water regimes in Israel and proposed that these sectors can co-evolve towards greater ecological sustainability. They make the point that technology is not developed in silos; instead, it is formed in broader social contexts. They propose *integrated transitions management* as a way to deal with multiple regimes in a transitions framework. Both Konrad et al. (2008) and Teschner et al. (2012) highlight the co-evolutionary nature of multi-regime change, a key feature of systemic governance.

These examples support the idea of embedding the nexus approach in a transition management framework, especially where regimes are in transition and there are opportunities for co-evolutionary change.

Social learning

Social learning can be understood as a form of policy that brings about changes in understanding and practice among a social group, leading to an improved situation through concerted action. The model I prefer is the SLIM heuristic (SLIM stands for 'Social Learning for the Integrated Managing and sustainable use of water at catchment scale', a European-based research project that ran from 2001 to 2004). The SLIM heuristic (see Box 15.3) draws on a large body of empirical work across several case studies and is a sense-making device for understanding transforming situations through learning.

Box 15.3: The SLIM social learning heuristic

The most recent version of the SLIM social learning heuristic appears in Collins and Ison (2009) and essentially comprises six variables:

(1) *History* is a key influencing variable on the state of the current situation. Technological lock-in and path-dependencies (eg as described in Brown et al., 2011) can structure the current socio-institutional situation, making it resistant to change. History can also provide lessons for future action.
(2) *Learning processes* are important in the transformation of a situation, and are based on existing social theories of single (reflection on action), double (reflection on framing) and triple-loop (reflection on context) learning.
(3) *Institutions* are taken to refer to the 'rules of the game', or the set of norms, regulations and policies that structure and shape what is currently done and what is possible.
(4) *Facilitation* is an enabling role in learning, and can be carried out by a leader or a 'mediating object' (in the sense of Latour's actor-network theory).
(5) *Stakeholding* is carried out by stakeholders who actively maintain and represent their stake in the situation and have different perspectives.
(6) *Epistemology* refers to the different ways of knowing that is carried by each person in the situation.

Various claims are made about the usefulness and nature of social learning. Ison et al. (2013) examine a variety of metaphors for social learning and contend that the concept, as fluid as it is, ought to be made real and articulated by those who use it. They also claim that it has not been taken up by policy makers relative to the amount of research interest (the same claim could probably be made for transition management). However, in a case study of the 1980s salinity program in the State of Victoria, Australia, Wallis and Ison (2013) drew attention to the role of novel institutional arrangements, facilitation and stakeholding in moving salinity management away from potentially being the sole responsibility of a single-focus, bureaucratic 'salinity bureau' to a community-led experiment in regional governance.

In an example that deals directly with climate, energy and water together, McEvoy and Wilder (2012) critically analysed the discourse of desalination (of water) as a 'solution' or technological climate change adaptation. As Barnett and O'Neill (2010) have highlighted, desalination has the potential to be a serious 'maladaptation' to climate change, as it is energy and emissions intensive and locks in a long-term reliance and path dependency on high technology, thus reducing adaptive capacity. McEvoy and Wilder (2012) propose, among other ideas, social learning as a way to build adaptive capacity to respond to climate change.

Social learning can fit into the framework of systemic governance by creating a set of conditions that encourage iterative experimentation and changes in practice and understanding. The nexus approach can be represented in social learning through facilitation, acting as a mediating object for changes in understanding about interconnected policy domains.

Responding to climate change: Mitigation and adaptation

Climate response is used here as a catch-all term for actions that a society takes to either mitigate or adapt to climate change. From a systemic perspective, many of these actions connect traditionally separate policy silos, and are time critical in that projected climate change is a ramp disturbance (ie becoming more severe over time). Thus, an effective climate response is in tension between 'acting now' and 'getting it right'. Opportunities to improve climate, energy and water governance together offer the potential to meet both of these criteria and set society on a trajectory towards sustainability. In this section, I examine some of the features of Australia's mitigation and adaptation policy context, and suggest some areas of beneficial integration.

Policy frameworks that deal with climate change mitigation rarely include consideration of impacts on other policy areas, such as the management of water resources (Pittock 2011). In particular, pricing mechanisms tend to be resource-specific and siloed: for instance, markets for water, markets for carbon and markets for biodiversity offsets. A preference for market-based instruments can be seen as a way of avoiding regulatory mechanisms, however as Peel (Chapter 12) points out, detailed regulation usually accompanies markets to manage their operation. Regulation and climate change legislation can be used to carefully define allowable mitigation activities and potentially limit any unintended consequences of a sole focus on emissions reduction.

Targets are another mechanism for setting governance goals for climate response. The national climate change policy response in Australia is the *Clean Energy Futures Plan*, which is designed to meet a commitment to an emissions reduction target of 5 per cent compared to 2000 levels by 2020 and 80 per cent by 2050 (DCCEE 2011). Australia's primary method for reducing carbon emissions is a carbon price mechanism that requires emitters of greenhouse gases (only the largest companies listed on the *Liable Entities Public Information Database* – approximately 350 in 2013) to purchase permits for emissions, starting at a fixed price and then moving to an emissions trading scheme. This combination of targets and markets is accompanied by a raft of new institutional arrangements.

A *Clean Energy Finance Corporation* (CEFC) was set up to fund renewable energy projects in the power sector. It is proposed that CEFC will invest in Australian renewable energy, take a commercial approach and invest in ways that overcome existing financial barriers (CEFC Expert Review Panel 2012). A *Carbon Farming Initiative* has been established in the land sector to provide incentives for land-based mitigation, as land-based emissions are not currently subject to the carbon pricing mechanism directly. Another tool in the governance toolbox is information, which is the focus of an *Energy Efficiency Information Grants Program*, intended to increase the uptake of efficiency measures in small businesses and communities.

Adapting to the effects of climate change will require managing uncertainty, interconnected issues, complexity and controversy. The policy discourse around adaptation in Australia is at a very early stage, but is already being framed by perspectives that preference market-based and 'value-free' rational scientific input to policy development (eg Productivity Commission 2012). To date, there is no national adaptation plan in Australia; however, many existing areas of policy have been reframed as adaptation. For instance, efforts to upgrade irrigation infrastructure in the Murray–Darling Basin have been recast as an adaptive response, as it will extend the life of the irrigation industry in a changing climate.

Climate responses, both mitigation and adaptation, are interconnected physically, socially and institutionally. For example, carbon farming can be framed as mitigation, as it can result in the sequestration of carbon in soils. It can also be cast as adaptation, as it can build soil health through increased soil organic carbon levels, buffer soils against drying out and maintain the viability of agriculture. Carbon farming has implications for several policy areas: agriculture, food security, mitigation, adaptation, carbon and water (and even energy, if carbon crops are used to produce biofuels). Changes to residential energy efficiency and building envelope can also be considered as both mitigation and adaptation. Improving the energy efficiency of appliances results in reduced energy demand and a lower carbon footprint. Activities that improve the whole energy envelope of a building can also be an adaptive measure, as it results in less energy required to cool and heat.

Reconceptualising climate response as a nexus of nexuses implies that efforts to find optimal outcomes will not work. However, it is worth searching for areas of nexus that can improve the situation overall, as long as it is done in a systemic framework that considers transformation and trajectory over time. This is the focus of the next section.

Systemic governance for climate response

How can we govern a transition from a society that is carbon-intensive and climate-exposed to one that is defined by low carbon intensity and high adaptive capacity? Aside from setting general targets of emissions reduction and relying on regulated market mechanisms to do the rest, there is no clear 'solution' to the *wicked problem* (APSC 2007; Rittel and Webber 1973) of climate change. Continuing as we have is not a viable option, so we need to find new trajectories that lead to overall improvement of the situation, and not just trade-off current interests. In this section, I explore the role of the two models of systemic governance outlined earlier (ie transition management and social learning) as ways of engaging with climate response situations that are capable of dealing with climate, energy and water policy together.

Transitioning to a low carbon-intensity society

Low carbon development, growth or transition plans are a means by which organisations or governments from a range of scales (eg nations, states or local regions) can identify lowest-cost options for carbon mitigation and abatement. Wiseman and Edwards (2012) reviewed eighteen low-carbon transition plans, from governmental and non-governmental organisations, and found a range of different targets (purposes), strategy types, assumptions and paradigm-challenging ambition. Generally speaking, such transition plans promote emissions targets combined with a range of technological measures in power, land and energy efficiency sectors.

These three policy areas (power, land and energy efficiency) represent a significant proportion of Australia's potential to mitigate greenhouse gas emissions. A synthesis of what are considered by one analysis to be the most cost-effective of these opportunities is presented in the ClimateWorks (2010) *Low Carbon Growth Plan for Australia* (see Box 15.4 for details), which was assessed in Wiseman and Edwards (2012).

Box 15.4 The Australian carbon abatement cost curve

ClimateWorks Australia (2010) produced a McKinsey carbon abatement cost curve for Australia that presents 74 lowest-cost opportunities to reduce greenhouse gas emissions in 2020 by 249 $MtCO_2$-e. This represents a 25 per cent reduction on Australian emissions in 2000. The cost curve classifies these activities into three broad categories: (1) cleaner power, including new sources of renewable energy as well as technologies to reduce transmission losses and make more efficient use of existing infrastructure; (2) land, which includes afforestation and changes to agricultural practices and land uses; and (3) energy efficiency, dealing with industrial and residential energy efficiency, as well as fuel efficiency in the transport sector. The cost curve is accompanied by an analysis of the enabling and constraining contextual factors that would be encountered in attempting to enact the cost curve, and forms a plan for overcoming them. The method for determining the marginal cost of abatement and abatement volumes used by McKinsey and Company is contested, and caution is warranted in their use (Kesicki and Ekins 2012).

An analysis of the water consequences of mitigation measures presented in the cost curve revealed several beneficial opportunities for climate, energy and water (Wallis et al., 2014). For example, most energy efficiency activities save energy (and emissions), as well as water, as they reduce the level of demand on water-cooled, coal-fired thermal power generation. Mitigation activities in the land sector have the potential to provide substantial ecosystem services, including biodiversity benefits through habitat provision and landscape connectivity, as well as benefits for salinity, erosion and flood control. Strategic reforestation around productive agricultural land can act as windbreaks and shade for livestock, and may improve agricultural productivity. However, activities such as large-scale afforestation are also likely to impact heavily on rural water resources.

Taking a transition management perspective on the Australian low carbon growth plan, I consider how such a plan can transform multiple regimes as a niche innovation. Turning analysis into a process of transitioning is probably the biggest challenge for any low carbon transition plan; in other words, how do you put the plan into practice? This challenge is enhanced when considering the multiple regimes that the plan seeks to influence energy, transport, construction, agriculture and forestry directly, and water, food security, planning, and so on indirectly.

In one example from the literature, Quitzau et al. (2012) examined the role of strategic niche management in the building sector to transition to more energy-efficient buildings. It recognised that conventional policy mechanisms had not worked, including economic incentives, minimum standards and demonstration projects. Taking an urban planning process as the basis of a niche innovation, they outlined how planners successfully challenged established regimes for heating and construction. For example, in their case study they described how the local authority and planners formulated an alternative strategy that utilised covenants on properties to create binding energy efficiency requirements for newly bought lots (Quitzau et al. 2012).

Following this example of transitions management, a low carbon transition plan could form the basis of a niche innovation, provided that they are supported by a network of motivated actors and institutional arrangements that are either novel, or as in the case cited by Quitzau et al. (2012) above, work through existing possibilities. From the theoretical perspective of transitions pathways (Geels and Schot 2007), such a niche could either substitute the existing multiple regimes, or more likely, lead to a reconfiguration. A reconfiguration might have already occurred in the Australian case, at least in the area of national mitigation policy, as many of the areas for action proposed in the ClimateWorks (2010) *Low Carbon Growth Plan for* Australia were reflected in the 2011 *Clean Energy Futures Plan.*

Learning to adapt in catchments

Australia's natural resources management (NRM) regions, managed by Catchment Management Authorities (CMAs), NRM Boards and similar authorities, are considered to be sites for learning and transformation in response to climate change (Walker et al. 2009; Marshall and Stafford Smith 2010; Robins and Kanowski 2011). A social learning approach has been proposed as a way to progress agendas for climate change adaptation in regional catchments (Collins and Ison 2009). Here, I briefly consider how social learning processes could guide such a transformation, incorporating a nexus approach to guide the viability of different climate response options.

Central to social learning for adaptation is reframing NRM regions, such as river catchments, as coupled social-ecological systems. Humans are systemically connected to their environment, and thus new institutional arrangements are needed that recognise and foster this connection. As discussed earlier, a characteristic of the silo approach is to separate 'things', such as water, from social processes and place (Gibbs 2013). Considering climate change adaptation as a process of adapting 'with' climate change, rather than 'to' climate change, governance that mediates the on-going connection of changing climate and changing society is needed. Maladaptive practices that seek to fix human activities in time and place relative to a changing climate may be resistant (or resilient), but not transformative.

Possible transformations in Australia's NRM regions include the conversion of large areas of agricultural land to carbon farming or afforestation, increasing the water efficiency of irrigation networks (including closing areas that have historically supported irrigated agriculture), repopulating rural and regional communities, investing in teleworking, reinventing monoculture agriculture as multifunctional agriculture, supporting distributed power generation, focusing on conservation and tourism, and the list goes on. This invokes a high level of complexity, or a nexus of nexuses.

As with the transition management example above, the process of transformation is challenging given that there are a wide range of viable trajectories and multiple perspectives and ways of valuing, and potential actions can be constrained by existing institutional arrangements. While I don't claim that a social learning performance can be designed and implemented deterministically, I do claim that the conditions from which social learning might emerge can be designed as 'learning systems'. For example, a systemic inquiry into the governance of water and energy in catchments (Ison and Wallis 2011) might reveal new trajectories that can improve the situation. A nexus approach – for instance, the collection and comparison of metrics for water and energy – would act as a facilitation or mediation device for learning.

In this scenario, the acts of 'knowing' and 'valuing' replace ideas of 'knowledge' and 'values' that can be produced, transmitted or consumed, and processes of social learning can underpin these trajectories. Stakeholders with different perspectives and ways of valuing continuously co-construct what is successfully adaptive.

Conclusion

The main point of this chapter is that a nexus approach to dealing with climate, energy and water together needs to be embedded in a systemic governance framework, especially in responding to climate change. Climate change is a systemic change that runs through all areas of policy: a nexus of nexuses. There is a danger that taking a nexus approach on its own can lead to some of the same unintended consequences of siloed governance.

The nexus approach can potentially be enhanced through this systemic governance framework, as it guides reflection on transformation and overall improvement of a situation. The key features of systemic governance include the ability to move between different levels of abstraction, an appreciation of the influence of history on future trajectories, the inclusion of people as an integral part of the situation that hold multiple perspectives and construct issues differently, and identification of an improved or transformed situation. In this chapter, I explored two models of systemic governance, transition management and social learning, and considered how they could handle the challenges of climate response. This is an area of inquiry that can be explored further through empirical research.

A valuable contribution of the nexus approach to this framework is in the analytical framing of multiple resource metrics. If government agencies were required to collect and publish data on energy, water and greenhouse gas emissions in compatible formats, it could inform the viability and practicality of options for governing for transformation. An international project called the Nexus Dialogue (www.waternexussolutions.org) is one example of a pathway to collecting evidence of the success of such projects.

References

Australian Public Service Commission (APSC) (2007). *Tackling Wicked Problems: A Public Policy Perspective*. Canberra: Australian Public Service Commission.

Barnett, J. and O'Neill, S. (2010). Maladaptation. *Global Environmental Change* 20: 211–213.

Bos, J. J. and Brown, R. R. (2013). Realising Sustainable Urban Water Management: Can Social Theory Help? *Water Science & Technology* 67: 109.

Brown, R., Ashley, R. and Farrelly, M. (2011). Political and Professional Agency Entrapment: An Agenda for Urban Water Research. *Water Resources Management* 25: 4037–50.

Clean Energy Finance Corporation (CEFC) (2012). Clean Energy Finance Corporation: Expert Review. Report to Government, Commonwealth of Australia.

Collins, K. and Ison, R. (2009). Jumping Off Arnstein's Ladder: Social Learning as a New Policy Paradigm for Climate Change Adaptation. *Environmental Policy and Governance* 19: 358–73.

ClimateWorks Australia (2010). *Low Carbon Growth Plan for Australia*. Melbourne: ClimateWorks Australia.

Department of Climate Change and Energy Efficiency (DCCEE) (2011). Securing a clean energy future. The Australian Government's Climate Change Plan. Department of Climate Change and Energy Efficiency, www.cleanenergyfuture.gov.au/wp-content/uploads/2011/07/Consolidated-Final.pdf.

De Haan, J. H. and Rotmans, J. (2011). Patterns in Transitions: Understanding Complex Chains of Change. *Technological Forecasting and Social Change* 78: 90–102.

Dovers, S. R. and Hezri, A. A. (2010). Institutions and Policy Processes: The Means to the Ends of Adaptation. *Wiley Interdisciplinary Reviews: Climate Change* 1: 212–31.

Geels, F. W. and Schot, J. (2007). Typology of Sociotechnical Transition Pathways. *Research Policy* 36: 399–417.

Gibbs, L. M. (2013). Bottles, Bores, and Boats: Agency of Water Assemblages in Post/colonial Inland Australia. *Environment and Planning A* 45: 467 – 84.

Hajer, M. and Wagenaar, H. eds. (2003). *Deliberative Policy Analysis. Understanding Governance in a Networked Society*. Cambridge University Press.

Ison, R. L. (2010). *Systems Practice: How to Act in a Climate Change World*. London: Springer.

Ison, R. L. and Wallis, P. J. (2011). *Planning as Performance: The Murray–Darling Basin Plan*. In *Basin Futures*, eds. Connell, D. and Grafton, R. Q., 399–411.

Kesicki, F. and Ekins, P. (2012). Marginal Abatement Cost Curves: A Call for Caution. *Climate Policy* 12: 219–36.

Koebel, J. W. (1995). An Historical Perspective on the Kissimmee River Restoration Project. *Restoration Ecology* 3: 149–59.

Konrad, K., Truffer, B. and Voß, J.-P. (2008). Multi-regime Dynamics in the Analysis of Sectoral Transformation Potentials: Evidence from German Utility Sectors. *Journal of Cleaner Production* 16: 1190–202.

Lindblom, C. (1959). The Science of "Muddling Through". *Public Administration Review* 19: 79–88.

Lindquist, E. (2012). Horizontal Management in Canada Ten Years Later. *Optimum Online* 42: 26–35.

Marshall, G. R. and Stafford Smith, D. M. (2010). Natural Resources Governance for the Drylands of the Murray–Darling Basin. *Rangeland J* 32: 267–82.

McEvoy, J. and Wilder, M. (2012). Discourse and Desalination: Potential Impacts of Proposed Climate Change Adaptation Interventions in the Arizona–Sonora Border Region. *Global Environmental Change* 22: 353–63.

Molle, F., Mollinga, P. P. and Wester, P. (2009). Hydraulic Bureaucracies: Flows of Water, Flows of Power. *Water Alternatives* 2: 328–49.

Olsson, P., Gunderson, L. H., Carpenter, S. R., et al. (2006). Shooting the Rapids: Navigating Transitions to Adaptive Governance of Social-Ecological Systems. *Ecology and Society* 11: 18.

Pittock, J. (2011). National Climate Change Policies and Sustainable Water Management: Conflicts and Synergies. *Ecology and Society* 16: 25.

Pittock, J., Hussey, K. and McGlennon, S. (2013). Australian Climate, Energy and Water Policies: Conflicts and Synergies. *Australian Geographer* 44: 3–22.

Productivity Commission (2012). *Barriers to Effective Climate Change Adaptation. Draft Report*. Canberra.

Quitzau, M.-B., Hoffmann, B. and Elle, M. (2012). Local Niche Planning and Its Strategic Implications for Implementation of Energy-Efficient Technology. *Technological Forecasting and Social Change* 79: 1049–58.

Rittel, H. W. J. and Webber, M. M. (1973). Dilemmas in a General Theory of Planning. *Policy Science* 4, 155–69.

Robins, L. and Kanowski, P. (2011). "Crying for our Country": Eight Ways in Which "Caring for our Country" Has Undermined Australia's Regional Model for Natural Resource Management. *Australasian Journal of Environmental Management* 18: 88–108.

Rotmans, J. and Loorbach, D. (2006). *Transition Management: Reflexive Governance of Societal Complexity Through Searching, Learning and Experimenting*. In *Reflexive Governance for Sustainable Development*, eds J. Voss, D. Bauknecht, R. Kemp. Cheltenham: Edward Elgar, p. 53.

Smith, A., Stirling, A. and Berkhout, F. (2005). The Governance of Sustainable Socio-technical Transitions. *Research Policy* 34: 1491–510.

Smith, A., Voß, J.-P. and Grin, J. (2010). Innovation Studies and Sustainability Transitions: The Allure of the Multi-level Perspective and Its Challenges. *Research Policy* 39: 435–48.

Teschner, N., McDonald, A., Foxon, T.J. and Paavola, J. (2012). Integrated Transitions Toward Sustainability: The Case of Water and Energy Policies in Israel. *Technol. Forecast. Soc. Chang.* 79: 457–68.
Verbong, G. and Geels, F. (2007). The Ongoing Energy Transition: Lessons from a Socio-technical, Multi-level Analysis of the Dutch Electricity System (1960–2004). *Energy Policy* 35: 1025–37.
Vörösmarty, C. J., McIntyre, P. B., Gessner, M. O., et al. (2010). Global Threats to Human Water Security and River Biodiversity. *Nature* 467: 555–61.
Walker, B. H., Abel, N., Anderies, J. M. and Ryan, P. (2009). Resilience, Adaptability, and Transformability in the Goulburn-Broken Catchment, Australia. *Ecology and Society* 14: 12.
Wallis, P. J. and Ison, R. (2011). Institutional Change in Multi-scalar Water Governance Regimes: A Case from Victoria, Australia. *Journal of Water Law* 22: 85–94.
Wallis, P. J., Ison, R. L. and Samson, K. (2013). Identifying the Conditions for Social Learning in Water Governance in Regional Australia. *Land Use Policy* 31: 412–21.
Wiseman, J. and Edwards, T. (2012). *Post Carbon Pathways: Reviewing Post Carbon Economy Transition Strategies*. Sydney: Centre for Policy Development, pp. 132.

16

Integrated modelling of the energy-water nexus in the American West

TOM ISEMAN[1] AND VINCE C. TIDWELL

Introduction

As demonstrated in previous chapters, energy development and water resource management are inextricably linked. However, all too often policy making has focused on energy or water issues without a full appreciation of their nexus.

For example, energy policy aimed at reducing greenhouse gas emissions could have the unintended consequence of increased water use; implementing carbon capture and sequestration (National Energy Technology Laboratory 2009) and increasing irrigation of corn for ethanol (Wu et al. 2009) are two such examples. Promotion of 'green' energy has similar implications for increased water use as wet-cooled solar thermal (Carter et al. 2009) and geothermal technologies can require significant quantities of water for operation. Also, the US Environmental Protection Agency's (EPA) proposed 316(b) rule, aimed at reducing impingement- and entrainment-induced mortality of aquatic wildlife related to thermoelectric power plant cooling water intakes, could have the unintended consequence of increased water consumption as power plants shift from open-loop to more water-intensive closed-loop cooling systems (Macknick et al. 2012).

Such narrowly focused policy choices, often targeting the energy side of an issue, suggest the need for accurate models of the energy-water nexus and integrated planning venues to examine the interactions of energy and water.

In the American West (west of the 100th meridian), the intersection of energy and water is particularly pressing, given the abundance of energy resources (US Energy Information Agency 2012) and relative dearth of water (Anderson and Woosley 2005).

The Western Governors' Association (WGA), the Western Interstate Energy Board (WIEB), the Western States Water Council (WSWC) and the Western Electricity Coordinating Council (WECC) are working together to gather information on the water-energy nexus to help facilitate regional planning for electric generation and transmission. As part of that effort, the WSWC and Sandia are examining the potential impacts of electric

[1] Tom Iseman is currently the Deputy Assistant Secretary for Water and Science at the Department of the Interior. Before joining Interior Mr. Iseman served as the Director of the Water Program at the Western Governors' Association. The co-authors started work on this chapter when Mr. Iseman was at WGA and completed it while employed at Interior. The views expressed herein are those of the authors and do not reflect the views of Sandia National Lab, the Department of the Interior or the Western Governors' Association or any of its individual members.

generation on water resources, as well as how water scarcity and drought could undermine electric reliability.

The project team, including Sandia National Laboratories and a consortium of National Laboratories, universities and non-profits, has spearheaded the development of a regional model of water availability that can be used to inform future electric planning and policy decisions. To date, the team has considered ten- and twenty-year planning horizons, with a focus on how different configurations of energy development and generation perform with respect to water consumption and drought vulnerability.

This chapter describes the development of an integrated regional model of energy and water and its use in planning and policy development in the Western United States. This chapter also addresses the public policy challenges of working across largely distinct sectors to develop integrated policy and planning strategies, and how regional models can inform discussions, illustrate resource trade-offs and advance dialogue across sectors. While this chapter focusses on a specific example of regional energy-water modelling in the Western United States, lessons learned from this effort can be instructive for other regions that require tools to frame policy making at the energy-water interface.

Given the breadth of this topic and the need to manage the length of the chapter, technical results are not included here. Rather, a number of journal articles that detail specific nexus issues are in preparation, and findings from this effort will be reported to WGA and the leadership of the US Department of Energy when this project is completed. Here, the focus is on the process toward integrated energy-water modelling and planning; specifically, the technical challenges associated with data availability, model compatibility and user-friendly interfaces.

Context

In December 2009, the US Department of Energy (DOE) announced a US$60 million program to support transmission planning for the three interconnections that cover the United States. Announcing the program, Secretary of Energy Stephen Chu said, "As we move the country toward a clean energy future, it is critical that we analyse the capacity of the country's transmission infrastructure and plan for future growth in this important industry . . . (and to) operate a reliable, efficient and secure electricity system" (USDOE 2009). The program dedicated nearly US$30 million to the Western Interconnection (WECC; Figure 16.1), which includes twelve Western states, for the purpose of developing collaborative long-term planning and the models and institutional forums it would require.

Notably, the original program guidelines did not explicitly include analysis of the energy-water nexus. Given the important connections between energy and water, the WSWC and Sandia recognised that water must be incorporated as a key input to a 'reliable, efficient and secure' electricity system. The operation of existing facilities and the siting of new projects will be influenced and sometimes dictated by the availability of a reliable water supply. As a result, WGA included a module on the energy-water nexus in its successful grant proposal to DOE.

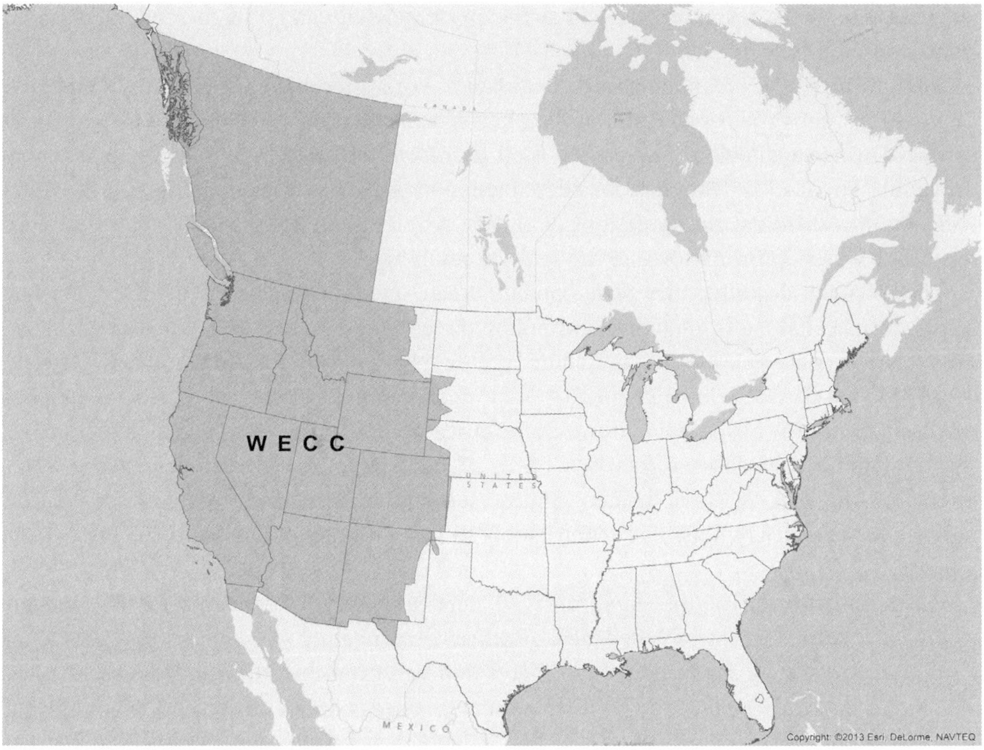

Figure 16.1 Geographic extent of the Western Electricity Coordinating Council (WECC).

Technical assistance was made possible by a grant from DOE's Office of Electricity that funded a coalition[2] of national laboratories and universities to undertake the quantitative assessment of energy and water, including the development of regional models and the assessment of energy scenarios. This work is nested within the broader regional transmission planning effort spearheaded by the WECC.

From the beginning, this project faced technical and political challenges while integrating water resource considerations into an electric-centric institution and planning process – largely because this is the first time integrated planning has been attempted at such a scale.

Approach

Fundamentally, the project team is seeking to develop a regional, integrated energy-water model to inform policy making – providing an unbiased, scientific basis for decision making about the siting and technology of electric generation and the trade-offs associated with water supplies. The foundation of this model is WECC's Long-Term Planning Tool

[2] Sandia National Laboratories serves as the lead for this effort, supported by Argonne National Laboratory, the Electric Power Research Institute, Idaho National Laboratory, the National Renewable Energy Laboratory, Pacific Northwest National Laboratory and the University of Texas.

(LTPT) that consists of two parts, the Study Case Development Tool (SCDT) and the Network Expansion Tool (NXT) (WECC 2013). The SCDT converts 'energy future' scenarios into cases (data models) that can be analytically studied. These cases include nodal details for each corresponding future scenario capturing load and generation profiles in addition to other decision factors encompassed within scenarios such as Renewable Portfolio Standards, resource mix, load growth, etc. The SCDT in turn provides the input data needed by the NXT, which optimally derives physically feasible network expansions necessary to meet the load and generation requirements of each scenario study case while minimising the net present value of capital costs.

WSWC and Sandia Labs have teamed to develop information on the availability and cost of water, in order to allow modellers to constrain power plant siting within the LTPT. The team focused on two primary components: (1) water availability, including surface and groundwater sources (both fresh and non-fresh) and current and future demands for water use; and (2) water variability, including inter-annual variability, extended drought and long-term climate change. This input will provide the LTPT and the regional planning team an important new screening capability to assess the adequacy of water supplies to sustain reliable electric generation.

The LTPT constrained by water availability and cost data forms the basis of the regional integrated energy-water model that was subsequently used in west-wide transmission planning. The goal of the regional integrated model was to identify potential water scarcity 'pinch points' or drought vulnerabilities and inform decisions about siting, fuel type and cooling technologies for future electric generation. In addition to characterising water availability and its variability, the team also developed cost estimates for the various sources of water supply in order to illuminate potential economic trade-offs. Further, these cost estimates play into the cost minimisation within the transmission modelling.

Integrated modelling and collaboration

This section addresses challenges and lessons learned in the process of developing an integrated energy-water planning model – not simply a model, but a model that will provide results that inform policy and management decisions. Although not an exhaustive list, four factors are explored: (1) learning to speak the same language; (2) integrating disciplinary databases and analysis tools; (3) fostering an environment of collaborative modeling; and (4) creating a visual, interactive and accessible analysis framework. A discussion of each factor is given below.

Learning to speak the same language

To achieve any real cooperation energy and water managers must learn to speak the same language (Cockerill et al. 2007 Mosey and White 1998; Nicolson et al. 2002; Wear 1999). While each often appreciates the influence of the other's domain on their own success, they lack a depth of knowledge of the constraints, uncertainties, regulations or cost points informing the other's decision making. For example, energy managers recognise that they are likely to need water for new power generation; however, they are unlikely to fully comprehend how resources in a given basin are managed, what the competing demands are, or how their decisions may affect surrounding communities and the environment.

Energy and water managers also often lack a shared vocabulary; that is, words with very specific meaning to one discipline fail to convey importance to the other. A simple example is the difference between withdrawal and consumption – water taken from a particular source, some of which may return to the hydrologic system, versus water that is removed and does not return but is lost to evaporation or incorporation into a specific product like a crop. These terms carry important distinctions to a water manager but likely no difference to an energy manager.

Finally, there is often a lack of context for integrated planning – integrated planning has not been done before, so no one knows what to expect.

Developing effective cross-sector communication requires an on-going commitment to education. Throughout this project, learning was accomplished through a series of face-to-face meetings where roughly half of each meeting was reserved for presentations aimed at informing each other about the challenges, opportunities, approaches and expectations unique to one's discipline. The remaining time was largely spent focusing on developing consensus around the specific question that had to be answered by the analysis. It was important to understand what part of the question each party was responsible for, how each was to answer their question and, ultimately, how all the pieces were to come together.

The unifying question for regional transmission planning was to identify the least cost, most reliable transmission expansion option subject to federal and state environmental protection standards. The energy managers took responsibility for evaluating the cost and reliability factors for various future transmission expansion scenarios. The evaluation of water, on the other hand, had several unique dimensions: it was recognised to be a constraint on siting new generation, a capital and operating expense, a vulnerability to grid reliability and a potential environmental impact. For the purposes of this analysis, the energy and water experts agreed that water would serve as one potential 'constraint' in the context of a multi-objective optimisation model.

Integrating disciplinary databases and analysis tools

Scientific and engineering analysis tools have largely grown up along disciplinary boundaries. Each tool is designed to answer a specific suite of questions, subject to prescribed physical dynamics, operating at unique time and space scales. For example, a groundwater model is formulated on a mass and energy balance solved on a continuous grid of cells that map the aquifer of concern. Spatial scales generally range from hundreds of metres to kilometres, while temporal scales are seasonal to annual.

By contrast, an electric transmission model balances power production with demand represented by a set of nodes interconnected by a network of transmission capacity. The models are spatially organised by physical generation facilities and service areas, with dispatch modelling treated on timescales of minutes.

Rarely is water explicitly considered in transmission modelling (except in the case of hydroelectric operations), or electric power production addressed in water resource modelling (except as a demand). The fact that both systems are bounded in very different ways – dimensions of a basin or aquifer for water modelling and that of the power grid for transmission modelling – adds significantly to the difficulty in rectifying both the acquired data and physical processes to a common footprint.

A fully integrated model would involve a tightly coupled electric power dispatch model and a highly resolved water resources model (eg river routing or groundwater model). Such modelling is necessary to evaluate the potential for periods of peak energy demand to overlap periods of low water availability. Such analyses are necessary for detailed siting and planning for a given power plant or cluster of plants. However, for planning at the scale of the western United States, such tight coupling is not computationally feasible nor is a consistent set of supporting data available for every watershed and aquifer in this region. Given the constraints on both the transmission and water resources modelling, the approach taken here has been to treat the interconnection-wide planning as a scoping analysis: that is, determine the locations most amenable to generation and transmission expansion given the available supply of water. Candidate locations would then be subject to a detailed siting analysis (tightly coupled energy-water modelling) at a later time.

Given the scoping level of analysis appropriate for interconnection-wide transmission planning, a loose coupling between the disciplinary energy and water models is adopted. Water availability and its associated cost are calculated independently and then passed as an input to the electric transmission Network Expansion Tool. Specifically, water availability and cost are a constraint to planned expansion of the power system, and as new generation comes on line, water availability is decreased to account for the new demand.

A complicating feature is the fact that there are multiple supplies of water that vary spatially due to differences in the local climate, physical hydrology, water use characteristics and management institutions. In most locations there are sources of freshwater and non-freshwater. Freshwater sources include surface water and groundwater that can be further distinguished by unappropriated water and appropriated water (both surface and groundwater) – the difference being that unappropriated water can be acquired by securing a water right directly from the state water management agency, while appropriated water must be transferred (sold or leased) out of its current use to a new use such as thermoelectric power generation, which also requires state approval. In terms of non-freshwater, there is both municipal wastewater and brackish groundwater that could be used for new development. Each source of water is characterised by a different availability and cost at any given location in order to evaluate the feasibility of power plant siting. Costs are associated with transferring or leasing a right to the water, lifting or transporting the water to the point of use, and treatment of the water as necessary.

The complexity in source and cost is captured and conveyed to the transmission planning model in the form of supply curves (Figure 16.2). These curves identify how much water is available at what price. One additional step is required to cast the constraint in terms of allowable future power generation. This involved a simple multiplication of the cost curve by the water consumption factor unique to the generation and cooling

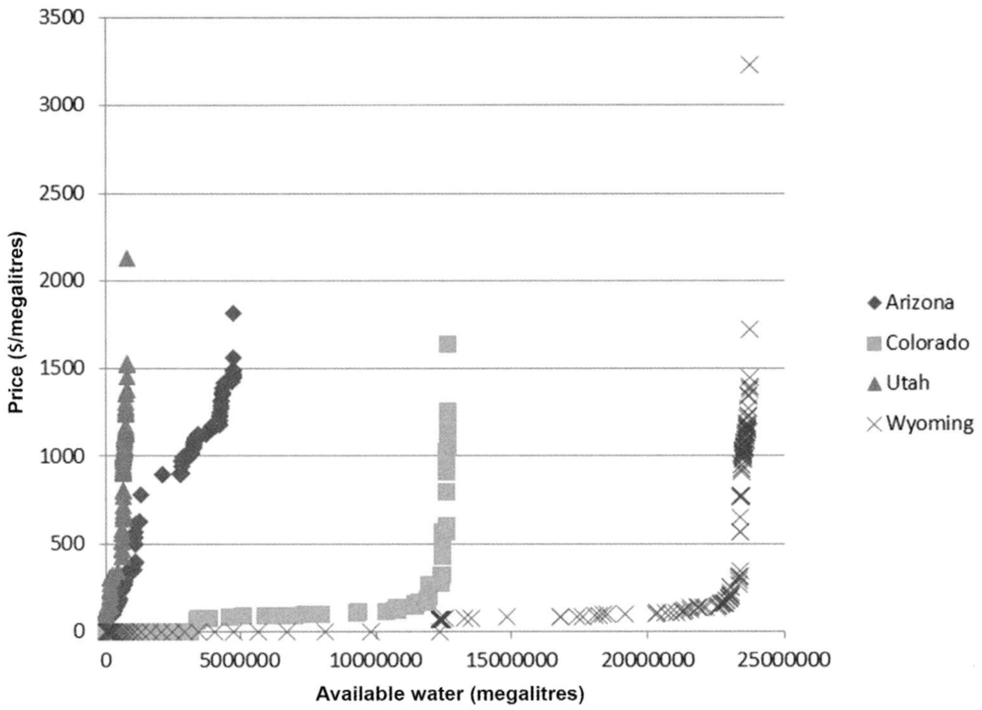

Figure 16.2 Cost curves showing the quantity of water available at a given price. Although developed at an eight-digit HUC level, curves shown here are aggregated at the state level for Arizona, Colorado, Utah and Wyoming.

technology pair (eg natural gas combined cycle generation with recirculating cooling towers) (Macknick et al. 2012). Additionally, these consumptive use factors were adjusted for the influence of environmental factors, such as ambient temperature, humidity and elevation. Through the simplicity of the cost curve, transmission planning could reflect constraints due to location, generation and cooling technology, water supply and cost.

Another important hurdle was the adoption of a consistent spatial reference system to accommodate information on both water availability and electric transmission planning. Water availability has a physical basis of reference at the watershed (catchment) level, while electric power is based on analysis at the nodal level, which corresponds to discrete service areas and power generation clusters. This issue was handled by first mapping water availability and cost at the highest resolution possible (in this project, at the eight-digit Hydrologic Unit Code (HUC) level, corresponding to over 1,000 watersheds in the conterminous Western United States). The watersheds were in turn aggregated to match the nodal service areas and generation clusters (approximately fifty individual hubs) used in the LTPT.

As described earlier, we found that loosely coupled energy and water models were adequate for our purposes and appropriate for regional integrated energy-water model. However, this is not the case for evaluating potential impacts of climate variability on individual power plant operations. For this analysis, a loose coupling between the water and energy models was still employed; however, more detailed hydrologic modelling was required. Specifically, downscaled climate projections to 2030 for a range of future emission scenarios were utilised to drive projections of water availability. Resulting precipitation, temperature and water-use projections were used as input to precipitation–runoff–reservoir routing modelling to simulate future stream flow and reservoir storage at an eight-digit HUC level. Additionally, power plant effluent temperatures were estimated. In this way, system vulnerabilities due to reservoir levels dropping below plant intake structures or plant effluent exceeding EPA discharge temperature limits could be identified (Yan et al. 2013).

Fostering an environment of collaborative modelling

While scientific and engineering analysis can contribute to a technically defensible basis for resource planning, results must be interpreted and applied in the context of their broader societal and environmental impacts. Given the complex and often contentious nature of such decisions, success requires a process for inclusive and transparent sharing of ideas complemented by tools to structure, quantify and visualise the collective understanding and data, providing an informed basis of dialogue, exploration and decision making (Koontz et al. 2004; Mullner et al. 2001; Selin and Chavez 1995; Selin et al. 2000). Actionable science is promoted where modeling and analysis step out of the 'black box' mentality to fully expose and vet the data, physical and social process dynamics, and assumptions on which the simulations are based (Cockerill et al. 2004; 2006). Broad participation is necessary to frame the analysis on the issues of principal concern to the stakeholders. These same stakeholders bring valuable perspectives and experiences that help quantify the linkage between human behavior and the natural environment.

Direct participation (eg interactive governance) by stakeholders in processes of collective decision making is reducing the distance between government and their citizenry. The value of collaboration, whereby various stakeholders work with policy makers to address a particular issue, has been well documented (Claussen 2001; Connick and Innes 2003; Splash 2001; Susskind et al. 2001). Such collaboration has increasingly included model building to inform the decision process. Here we have invoked the idea of collaborative modelling, which involves engaging stakeholders in the conceptualisation, specification and synthesis of knowledge and experience into useable information (ie a numerical computer model) for the express purpose of addressing a complex problem.

Collaborative modelling was implemented within the context of the project. Stakeholder engagement was managed with a three-tiered system, including a water availability team, an energy-water team and a steering committee, each with an overlapping set of participants (Figure 16.3). The water availability team largely focused on issues concerning water resources. Team members included the WSWC, WGA staff and state water managers along with representatives from the integrated modelling and energy-water teams.

Collaboration was accomplished at three different levels. First, meetings with individual states were conducted to gather and vet the data to be used in the transmission planning effort. Second, ad hoc working groups with volunteer members were formed that assisted

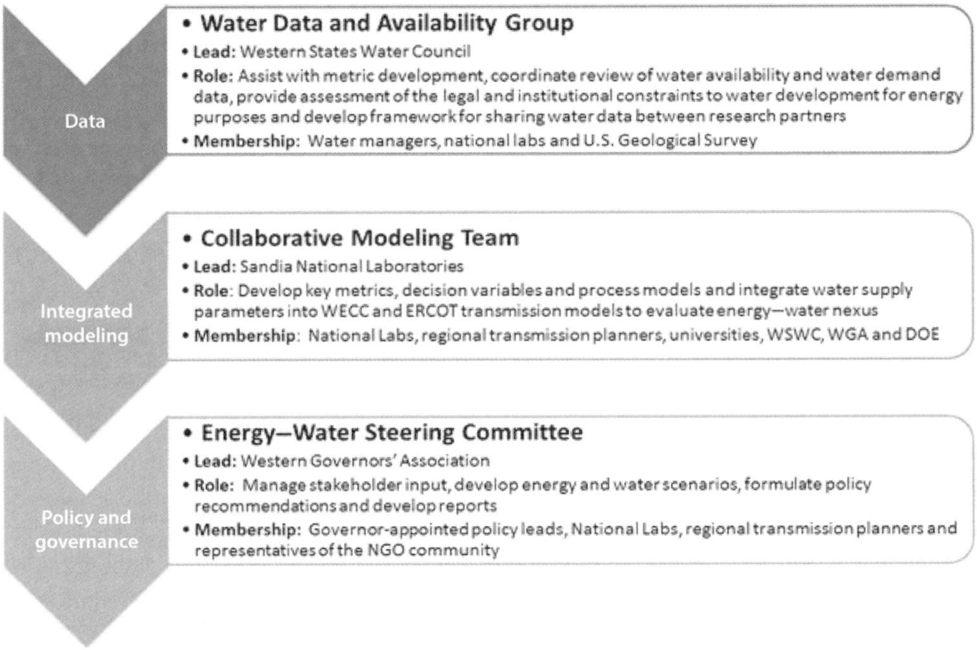

Figure 16.3 Organisation of stakeholder teams in the collaborative modeling process supporting long-term transmission planning for the Western Electric Reliability Coordinating Council (WECC).

in developing the water availability, water cost and environmental metrics used to support WECC transmission planning. These metrics were developed through a series of weekly webinars conducted over two to three months. Third, WSWC meetings were attended, where updates on the project were made to the full council.

The focus of the integrated modelling team was management of the technical aspects of the scientific and engineering analysis. Team members included the formal project team members and representatives from the WGA, WSWC, DOE and WECC. The team met roughly twice a month by webinar. The webinars were used to co-ordinate activities, address issues and keep team members informed on project progress.

The steering committee's focus was to manage stakeholder input and evaluate technical and planning options to address the energy-water nexus. Primary team membership included WGA, WSWC, DOE and WECC, and select members of the technical project team. Co-ordination was accomplished primarily by face-to-face meetings held on an as-needed basis. The subject of these meetings tended to be around details of how the water-related metrics would be included into WECC's long-range transmission-planning modelling. Also developed were water-centric (eg climate variability) scenarios for inclusion in WECC's ten-year planning analysis.

While this project has been successful in bringing consideration of water demand and supply to the attention of energy managers, it has also led to the realisation that a number of stakeholders and decision makers are absent from the process; in particular, the lay public. This gap is natural, given the very large geographic and sectorial (energy and water) footprint of the project. The expectation is that broader participation is best suited to decisions concerning local siting of new generation and transmission, which will occur in subsequent steps of planning. Nevertheless, there is a need to extend the reach of project participation, which is addressed in the conclusion to this chapter.

Creating a visual, interactive and accessible analysis framework

Providing a visualisation of the intersection of energy development and water availability can help to identify data gaps and potential next steps in addressing the water-energy nexus in the Western United States. It provides a big-picture, high-level, comparative snapshot of available supplies and demands as they apply to power plant siting and energy transmission planning. The visual dimension also helps convey interconnection between different sectors, such as the locations of energy hubs superimposed over areas of water abundance or relative scarcity. That is, visualisation helps comprehend the non-local aspects of resource allocation problems, such as the juxtaposition of watersheds with smaller water supply portfolios with locations of high water demand growth.

To help visualise and analyse the breadth of water and energy data collected through the project a web-served, interactive decision support system (DSS) has been developed. The DSS is created using ESRI's proprietary ArcGIS software, a Geographic Information System (GIS) management platform. Data are imported as unique data layers in point or areal format. Broad data types include individual power plant attributes (eg type, capacity, water source, water use), water demand (current or future, source, withdrawal and consumption, sector), water supply (gauged flows, groundwater recharge, reservoir storage), institutional controls (eg unappropriated water, closed basins, compact deliveries) and planning metrics (water availability, cost, environmental). All data are rendered in consistent units for the fourteen WECC states. Data can be viewed over a range of different reference systems, including eight-digit HUC, county, state and interconnection. Data can be viewed, overlaid and displayed in bar and pie chart format (Figure 16.4).

The DSS also supports analysis of the raw data. A simple example would be subtracting future water demand from available water supply to identify regions potentially facing water stress conditions. As there are no universal measures of water availability, the analyst might want to construct their own water availability metric – the DSS will support such personal exploration of the data. Ultimately, the breadth of data and flexibility of analysis should promote use of the data set well beyond the bounds of the current long-range transmission project.

While the water availability teams were gathering the initial round of data to support metric generation, it became apparent that, as water availability itself is in constant flux, so are the data that support the study metrics. In order to establish a basis for reproducible water modelling that can meet long-term western water-energy needs and requisite DSS accuracy, the WSWC began a separate effort to create a framework for sharing and publishing relevant water planning and water use information generated by state, federal and local governmental agencies. The Water Data Exchange (WaDE) project is a collaborative effort between the WSWC, the Western States Federal Agency Support Team (WestFAST), the WGA and the Department of Energy Labs. The purpose of WaDE is to better enable the western states and

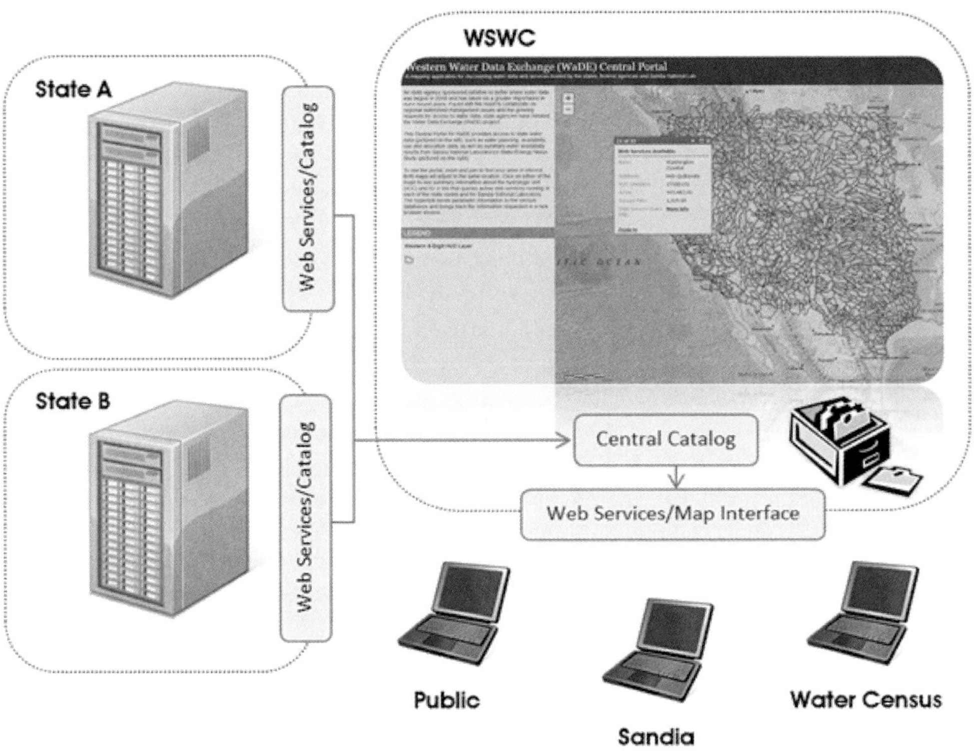

Figure 16.4 Conceptual diagram of the Water Data Exchange distributed architecture.

others to share water information with each other and support regional-scale analyses of both physical and legal/institutional water availability. These data would also support modelling efforts within federal agencies that have an interest in water quantity and use, such as the USGS's Water Census and National Water Assessment initiative. It also seeks to improve the sharing of federal data that supports state water-planning efforts.

WaDE is a long-term project that uses a web services–based approach. This allows each of the states to maintain its current data systems, while allowing common datasets among the states to be mapped to a standard format. Using automated processes, these data are published over the web using eXtensible Markup Language (XML) and are discoverable via a centralised data catalogue and web portal, maintained by the WSWC. In the future, the DSS may be updated using data shared by water agencies, planners and managers via WaDE.

Modelling applications

Despite the challenges identified above, the project has generated initial modeling results that illuminate the interplay of energy and water in the Western United States. Results were summarised in WECC's initial ten-year plan (WECC 2011) under the regional transmission expansion-planning project, released in September 2011. Key findings included:

1. While thermoelectric water demands are relatively small compared with water demands for agricultural or municipal water use, thermoelectric generation is likely to make up a significant portion of the *increase* in water consumption between now and 2020.
2. The projected growth of thermoelectric water demand occurs predominantly in basins that may be prone to water-related stress and where there will be competition for new water supplies from the municipal and industrial sectors.
3. WECC study cases perform differently with respect to water withdrawal and consumption, indicating that choices about the fuel type, cooling technology and location of new facilities should be considered in the context of the availability of water in a given location.

The national lab team has also analysed the risk posed by drought to electric reliability. Initial results suggest the Pacific Northwest will face the highest vulnerability due to the extensive use of hydropower (Harto and Yan 2011). Thermoelectric generation in the WECC region has largely taken steps to insulate itself from the common occurrence of drought. Many plants have limited their exposure through use of dry cooling, or use water sources that are largely insulated from the effects of drought (wastewater, groundwater). Those plants that do depend on surface water generally have elaborate contingency plans to secure water in times of drought.

Recently WECC completed its 'long-term' transmission planning exercise that considers a twenty-year time horizon. WECC's analysis utilised the regional integrated energy-water model. Considered were a reference or business-as-usual case and four scenarios that differ according to assumptions on projected demand schedules and technology evolution and adoption. For each of these test cases, assumed economics, emission standards and renewable portfolio standards dictated that most new generation would come from renewables (largely photovoltaic and wind) and natural gas – all of which have favourable water-use characteristics when compared with other electric-generating technologies. Additionally, scenarios involving progressive emission standards resulted in significant water savings due to the de-rating of much of the available coal-fired generation capacity. Finally, the regional integrated energy-water model identified opportunities for distributing new load to watersheds where water was available for development.

Conclusion

The project has seen some early successes. For example, the project has begun a dialogue between the energy planners and water supply managers in the Western United States. The project has provided resources for extensive regional data collection and the preliminary analysis of the intersection of energy and water. And finally, initial analysis has confirmed the intuition of project proponents that energy and water interactions matter as we consider the configurations and vulnerabilities of energy generation and transmission networks in the Western United States.

However, clear challenges remain, particularly as we seek to apply model results to planning and policy making. While state water managers understood the need for a uniform set of water availability and cost data, they also expressed concern over the final data because of recognised gaps in information, the dynamic nature of water use and the generally contentious nature of water rights and water allocation in the West. This concern illustrates the need for more uniform data at more refined geographic scales to assist future planning efforts. Projects such as this one as well as the WaDE tool development are necessary steps in exposing critical data gaps as well as to provide a vehicle for collecting and vetting the next generation of water and energy data.

Also, project partners realize the need to reach other audiences and at other scales. Additional outreach should focus on state water resource managers and state water planning, state Public Utility Commissions, and individual utilities and water resource managers. Regional models will need to be deployed to a distributed array of decision-making venues, including regional and local, public and private, in order to have the intended impact. At the same time, new venues should be considered to promote smart and robust planning and policy around this critical issue. The potential audiences and different applications will have implications for the design of models.

Further, those working at the intersection of energy and water will need to be cognizant of how energy and water differ in terms of policy development and management. Energy is amenable to national and even global policy; Congress often legislates on energy, and the Federal Energy Regulatory Commission provides a federal regulatory presence. On the other hand, states are primarily responsible for allocating, administering and protecting water resources within their borders. These differences carry through to how the resources are managed. The need to ensure electric reliability has driven the development of an extensive, nested network of subregional, regional and national planning entities, under the auspices of the North American Electric Reliability Corporation. Conversely, the federal government has historically deferred to state authority and control of water resources because states are best positioned to determine how to allocate their water resources. So not only are these issues addressed in different venues and through different constituencies, but they also have a fundamentally different approach to policy and management.

Ultimately, this project and other similar efforts indicate the need to engage high-level policy makers, including governors and the US Congress, in developing policy and technical solutions to address the intersection of energy and water. While we see opportunities to better integrate energy-water considerations into regional and national policy[3], we also foresee challenges as we move from model integration to policy integration. In addition to the issues identified above, we note that regional energy and water planning is often done at a decadal or longer time scale, which can undermine immediacy or urgency for policy makers. The energy-water nexus will need to compete for limited bandwidth of elected officials, and while energy at least is certainly a national and gubernatorial priority, budgets, jobs, health care and education remain at the top of the public polls (PollingReport.com 2013).

[3] For example, Senator Jeff Bingaman had proposed the *Energy-Water Integration Act* (S.1343) in 2011 and companion bills had been introduced in the House. The US DOI's *Water Smart* program has a module on energy-water, and a variety of national NGOs are working on the energy-water nexus.

Regional and integrated modelling has provided an objective entry point and served to confirm and highlight important planning and policy considerations at the energy-water nexus. The model demonstrates that some parts of the American West may experience imbalances between supply and demand that could impact energy transmission efforts, while also highlighting potential impacts to other sectors such as disruptions to agriculture and the environment. This effort to develop an integrated model of energy and water demand and availability also provides an in-depth analysis of the best available data regarding water supply and demand. Ideally, these combined results will inform policy makers working to address the energy-water nexus.

Acknowledgements

The work described in this article was funded by the US Department of Energy's Office of Electricity Delivery and Energy Reliability through the American Recovery and Reinvestment Act of 2009, under Contract No. M610000581.

Sandia National Laboratories is a multi-program laboratory managed and operated by Sandia Corporation, a wholly owned subsidiary of Lockheed Martin Corporation, for the US Department of Energy's National Nuclear Security Administration under contract DE-AC04–94AL85000.

The views expressed in this chapter are those of the authors and do not represent the views of WGA, WIEB, WSWC, WECC or any of its individual members.

References

Anderson, M. T. and Woolsey, L. H. Jr (2005). Water Availability for the Western United States – Key Scientific Challenges. *US Geological Survey Circular* 1261: 85.

Carter, N. T. and Cambell, R. J. (2009). Water Issues of Concentrating Solar Power (CSP) Electricity in the US Southwest. *Congressional Research Service* R40631.

Claussen, E. (2001). Making Collaboration a Matter of Course: A New Approach to Environmental Policy Making. *Environmental Practice* 3(4): 202–5.

Cockerill, K., Tidwell, V. and Passell, H. (2004). Assessing Public Perceptions of Computer-based Models. *Environmental Management* 34(5): 609–19.

Cockerill, K. C., Passell, H. and Tidwell, V. C. (2006). Cooperative Modeling: Building Bridges between Science and the Public. *Journal of American Water Resources Association* 42(2): 457–71.

Cockerill, K. C., Tidwell, V. C., Passell, H. and Malczynski, L. (2007). Collaborative Modeling Lessons for Environmental Management. *Environmental Practice* 9(1): 28–41.

Connick, S. and Innes, J. E. (2003). Outcomes of Collaborative Water Policy Making: Applying Complexity Thinking to Evaluation. *Journal of Environmental Planning and Management* 46(2): 177–97.

US Department of Energy (2009). *Secretary Chu Announces Efforts to Strengthen US Electric Transmission Networks*, Press Release, 18 December.

Harto, C. B. and Yan, Y. E. (2011). *Analysis of Drought Impacts on Electricity Production in the Western and Texas Interconnections of the United States*. ANL/EVS/R-11/14, Argonne, IL: Argonne National Laboratory.

Koontz, T. M., Steelman, T. A., Carmin, J., Korfmacher, K. S., Moseley, C., and Thomas, C. W. (2004). *Collaborative Environmental Management What Roles for Government?* Washington, DC: Resources for the Future.

Macknick, J., Newmark, R., Heath, G., and Hallett, K. C. (2012). Operational Water Consumption and Withdrawal Factors for Electricity Generating Technologies: A Review of Existing Literature. *Environmental Research Letters* 7(4): 045802.

Moxey, A. and White, B. (1998). NELUP: Some Reflections on Undertaking and Reporting Interdisciplinary River Catchment Modelling. *Journal of Environmental Planning and Management* 41(3): 397–402.

Mullner, S. A., Hubert, W. A. and Wesche, T. A. (2001). Evolving Paradigms for Landscape-Scale Renewable Resource Management in the United States. *Environmental Science & Policy* 4(1): 39–49.

National Energy Technology Laboratory (2009). Estimating Freshwater Needs to Meet Future Thermoelectric Generation Requirements, 2009. DOE/NETL-400/2009/1339, Pittsburg, PA.

Nicolson, C. R., Starfield, A. M., Kofinas, G. P., and Kruse, J. A. (2002). Ten Heuristics for Interdisciplinary Modeling projects. *Ecosystems* 5(4): 376–84.

PollingReport.com (2013). Problems and Priorities, www.pollingreport.com/prioriti.htm.

Selin, S. and Chavez, C. (1995). Developing a Collaborative Model for Environmental Planning and Management. *Environmental Management* 19(2): 189–95.

Selin, S. W., Schuett, M. A. and Carr, D. (2000). Modeling Stakeholder Perceptions of Collaborative Initiative Effectiveness. *Society and Natural Resources* 13(8): 735–45.

Spash, C. L. (2001). Broadening Democracy in Environmental Policy Processes. *Environment and Planning C: Government and Policy* 19(4): 475–81.

Susskind, L. E., Jain, R. K. and Martyniuk, A. O. (2001). *Better Environmental Policy Studies: How to Design and Conduct More Effective Analyses.* Washington, DC: Island Press.

US Energy Information Agency (2012). Annual Energy Outlook 2012 with Projections to 2035, www.eia.gov/forecasts/aeo/pdf/0383(2012).pdf.

Wear, D. N. (1999). Challenges to Interdisciplinary Discourse. *Ecosystems* 2(4): 299–301.

Western Electricity Coordinating Council (WECC) (2011). *10-Year Regional Transmission Plan, Plan Summary.* Western Electric Coordinating Council, www.wecc.biz/library/StudyReport/Documents/Plan_Summary.pdf

Western Electricity Coordinating Council (WECC) (2013). *Long Term Planning Tool,* www.wecc.biz/Planning/TransmissionExpansion/RTEP/LTPT/default.aspx

Wu, M., Mintz, M., Wang, M. and Arora, S., (2009). *Consumptive Water Use in the Production of Ethanol and Petroleum Gasoline.* ANL/ESD/09-1, Argonne, IL: Argonne National Laboratory, www.transportation.anl.gov/pdfs/AF/557.pdf

Yan, Y. E., Tidwell, V. C., King, C. W., and Cook M. A. (2013). *Impact of Future Climate Variability on ERCOT Thermoelectric Power Generation,* ANL/EVS/R-13/2, Argonne, IL: Argonne National Laboratory.

17

Biodiversity and the climate, energy and water nexus

JAMIE PITTOCK

Introduction

Climate change driven by the use of carbon-intensive energy is a proposed planetary boundary: a threshold that demarcates a safe living space for humanity (Rockström et al. 2009). Two other planetary boundaries are water consumption and biodiversity loss (Rockström et al. 2009).

The extensive loss of biodiversity impacts on humanity in two main ways. First, ecosystems provide humanity with a range of services, namely provisioning services such as fish and clean water, regulating services such as clean air and flood mitigation, cultural services such as recreational opportunities and spiritual succour, and supporting services like soil formation (UNEP 2007). Second, many argue that humanity has a moral imperative to conserve the diversity of life. This concern for biodiversity conservation has been codified by our national governments in international law, as discussed further in this chapter.

Demands to supply energy with lower carbon emissions and to secure water supplies under a changing climate are driving rapid decision making and deployment of new technologies, or old technologies at a greater scale than before, risking perverse outcomes (Pittock et al. 2013). In this volume, key aspects of the linkages between climate, energy and water are discussed, with a view to identifying management interventions by which positive interactions may be favoured and perverse impacts avoided. This chapter examines how management of the links between climate, energy and water can hinder or help progress in another sector, the conservation of biodiversity. This discussion begins with an overview of the conservation of biodiversity and a summary of the impacts of water use and energy supply on biodiversity.

Biodiversity conservation

The Millennium Ecosystem Assessment attributes the major causes of biodiversity loss as being driven by infrastructure development, land (habitat) conversion, pollution, overharvesting and overexploitation, the introduction of invasive species and global climate change (MEA 2005a). The drivers of biodiversity loss have cumulative impacts and are likely to be exacerbated by climate change.

This chapter focusses on the conservation of the freshwater portion of the world's biodiversity as it is of great concern in considering the climate, energy and water nexus. Freshwater biodiversity is often overlooked in simplistic assessments that divide the world's biodiversity between 'terrestrial' and 'marine', without appreciating the distinctive ecological processes that make conservation of the freshwater biome a unique challenge. For example, the United Nations List of Protected Areas does not distinguish freshwater wetlands from terrestrial ecosystems (other than lakes, which it identifies as under-represented; Chape et al. 2003). In another example, a recent assessment of biodiversity risks from fossil fuel extraction assesses marine and terrestrial but not freshwater eco-regions (Butt et al. 2013). Terrestrial-based ecosystems are largely visible and can often be conserved within a set area of land. By contrast, for surface and groundwater-dependent ecosystems, key processes are not readily observed, data availability is poorer and ecosystem health depends on managing processes across entire hydrological catchments. This poses greater challenges in managing the trade-offs among freshwater, climate and energy policies.

Freshwater biodiversity and water use

Freshwater ecosystems occupy around 0.8 per cent of the Earth's surface yet are the most species rich per unit area with nearly 6 per cent of the world's described species (MEA 2005b; Dudgeon et al. 2006). Freshwater ecosystems formed and species evolved to thrive with all of the available water in flows of specific qualities and volumes at particular times (Postel and Richter 2003; Richter et al. 2011). Thus, the health of this biodiversity depends on maintenance of the physical water cycle. Whenever people divert water for their use, there are knock-on effects along the water cycle and on the biodiversity that depends on this water. Even seemingly innocuous activities are likely to have perverse impacts if not carefully managed. For instance, planting trees on former pastures to sequester carbon in large quantities is likely done at a local scale to increase evapotranspiration (green water) and diminish runoff and inflows into freshwater ecosystems (blue water) (van Dijk and Keenan 2007). Freshwater and estuarine species and ecosystems are particularly impacted due to their dependence on vulnerable water flows, susceptibility to fragmentation by in-stream infrastructure and because these habitats are often a focus of human settlement (MEA 2005b). Hence, most human activities involve some sort of direct or indirect trade-off between desired services and freshwater biodiversity conservation. The policy challenge is to be sufficiently cognisant of the links to be able to make transparent decisions on the trade-offs that increase the benefits and minimise the impacts (Pittock 2011; Hussey and Pittock 2012). This chapter attempts to identify what some of those measures may involve.

One principle of conservation is that adequate and representative areas of each type of ecosystem should be designated for sustaining biodiversity. Limited agreement on how to categorise and measure conservation of freshwater ecosystems makes it harder to assess progress in conserving freshwater ecosystems, as indicated by the earlier example of the United Nations List of Protected Areas (Chape et al. 2003). In terms of conservation of representative areas of each type of habitat, the Ramsar Convention on Wetlands provides a framework for categorising wetland types and assessing the extent to which they are incorporated in the global network of protected areas (Ramsar 2012a). At the time of writing, the Ramsar Convention reports that it had designated 'wetlands of international importance' or Ramsar sites covering 2,168 sites and nearly 207 million hectares (see www.ramsar.org), possibly the world's largest protected-area system. Ramsar Convention data suggests that some types of wetlands are significantly under-represented, including peat lands, wet grasslands, mangroves and coral reefs, karst and other subterranean wetland types, temporary pools and bivalve (shellfish) reefs (Ramsar 2012a). What is clear is that many freshwater ecosystems are poorly represented within conservation areas, and those that are in reserves are often degraded by water infrastructure or impacts emanating from outside park boundaries (Pittock 2008).

Protection and recovery of threatened species is another measure of conservation. Wetland ecosystems contain the greatest proportion of species threatened with extinction (MEA 2005b). Based on 2,849 populations of 737 species of fish, birds, reptiles, amphibians and mammals found in temperate and tropical freshwater lakes, rivers and wetlands, a global freshwater biodiversity index declined by 37 per cent between 1970 and 2008 (WWF et al. 2012). One reason that such a large portion of freshwater species are threatened is their vulnerability to changes in water quantity, quality and timing.

Human civilisations have grown up alongside and impacted major freshwater ecosystems as the places where the best agricultural lands, transport routes and access to freshwater for industrial and domestic uses may be found (Vorosmarty and Sahagian 2000; Vorosmarty et al. 2010). Many freshwater species depend on migration along rivers to breed and complete their life cycles (often triggered by seasonal floods), and on the movement of sediments and nutrients down river, movement that is increasingly blocked by dams. Organic matter that drives the food chain in freshwater ecosystems is derived from riverside forests and the seasonal rise and fall of waters up onto floodplains. This dependence on longitudinal, vertical and latitudinal hydrological connectivity is directly challenged by developments to increase water security for people, for instance, by storing more water in dams to boost supplies in dry seasons and years, evening out the natural variability the freshwater biota depend upon (Postel and Richter 2003). Infrastructure proponents are calling upon governments to increase such water storages as a strategy to adapt to climate change (Grey and Sadhoff 2007). Freshwater species have also evolved to thrive in water of particular temperature and other chemical qualities, which are readily altered by dams, loss of river-side forests, and through pollution, including from agricultural and other land-use changes in the catchment (Olden and Naiman 2009; Davies 2010).

Groundwater systems are also important for conservation of freshwater biota as well as being exploited as a water supply for people (Barquín and Scarsbrook 2008). Stygofauna – animals living in groundwater – have been found living in aquifers several kilometres underground (Tomlinson and Boulton 2010). Aquifers may discharge into surface water ecosystems providing water that is buffered from seasonal and inter-annual variability, as well as being of a consistent quality (Boulton and Hancock 2006). Endemic species have often evolved in spring habitats, and these may comprise the only perennial freshwater ecosystems in arid areas (Strayer and Dudgeon 2010). High water tables may support floodplain forests and similar biota. Aquifer discharge into channels is often important for sustaining base river flows, and discharge points may form refugia for species requiring water of a particular temperature and other qualities during seasonably unfavourable times of the year (Olden and Naiman 2009). The characteristics of groundwaters that make them important for conservation of freshwater biota – namely, presence in arid areas of consistent volumes and quality of water – are also valued for supplying water to people. There is greater use of groundwater in adaptation to climate change as it is buffered from seasonal and inter-annual hydrological variability, leading to consideration of how to boost aquifer recharge (Kabat and van Schaik 2003; Spies and Dandy 2012).

Terrestrial biodiversity

Despite the freshwater focus of this chapter, there are many links between the conservation of distinctive freshwater and terrestrial biota. For example, the plains wildlife of the Maasai Mara region of East Africa depends on the waters of the Mara River (Negussie et al. 2011). In the past, proposals to increase water diversions from its Amala River tributary for irrigation and power generation threatened this ecosystem (Gereta et al. 2002). Hypothetically, such water diversion plans could be revisited if a low-carbon emissions policy decision was made to grow biofuels or increase hydropower production.

Terrestrial biota may also be threatened by climate and energy policies. An assessment of species-rich terrestrial and marine ecoregions identifies the risks posed to biodiversity by extraction of overlapping fossil fuel reserves (Butt et al. 2013). Renewable energy may also have impacts: for example, poorly sited wind turbines to strike particular bat or bird species (Kuvlesky et al. 2007).

Marine biodiversity

Climate change severely impacts marine ecosystems by changing the acidity and temperature of water, and through sea-level rise (IPCC 2007). Climate mitigation and energy policies may also have modest impacts. Some coastal marine habitats may be converted to tidal power production (Wolf et al. 2009). Damming parts of the coast is likely to impede aquatic wildlife passage, and by changing tidal water flows, impact on near-shore wildlife recruitment and feeding (Clark 2006; Wolf et al. 2009). As an example, the Severn Estuary in the United Kingdom was proposed to be dammed for electricity generation, trading off internationally recognised protected areas for waterbirds and other aquatic wildlife for low carbon energy (DECC et al. 2010). Although the most recent proposal was not supported by the UK Government, similar proposals are likely in that estuary and elsewhere around the world.

The impact of generating electricity from tides or waves is lowered by placing turbines or other generating equipment in open water, although there may still be questions of habitat loss, collision risks, noise and electromagnetic fields impacting on biodiversity (Drewitt and Langston 2006; Inger et al. 2009; Witt et al. 2012). Energy generation already involves placement of wind turbine infrastructure in open waters, and this practice may be expanded with various wave and tidal power technologies currently under development. This infrastructure changes the marine substrate and structure, creating new niches for different species of wildlife to live in these localities (Ashley et al. in press). It is conceivable that over-exploited fisheries may benefit in areas where infrastructure is placed in open waters if fishing operations are then restricted, creating de facto wildlife reserves (Inger et al. 2009).

Impacts of the energy sector on biodiversity

The preceding discussion highlights the wide range of impacts that the energy sector has on biodiversity. The Ramsar Convention summarises the impacts by stating that: "Energy sector activities in all phases of the energy supply chain can potentially have negative impacts on the ecological character of wetlands" (Ramsar 2012b, Annex cl.5). They go on to say that the impacts are primarily expressed through (but not limited to) the following aspects, which are summarised and modified in Table 17.1 to apply to all biodiversity.

Table 17.1 Impacts of energy activities on biodiversity (Ramsar 2012b).

Energy activity	Impact on biodiversity
Consumptive use of surface water or groundwater or alterations of natural flow regimes or drainage	Changes in water flows that change ecosystems or eliminates habitat of some species
Emission of chemical, thermal, radioactive and organic pollutants	Degradation of water, soil and air quality that changes ecosystems or eliminates habitat of some species
Construction of energy generation facilities such as tidal power plants	Changes in natural coastal water-level fluctuations that change ecosystems or eliminates habitat of some species
Construction and operation of energy generation facilities and infrastructure and disruption in sediment flow and ecosystem connectivity of energy-related activities or infrastructure	Direct impacts on habitats
Construction and operation of energy generation and transmission infrastructure such as wind turbines	Collision of fauna, especially birds and bats
Construction and operation of transport or water supply facilities, such as inter-basin transfer of water for hydropower generation	Habitat fragmentation and connectivity between hydrographic basins, wetlands and forests that prevent species movement and may reduce populations of species below viable sizes.
Atmospheric emissions	Indirect impacts on water or soil quality from emissions and those due to climate change that changes ecosystems or eliminates habitat of some species.

This chapter began by focussing on biodiversity, climate and energy, and water use as three key planetary boundaries. This section has outlined the threats to biodiversity with particular reference to current water use and energy supply. The responses of global institutions for conservation of biodiversity are now considered in relation to climate change and energy policies.

Global institutions

Biodiversity conservation has been codified by our national governments in international law through a number of environmental agreements. Among them, and having the broadest scope, is the Convention on Biological Diversity (CBD) (UN 1992). More recently under the CBD, the world's governments agreed to the 2020 Aichi biodiversity targets. Summarising the twenty targets are five overarching 'strategic goals' (CBD 2010), namely:

1. Mainstreaming biodiversity across government and society
2. Reduce the pressures on biodiversity and promote sustainable use
3. Improve the status of biodiversity by safeguarding ecosystems, species and genetic diversity
4. Enhance the benefits to all from biodiversity and ecosystem services
5. Enhance implementation through participatory planning, knowledge management and capacity building.

These goals clearly demand of policy makers to consider biodiversity conservation when crafting policies to respond to climate change or to supply energy and water to communities and industry. The measures in the CBD are complemented by a range of other thematic and regional biodiversity conservation agreements, including the Ramsar Convention for the conservation of wetlands, treaties conserving migratory species and their habitats, and regulation of trade in particular species. For example, in the 2009–15 Strategic Plan of the Ramsar Convention, 168 national governments have committed to "work towards achieving the wise use of all wetlands" (Goal 1) and to "develop and maintain an international network of wetlands that are important for the conservation of global biological diversity, including water bird flyways and fish populations and for sustaining human life" (Goal 2) (Ramsar 2008a).

In 2012, through both the CBD and the Ramsar Convention, the member national governments expressed concern about the impacts of climate change and various types of energy supply on biodiversity and made commitments to manage these conflicts. The CBD felt the need to reiterate "the importance of activities to integrate biodiversity into relevant climate change activities and to ensure coherence in national implementation of both the United Nations Framework Convention on Climate Change and the Convention on Biological Diversity" (Decision XI/21 cl.2) and adopted decisions on integrating biodiversity and climate change activities, emphasising the need to better govern biofuel production, geoengineering and forest use in carbon sequestration (CBD 2012).

Through the Ramsar Convention, the member governments committed to maintain or improve the ecological character of wetlands, including their ecosystem services, to enhance the resilience of wetlands as far as possible in the face of climate-driven ecological changes including, where necessary, to promote the restoration of degraded wetlands, and further to promote the ability of wetlands to contribute to nature-based climate change adaptation, particularly the roles of wetlands in regulating water, including reducing risks from water-related disasters, and to sequester and store carbon as important responses for climate change mitigation through the maintenance and enhancement of their ecological functions, and to reduce or halt the release of stored carbon that can result from the degradation and loss of wetlands. (Ramsar 2012c, cl.26)

Further, the Ramsar member authorities expressed concern "about the potential for certain activities related to the extraction of non-renewable energy resources, the production or harvesting of renewable energy resources including biofuels . . ., and the processing, distribution and utilization of energy resources and generation of electricity, if they are not appropriately managed and regulated, to have direct and indirect adverse impacts on the ecological character of wetlands" (Ramsar 2012b, cl.6). The Convention adopted "[g]uidance for addressing the implications for wetlands of policies, plans and activities in the energy sector" (Ramsar 2012b).

The blunt decisions of these biodiversity treaties emphasise concern that biodiversity is being threatened or lost from poorly considered activities in the climate change mitigation and energy sectors. They express concern that institutions for climate and energy policies, including the UN Framework Convention on Climate Change, give insufficient attention to avoiding perverse impacts of sectoral policies and seizing opportunities to achieve multiple objectives. The decisions also highlight that parts of most national governments understand the potential for conflicts and synergies between these sectoral policies sufficiently well to reach consensus on the problems and potential solutions. Normalising the issues has commenced leading to the prospect that solutions could be found to minimise impacts and maximise benefits. In this context, the impacts on biodiversity from climate change and related policies are now examined.

Direct impacts of climate change on biodiversity

Earth's climate and hydrological systems are intimately linked so that climate change will significantly impact on water resources and ecosystems. The expected changes are detailed by the Intergovernmental Panel on Climate Change (Bates et al. 2008). A warmer world increases the energy that drives the climate system, including intensifying evaporation, the volume of water in the atmosphere which subsequently falls as precipitation. One consequence is the movement of climatic zones, including expansion of the tropics and shift of the mid-latitude arid zones towards the poles. Consequently, many areas will get wetter and some drier. More frequent extreme events are anticipated in the form of droughts and floods. While tropical storm intensity is expected to increase, any change in frequency is uncertain. In mid-to-low rainfall regions, greater evapotranspiration may reduce inflows into surface water bodies. Higher temperatures, greater aridity and more frequent floods are expected to diminish water quality through increased sedimentation, diffuse pollution and eutrophication. While aquifers are expected to be partially buffered against seasonal and inter-annual variation in precipitation, the implications for aquifer recharge are uncertain.

These physical hydrological changes will impact freshwater biodiversity. Aquatic flora and fauna is adapted to specific temperature ranges such that cold-water-dependent species have already been observed moving to higher elevations and closer to the poles (Daufresne and Boet 2007). Many species will be unable to change their distributions, where natural barriers (such as unconnected east–west flowing rivers) or human barriers (such as dams) prevent movement. Wetland flora and fauna often depend on variations in flow to thrive – flow patterns that may change with the climate.

Aquatic fauna often depend on high flows of a particular temperature or other quality as a trigger to migrate or breed, which may change (Olden and Naiman 2009). Higher or lower water flows will result in changes to geomorphology that may expand or decrease habitat for particular species – for instance, by reducing the flows that flush silt from gravel beds needed for egg laying by salmonoid fish (Postel and Richter 2003). Wetland species require inundation at a certain frequency to survive and thrive, and this hydro-period is likely to change with the climate (Poff et al. 2002). The added risk from degradation of water quality from greater erosion, eutrophication and diffuse pollution is anticipated to change the distribution of freshwater species and ecosystems.

Many assessments of climate change on biodiversity stop at the direct impacts. However, no assessment is complete without considering how societal responses to climate change may ameliorate, or more likely exacerbate, the direct impacts (Pittock 2011). Focussing on energy and water, the impacts on biodiversity are now considered in terms of (a) changes to energy supply, (b) carbon sequestration to offset greenhouse gas emissions from energy generation and (c) changes to water supply.

Climate mitigation policies for energy supply

Climate change mitigation policies focus on reducing greenhouse gas emissions from energy supply. Although administrative mechanisms and regulatory interventions are possible, economists favour central policies that price greenhouse gas emissions, such as carbon taxes or emission trading schemes (Dovers and Hezri 2010). Increasing the price of polluting energy production from fossil fuels, if sufficiently high, reduces energy demand and favours less-polluting energy supply technologies. There are three major outcomes likely for water and biodiversity that are discussed here, namely energy demand management, supply from renewable technologies and use of lower-emission fossil fuel technologies.

Demand management seeks to ensure that the same services are provided with less energy. To the extent that demand management obviates the need to use more fuel and water to generate energy, biodiversity benefits from avoided impacts. Potentially, the reduced cost of water per unit of services delivered may add to the economic benefits obtained from demand management, but there are many institutional barriers to realising these benefits (Wallis et al. 2014).

Energy demand could be met from renewable technologies that vary greatly in terms of the required land and water, and concomitant impact on biodiversity (Carter 2011; Pittock 2011). At one end of a scale of impacts, first-generation biofuel crop production requires large areas of land and an order of magnitude increase in water consumption, and thus impact on biodiversity (Chapter 7; Gerbens-Leenes et al. 2008). At the other end, technologies such as wind and solar photovoltaics require very little water, and land used may be modest or located at sites of low biodiversity value. While intermittent energy generation technologies may have modest demands on land and water, there is a need to match generation to demand. This may be done through complementary fast-start generators, such as hydro or natural gas. Alternatively energy may be stored for later use, as with pumped storage hydropower (Pittock 2010a). These complementary energy generators may add to the impacts on biodiversity, land and water of solar and wind generators.

Fossil fuels with lower emissions may also be deployed. In many parts of the world, natural gas is displacing coal-fired electricity, in part because gas generators emit only 40 per cent of the greenhouse gas emissions at the point of generation as coal (although fugitive emissions from gas production are contested; IEA 2012; Cook et al. 2013). As discussed earlier in this chapter, expansion of unconventional natural gas production has extensive implications for land and groundwater use and management, with deleterious impacts on biodiversity. Surface water biota and threatened species are protected in developed countries through such measures as regulations on the temperature on water discharges that can be tolerated by aquatic fauna (Thomas 1996; Laws 2000). Developed countries like Australia and the United States have taken little action with endangered species laws to protect groundwater-dependant biota threatened by gas extraction. These biodiversity laws are powerful, and public interest test cases can be anticipated that will impact on government policies. There are further impacts on land and water with options for sequestration to offset emissions from energy generation.

Indirect impacts of climate change mitigation policies for carbon sequestration

The long-term future of fossil fuel energy generation is considered by many governments to be viable if linked to some degree of carbon sequestration. Two main options are being promoted, namely carbon capture and storage (CCS) from a point-source emitter, and carbon storage in soils and vegetation in the landscape (Pittock 2011). A number of governments have been promoting CCS from fossil fuel production and generation facilities (for example, under the name 'clean coal'), with the waste stream injected into deep aquifers and other geologic structures for long-term storage. The main environmental impact identified thus far is an increase in water consumption associated with cooling solvents used to capture waste gases as well as the water required in generation to replace what is used in the CCS process (NETL 2008). The additional water consumption may impact on biodiversity, especially as water resources are often fully exploited in the regions where coal mining and generation are clustered, such as the La Trobe and Hunter valley regions of Australia (Smart and Aspinall 2009). Thus far, the high costs of CSS and lack of demonstration plants are major barriers to deploying this technology.

Programs are being implemented to store more carbon in the landscape. At the international scale, under the auspices of the UN Framework Convention on Climate Change, REDD+ (reduced emission from degradation and deforestation plus) and the Clean Development Mechanism both have the potential to allocate credits for projects that sequester additional carbon in the landscape (UNFCCC 2007; CDM Executive Board 2010). National policies for catchment afforestation for adaptation and mitigation have been adopted in the climate change policies of a number of countries, including China, India and Mexico (Pittock 2011). Australia has gone further than other countries with its Climate Change (Carbon Farming) Act of 2011 in establishing mechanisms by which landholders who sequester carbon by applying additional, approved methodologies could earn carbon credits for sale in the domestic market (Australian Government 2011).

Afforestation is the primary landscape-scale carbon storage methodology being proposed globally. While superficially attractive, there are associated costs and benefits for biodiversity. Afforestation can be beneficial for biodiversity conservation if undertaken with indigenous species; however, this has not always been the case historically, for example, in the extensive use of at least one exotic species in China's 'Grain for Green' program (Chen et al. 2014). Australia's program mandates use of indigenous species on the basis that commercial plantings of non-indigenous species would not be additional and lack co-benefits (DCCEE 2011). However, replacing pastures through afforestation increases evapotranspiration and can significantly diminish stream inflows and dependent freshwater biodiversity (Jackson et al. 2005; van Dijk and Keenan 2007). One model of the Macquarie River catchment in Australia projected that afforestation of 10 per cent of the catchment would reduce flows in the over-allocated river by 17 per cent, impacting the Ramsar-listed Macquarie Marshes wetlands down river (Herron et al. 2002).

A number of mechanisms could be used to manage the risk of stream flow reductions (Pittock et al. 2013). Afforestation could be limited by regulation to those areas that are so wet that increased evaporation is insignificant, such as in the tropics or dry regions that do not regularly contribute to surface water flows. However, in dry regions, carbon sequestration may be limited by slower plant growth (Polglase et al. 2011). Spacing such plantings by time and location may also limit impacts on water to acceptable levels. To apply market influence, water consumption by carbon sequestration activities could be subject to usage fees, as in the stream flow reduction management provisions in South Africa (Mackay 2003), or incorporated into cap-and-trade water markets. The carbon farming program in Australia includes a complex set of national regulatory and market-based measures intended to limit impacts on water resources that to date has been criticised as ineffectual due to a lack of state-level implementation (NWC 2011; NWC 2013). Key criticisms include the failure to extend cap-and-trade water markets to plantations (other than in South Australia) and ineffectual regional planning by the state government's catchment management authorities for managing the water use in afforestation. The lessons are that there are technical means for managing the nexus between carbon sequestration, water and biodiversity conservation, but they are complex, and crafting institutions to apply them well is proving difficult.

Globally, there are many proponents of sequestering more carbon in the soil. One technique proposed is biochar, the incorporation of charcoal into the soil (Kleiner 2009).

Thus far, the focus of research on the method has been on the longevity of the carbon sequestration, with little or no consideration of the impacts on water or biodiversity. The source of biomass for charcoal production will influence the type and extent of land and water required and concomitant impacts on biodiversity. Once biochar is incorporated into the soil, presumably on existing agricultural lands, the soil properties will change. Biochar proponents argue that more less-compacted and carbon-rich soils will hold more precipitation and make this available for greater crop or pasture growth (Kleiner 2009). While there are no assessments, changes in hydrology would be expected. Potential impacts that require investigation include reduced river inflows, greater aquifer recharge and changes in water quality. These changes may affect freshwater biodiversity.

Many of the policy and market changes in energy supply and carbon sequestration described above would have the perverse outcome of increasing water consumption and scarcity, and thus impacting on freshwater biodiversity. They may also compound the challenge to adapt water supplies under a changing climate.

Indirect impacts of climate change adaptation policies for water supply

In addition to the increasing water demand from growing global populations and water-intensive, wealthier lifestyles, climate change is likely to increase water demand directly (Bates et al. 2008). A warmer climate is expected to increase evapotranspiration in agriculture and cooling requirements in domestic and industrial processes, increasing water demand. The challenges in meeting this increasing demand are expected to be complicated by the changes in precipitation patterns, more frequent floods and droughts, and declining water quality. Four main approaches to meeting these water supply challenges have been proposed: (a) water demand management, (b) greater reuse of waste waters, (c) diversifying water sources and (d) storing more water (Kabat and van Schaik 2003).

Water demand management, as detailed in Chapter 9, often has a double benefit in saving both energy and water, for example, with reduced use of hot water. To the extent that these savings obviate the need for additional resource withdrawals from the environment, then biodiversity will benefit. Water demand management measures on their own can increase energy demand and impacts on greenhouse gas emission and biodiversity, as indicated in the example of switching from gravity-fed to reticulated irrigation production outlined Chapter 8. These trade-offs need to be assessed for each intervention to inform decisions and to avoid perverse outcomes and maximise benefits.

Greater reuse of wastewater has a number of benefits, including reducing demand for waters from the natural environment, diversifying water supply by drawing on a reliable source and reducing discharge of polluted waters (Toze 2006). Nutrients may also be reused in agricultural fertilisers. In these contexts, there are benefits for biodiversity conservation. Treating wastewater to potable standards is energy intensive, but less so than desalination and with the added benefit of lower transmission costs due to co-location of supply and demand (Kenway et al. 2011). However, in developed countries, most communities have resisted the reuse of wastewater.

Diversifying water sources spreads the risk of any one source failing due to climate change and variability. In Australia, for example, most capital cities relied primarily on surface waters in the twentieth century, which made them vulnerable during the 2002–10 Millennium drought. The five largest state governments then embarked on extensive programs for diversification of water sources through linking previous separate water supply systems, increasing use of groundwater and wastewater reuse, and building major desalination plants, as well as managing demand (PC 2011). The new water resources involve extensive treatment or pumping and are energy intensive, exacerbating greenhouse gas emissions (Hall et al. 2009). In terms of biodiversity, linking previously separate river systems carries the risk of facilitating movement of invasive species and diseases, or conversely, migration of indigenous species to new habitats under climate change. Groundwater supplies may be buffered against short-term climatic variability, but overuse will impact on groundwater-dependent ecosystems. Desalination is particularly energy intensive and may have modest impacts on biodiversity, involving seawater withdrawal that entrains some marine species and the discharge of brine and chemicals (Meerganz von Medeazza 2005; Elimelech and Phillip 2011).

Storing more water enables supply to continue in areas of variable climates in dry times. For agriculture and hydropower, this is in large part about 'overturning the hydrograph' – that is, holding water in dams from wet seasons for release during dry seasons when river flows would be naturally low. It is this overturning of the ecological flow processes that causes maximum impacts on biodiversity (Postel and Richter 2003). Nevertheless, a number of storage proponents argue that in countries with 'difficult hydrologies' – namely, those with great seasonal variation in water availability and inter-annual variability with floods and droughts – economic development and social welfare are linked to storage, especially in dams (Grey and Sadhoff 2007). Surface water storages are controversial, expensive to build, heavily subsidised in construction and difficult to maintain as a self-funding business (WCD 2000). In developed countries, the most effective dam sites are occupied, and attention is turning to the difficulties of optimising dam operations. In the context of biodiversity conservation and climate change adaptation, this periodic reoperation of old dams provides opportunities to retrofit hardware and operating procedures to minimise environmental impacts (eg thermal pollution control and fish passages), and to enhance socioeconomic benefits (Pittock and Hartmann 2011). Many redundant or underperforming dams have been removed in this process.

In developing countries, small-scale rainwater harvesting and conservation agriculture techniques have been proposed as alternatives to new dams, although these too have costs and benefits (Makurira 2014). Rainwater harvesting structures have similar challenges as dams in raising capital for construction, although their use of decentralised small-scale and appropriate technologies means that these systems may be better maintained than larger irrigation schemes (Makurira 2014). The decentralised deployment and more effective operations of these methods may limit energy requirements for pumping, and spread out and minimise the environmental impacts (Stirzaker and Pittock 2014). However, rainwater harvesting and conservation agriculture may also significantly reduce stream flows if deployed widely in susceptible regions (van Dijk 2006).

Use of aquifers for water storage is another option for reducing the impact of short-term climatic variability. A number of developed countries are actively recharging aquifers for urban supply from treated wastewaters, and in developing countries, rainwater tanks (village or farm dams) are often used to replenish shallow aquifers (Dillon 2005; Gujja et al. 2009). If aquifers are not over-exploited, then the impacts on biodiversity may be modest compared to surface-water storages, although there are additional energy requirements in pumping. Governing use of shared aquifers to avoid the over-exploitation seen in many regions of the world is a huge challenge. Rather than relying on blunt regulatory and user-pays power pricing, recent programs of the International Water Management Institute, in partnership with Indian state governments, illustrate that politically tractable solutions are possible, including separate and rationed energy supply grids for groundwater pumps and subsidised access to low power pumps for withdrawal of annually recharged aquifers rather than deeper groundwater (IWMI 2011; IWMI 2012). These unconventional policy solutions illustrate the challenges and opportunities for governing the nexus well.

This discussion of water demand management, reuse of wastewaters, diversifying and storing more water highlights how adapting water supplies to climate change involves technology and policy choices. Some technologies consume more energy and diminish biodiversity, while others may conserve these resources. The remainder of this chapter describes the means of making better choices.

Governing the nexus for biodiversity conservation

Minimising the conflicts and seizing the positive synergies between biodiversity conservation, climate policy, energy and water supply described in this chapter are a tremendous challenge for societal institutions. There are four approaches that may favour better decision making by businesses, communities and governments to maximise benefits (Pittock et al. 2013).

1. Better information to inform decisions

Too often, the information needed to make decisions across sectors like biodiversity, energy and water is incompatible or not accessible in formats that enable comparisons and better choices. Quantifying the relationships across these sectors is needed (Hussey and Pittock 2012). Concepts such as the limits of acceptable change in the ecological character description for each Ramsar wetland site could be used to inform development decisions that may not cross thresholds for adequate biodiversity conservation (Ramsar 2008b). Simple comparative tools such as two-way tables or traffic light analyses can prompt information needed from different sectors to enable more integrated decisions (Pittock et al. 2013). A more complex Catchment Assessment Framework tool developed for freshwater biodiversity conservation can be adapted to multiple sectors for gathering information on climate change adaptation options, risks, costs and benefits (Lukasiewicz et al. 2013). Better and more structured use of information can help to avoid perverse impacts and maximise benefits.

2. Technology choice

A number of technologies favoured by climate mitigation policies have large impacts on biodiversity and water resources. Examples include first-generation biofuels (Chapter 7; Gerbens-Leenes et al. 2008) and hydropower (Pittock 2010b). Alternate technologies can meet the mutual goals of low-carbon energy supply and conservation of water and dependent freshwater biodiversity. Examples include dry or hybrid cooling for thermal power stations (Stillwell et al. 2010; Spies and Dandy 2012) and off-river pumped hydropower (Pittock 2010a). In the water sector, desalination is an example of a technology used in a climate adaptation context that has significant, perverse impacts (Meerganz von Medeazza 2005). Better financial and governance processes that favour technologies with lower impacts are needed.

3. Market incentives

While many countries have established markets that put a price on carbon, energy and water, these markets are rarely well integrated. Adequate or full-cost of supply prices for water and energy can provide incentives against over-use. For example, Australian governments committed in 2004 to ensure that all significant water use occurred within a cap-and-trade water market, but a number of state governments have not included use by extractive industries or power generators as yet (NWC 2011; RET 2011). Further, Australia's Carbon Farming Initiative provides a financial incentive for sequestration in the landscape, yet these businesses do not yet have to pay for resulting water consumption that may diminish river inflows (NWC 2011; Pittock et al. 2013). While such market mechanisms do not directly price biodiversity, internalising the cost of water may reduce impacts on freshwater ecosystems.

4. Better governance

Finally, there are examples of better integrating cross-sector decisions by applying leadership, legal mandates for work across sectors, vertical and horizontal integration mechanisms, and accountability mechanisms that monitor, identify problems and opportunities to enhance choices (Ross and Dovers 2008; Pittock 2011). One example is the periodic relicensing of many hydropower dams in the United States, where every thirty to fifty years, the performance of these dams is publicly re-evaluated to enable changes to meet current social, economic and environmental standards (Pittock and Hartmann 2011). This institutional policy reform window may more readily enable biodiversity conservation measures to be enacted in response to climate change. Another example concerns the potential for strategic environmental assessment to consider cumulative impacts on biodiversity and other attributes more flexibly, earlier in time and at a larger scale than project-by-project assessment (Pittock and Hartmann 2011; Pittock et al. 2013). This approach has great potential to minimise impacts as challenges of energy and water security in a changing climate induces rapid technological and policy change, such as with exploitation of unconventional gas resources.

Conclusion

Biodiversity, climate and energy, and water use are among the planetary boundaries that define a safe living space for humanity. Biodiversity is being lost at an excessive rate, exacerbated in large part by current water use and energy supply practices. While the direct impacts of climate change increase the threats to biodiversity, this chapter has focussed on the added, indirect impacts of climate policies in energy supply, sequestering carbon in the landscape and adapting water systems. Many of the resulting policy and market changes in energy supply and carbon sequestration would have the perverse outcome of increasing water consumption and thus impact freshwater biodiversity. They may also compound the challenge to adapt water supplies under a changing climate.

The Convention on Biological Diversity and Ramsar Convention on Wetlands have each adopted decisions that describe how biodiversity may be threatened or lost from poorly considered activities in the climate change mitigation and energy sectors. These statements also highlight that parts of their member national governments understand the conflicts and synergies across the sectors well enough to agree on the problems, leading to solutions that minimise impacts and maximise benefits. Four approaches that may favour improved cross-sectoral decision making by businesses, communities and governments are described in this chapter: better information, technological choices, market incentives and enhanced governance process.

Minimising the conflicts and seizing the positive synergies between biodiversity conservation, climate policy, energy and water supply are enormous challenges for our societies, but there are ways in which better outcomes for biodiversity and people can be achieved.

References

Ashley, M. C., Mangi, S. C. and Rodwell, L. D. (in press). The Potential of Offshore Windfarms to Act as Marine Protected Areas – A Systematic Review of Current Evidence. *Marine Policy*. Doi: http://dx.doi.org/10.1016/j.marpol.2013.09.002.

Australian Government (2011). Carbon Credits (Carbon Farming Initiative) Act 2011. No. 101, 2011. An Act about projects to remove carbon dioxide from the atmosphere and projects to avoid emissions of greenhouse gases, and for other purposes. Canberra: Commonwealth of Australia.

Barquín, J. and Scarsbrook, M. (2008). Management and Conservation Strategies for Coldwater Springs. *Aquatic Conservation: Marine and Freshwater Ecosystems* 18(5): 580–91.

Bates, B. C., Kundzewicz, Z. W., Wu, S. and Palutikof, J. P. eds. (2008). *Climate Change and Water. Technical Paper of the Intergovernmental Panel on Climate Change*. Geneva: IPCC Secretariat.

Boulton, A. J. and Hancock, P.J. (2006). Rivers as Groundwater-Dependent Ecosystems: A Review of Degrees of Dependency, Riverine Processes and Management Implications. *Australian Journal of Botany* 54(2): 133–44.

Butt, N., Beyer, H. L., Bennett, J. R., Biggs, D., Maggini, R., Mills, M., Renwick, A. R., Seabrook L. M. and Possingham. H. P. (2013). Biodiversity Risks from Fossil Fuel Extraction. *Science* 342(6157): 425–6.

Carter, N. T. (2011). *Energy's Water Demand: Trends, Vulnerabilities, and Management*. Washington, DC: Congressional Research Service.

Convention on Biological Diversity (2010). *Aichi Biodiversity Targets*. Montreal: Convention on Biological Diversity.

Convention on Biological Diversity (2012). Full Decision Report. *Eleventh meeting of the Conference of the Parties to the Convention on Biological Diversity, 8 – 19 October 2012 – Hyderabad, India*. Montreal: Convention on Biological Diversity.

Clean Development Mechanism Executive Board (2010). Project Design Document Form for Afforestation and Reforestation Project Aactivities (CDM-AR-PDD), Version 05. Bonn, UN Framework Convention on Climate Change.

Chape, S., Blyth, S., Fish, L., Fox P. and Spalding. M. (2003). 2003 United Nations List of Protected Areas. Gland and Cambridge: International Union for the Conservation of Nature and UN Environment Programme – World Conservation Monitoring Centre.

Chen, Y., Jessel, B., Fu, B., Yu, X. and Pittock, J. eds. (2014). *Ecosystem Services and Management Strategy in China*. Springer Earth System Sciences. Berlin Heidelberg: Springer-Verlag.

Clark, N. A. (2006). Tidal Barrages and Birds. *Ibis* 148: 152–7.

Cook, P., Beck, V., Brereton, D., Clark, R., Fisher, B., Kentish, S., Toomey, J. and Williams, J. (2013). *Engineering Energy: Unconventional Gas Production*. Melbourne: Australian Council of Learned Academies.

Daufresne, M. and Boet, P. (2007). Climate Change Impacts on Structure and Diversity of Fish Communities in Rivers. *Global Change Biology* 13(12): 2467–78.

Davies, P. M. (2010). Climate Change Implications for River Restoration in Global Biodiversity Hotspots. *Restoration Ecology* 18(3): 261–8.

Department of Climate Change and Energy Efficiency (DCCEE) (2011). Carbon Credits (Carbon Farming Initiative) Regulations 2011. Canberra: Department of Climate Change and Energy Efficiency.

DECC, SWRDA and WAG. (2010). Severn Tidal Power Feasibility Study: Conclusions and Summary Report. London: Department of Energy and Climate Change, South West Regional Development Authority, and Welsh Assembly Government.

Dillon, P. (2005). Future Management of Aquifer Recharge. *Hydrogeology Journal* 13(1): 313–16.

Dovers, S. R. and Hezri A. A. (2010). Institutions and Policy Processes: The Means to the Ends of Adaptation. *Wiley Interdisciplinary Reviews: Climate Change* 1(2): 212–31.

Drewitt, A. L. and Langston, R. H. W. (2006). Assessing the Impacts of Wind Farms on Birds. *Ibis* 148: 29–42.

Dudgeon, D., Arthington, A. H., Gessner, M. O., Kawabata, Z-I., Knowler, D. J., Lévêque, C., Naiman, R. J., Prieur-Richard, A-H., Soto, D., Stiassny M. L. J. and Sullivan. S. (2006). Freshwater Biodiversity: Importance, Threats, Status and Conservation Challenges. *Biological Reviews* 81(2): 163–82.

Elimelech, M. and Phillip W. A. (2011). The Future of Seawater Desalination: Energy, Technology, and the Environment. *Science* 333(6043): 712–17.

Gerbens-Leenes, P. W., Hoekstra, A. Y. and van der Meer, T. H. (2008). Water Footprint of Bio-energy and Other Primary Energy Carriers. *The Value of Water Research Report Series No. 29*. Delft: UNESCO-IHE Institute for Water Education.

Gereta, E., Wolanski, E., Borner M. and Serneels. S. (2002).Use of an Ecohydrology Model to Predict the Impact on the Serengeti Ecosystem of Deforestation, Irrigation and the Proposed Amala Weir Water Diversion Project in Kenya. *International Journal of Ecohydrology & Hydrobiology* 2(1): 135–42.

Grey, D. and Sadhoff, C. (2007). Sink or Swim? Water Security for Growth and Development. *Water Policy* 9: 545–71.

Gujja, B., Dalai, S. Shaik, H. and Goud. V. (2009). Adapting to Climate Change in the Godavari River Basin of India by Restoring Traditional Water Storage Systems. *Climate and Development* 1(3): 229–40.

Hall, M., West, J., Lane, J., de Haas, D. and Sherman, B. (2009). *Energy and Greenhouse Gas Emissions for the SEQ Water Strategy*. Brisbane: Urban Water Security Research Alliance.

Herron, N., Davis, R. and Jones, R. (2002). The Effects of Large-Scale Afforestation and Climate Change on Water Allocation in the Macquarie River Catchment, NSW, Australia. *Journal of Environmental Management* 65(4): 369–81.

Hussey, K. and Pittock, J. (2012). The Energy-Water Nexus: Managing the Links between Energy and Water for a Sustainable Future. *Ecology and Society* 17(1), 31 [online].

International Energy Agency (IEA) (2012). *World Energy Outlook 2012*. Paris: International Energy Agency.

Inger, R., Attrill, M. J., Bearhop, S., Broderick, A. C., Grecian, W. J., Hodgson, D. J., Mills, C., Sheehan, E., Votier, S. C., Witt, M. J. and Godley, B. J. (2009). Marine Renewable Energy: Potential Benefits to Biodiversity? An Urgent Call for Research. *Journal of Applied Ecology* 46(6): 1145–53.

Intergovernmental Panel on Climate Change (IPCC) (2007). *Impacts, Adaptation and Vulnerability. Intergovernmental Panel on Climate Change Climate Change 2007: Working Group II. Contribution to the Intergovernmental Panel on Climate Change Fourth Assessment Report*. Geneva: International Panel on Climate Change.

International Water Management Institute (2011). Innovative Electricity Scheme Sparks Rural Development in India's Gujarat State. *Success Stories* 9.

International Water Management Institute (2012). Boosting Water Benefits in West Bengal. *Sucess Stories* 14.

Jackson, R. B., Jobbágy, E. G., Avissar, R., Roy, S. B., Barrett, D. J., Cook, C. W., Farley, K. A., le Maitre, D. C., McCarl B. A. and Murray, B. C. (2005). Trading Water for Carbon with Biological Carbon Sequestration. *Science* 310(5756): 1944–7.

Kabat, P. and van Schaik H. (2003). *Climate Changes the Water Rules: How Water Managers Can Cope with Today's Climate Variability and Tomorrow's Climate Change*. Delft: Dialogue on Water & Climate.

Kenway, S. J., Lant, P., Priestley A. and Daniels. P. (2011). The Connection between Water and Energy in Cities – A Review. *Water Science and Technology* 63(9): 1983–90.

Kleiner, K. (2009). The Bright Prospect of Biochar. *Nature* (0906): 72–4.

Kuvlesky, W. P., Brennan, L. A. Morrison, M. L. Boydston, K. K. Ballard, B. M. and Bryant, F. C. (2007). Wind Energy Development and Wildlife Conservation: Challenges and Opportunities. *The Journal of Wildlife Management* 71(8): 2487–98.

Laws, E. A. (2000). *Aquatic Pollution: An Introductory Text*. John Wiley & Sons.

Lukasiewicz, A., Finlayson, C. M. and Pittock, J. (2013). Identifying Low Risk Climate Change Adaptation in Catchment Management while Avoiding Unintended Consequences. Gold Coast: National Climate Change Adaptation Research Facility.

Mackay, H. (2003). Water Policies and Practices. *Towards a just South Africa: The Political Economy of Natural Wealth*.

Makurira, H. (2014). Rainwater Harvesting and Conservation Agriculture in the Southern African Region. In *Water, Food and Agricultural Sustainability in Southern Africa*. Prahran: Tilde University Press, pp. 143–75.

Millennium Ecosystem Assessment (MEA) (2005a). *Ecosystems and Human Well-being: Synthesis*. Washington, DC: Island Press.

Millennium Ecosystem Assessment (MEA) (2005b). *Ecosystems and Human Well-being: Wetlands and Water Synthesis*. Washington, DC: World Resources Institute.

Meerganz von Medeazza, G. L. (2005). "Direct" and Socially-Induced Environmental Impacts of Desalination. *Desalination* 185(1–3): 57–70.

Negussie, Y. Z., Urbaniak, M. and Zalewski, M. (2011). Ecohydrology for a Sustainable Future in Africa – The Cases of Ethiopia, Kenya and Tanzania. *Ecohydrology & Hydrobiology* 11(3–4): 223–30.

National Energy Technology Laboratory (NETL) (2008). Estimating Freshwater Needs to Meet Future Thermoelectric Generation Requirements. 2008 Update. DOE/NETL-400/2008/1339. 30 September. Pittsburgh, PA: US Department of Energy, National Energy Technology Laboratory.

National Water Commission (NWC) (2011). The National Water Initiative – Securing Australia's Water Future: 2011 Assessment. Canberra: National Water Commission.

National Water Commission (NWC) (2013). Carbon Farming Initiative. Canberra: National Water Initiative.

Olden, J. D. and Naiman, R. J. (2009). Incorporating Thermal Regimes into Environmental Flows Assessments: Modifying Dam Operations to Restore Freshwater Ecosystem Integrity. *Freshwater Biology* 55: 86–107.

Pittock, J. (2010a). Viewpoint – Better Management of Hydropower in an Era of Climate Change. *Water Alternatives* 3(2): 444–52.

Pittock, J. (2010b). A Pale Reflection of Political Reality: Integration of Global Climate, Wetland, and Biodiversity Agreements. *Climate Law* 1(3): 343–73.

Pittock, J. (2011). National Climate Change Policies and Sustainable Water Management: Conflicts and Synergies. *Ecology and Society* 16(2): 25. [online].

Pittock, J., Hansen, L. J. and Abell, R. (2008). Running Dry: Freshwater Biodiversity, Protected Areas and Climate Change. *Biodiversity* 9(3 & 4): 30–8.

Pittock, J. and Hartmann, J. (2011). Taking a Second Look: Climate Change, Periodic Re-licensing and Better Management of Old Dams. *Marine and Freshwater Research* 62: 312–20.

Pittock, J., Hussey, K. and McGlennon, S. (2013). Australian Climate, Energy and Water Policies: Conflicts and Synergies. *Australian Geographer* 44(1): 3–22.

Poff, N., Brinson M. M. and Day Jr, J. W. (2002). *Aquatic Ecosystems and Global Climate Change*. Arlington, VA: Pew Center on Global Climate Change, 44.

Polglase, P., Reeson, A., Hawkins, C., Paul, K., Siggins, A., Turner, J., Crawford, D., Jovanovic, T., Hobbs, T., Opie, K., Carwardine, J. and Almeida, A. (2011). *Opportunities for Carbon Forestry in Australia: Economic Assessment and Constraints to Implementation*. Canberra: CSIRO.

Postel, S. and Richter, B. (2003). *Rivers for Life: Managing Water for People and Nature*. Washington, DC: Island Press.

Productivity Commission (2011). Australia's Urban Water Sector, Report No. 55, Final Inquiry Report. Canberra: Productivity Commission.

Ramsar (2008a). The Ramsar Strategic Plan 2009–2015 (adopted by Resolution X.1). Gland: Ramsar Convention on Wetlands.

Ramsar (2008b). Resolution X.15 Describing the Ecological Character of Wetlands, and Data Needs and Formats for Core Inventory: Harmonized Scientific and Technical Guidance. Gland: Ramsar Convention.

Ramsar (2012a). Resolution XI.8 Annex 2 Strategic Framework and Guidelines for the Future Development of the List of Wetlands of International Importance of the Convention on Wetlands (Ramsar, Iran, 1971) – 2012 revision. *11th Meeting of the Conference of the Parties to the Convention on Wetlands (Ramsar, Iran, 1971)*. Gland: Ramsar Convention on Wetlands.

Ramsar (2012b). Resolution XI.10 Wetlands and energy issues. *11th Meeting of the Conference of the Parties to the Convention on Wetlands (Ramsar, Iran, 1971).* Gland, Ramsar Convention on Wetlands.

Ramsar (2012c). Resolution XI.14 Climate Change and Wetlands: Implications for the Ramsar Convention on Wetlands. *11th Meeting of the Conference of the Parties to the Convention on Wetlands (Ramsar, Iran, 1971).* Gland: Ramsar Convention on Wetlands.

Department of Resources, Energy and Tourism (RET) (2011). Draft Energy White Paper 2011: Strengthening the Foundations for Australia's Energy Future. Canberra: Department of Resources, Energy and Tourism.

Richter, B. D., Davis, M. M., Apse, C. and Konrad C. (2011). A Presumptive Standard for Environmental Flow Protection. *River Research and Applications.*

Rockström, J., Steffen, W., Noone, K., Persson, A., Chapin III, F. S., Lambin, E. F., Lenton, T. M., Scheffer, M., Folke, C., Schellnhuber, H. J., Nykvist, B., de Wit, C. A., Hughes, T., van der Leeuw, S., Rodhe, H., Sörlin, S., Snyder, P. K., Costanza, R., Svedin, U., Falkenmark, M., Karlberg, L., Corell, R. W., Fabry, V.J., Hansen, J., Walker, B., Liverman, D., Richardson, K., Crutzen, P. and Foley, J. A. (2009). A Safe Operating Space for Humanity. *Nature* 461(24 September 2009): 472–5.

Ross, A. and Dovers, S. (2008). Making the Harder Yards: Environmental Policy Integration in Australia. *Australian Journal of Public Administration* 67(3): 245–60.

Smart, A. and Aspinall, A. (2009). *Water and the Electricity Generation Industry. Implications of Use.* Canberra: National Water Commission.

Spies, B. and Dandy, G. (2012). *Sustainable Water Management: Securing Australia's Future in a Green Economy.* Melbourne Australian Academy of Technological Sciences and Engineering.

Stirzaker, R. and Pittock, J. (2014). The Case for a New Irrigation Research Agenda for Sub-Saharan Africa. In *Agriculture and Water in Southern Africa*, ed J. Pittock, R. Q. Grafton and C. White. Prahran: Tilde University Press.

Strayer, D. L. and Dudgeon, D. (2010). Freshwater Biodiversity Conservation: Recent Progress and Future Challenges. *Journal of the North American Benthological Society* 29(1): 344–58.

Thomas, G. A. (1996). Conserving Aquatic Biodiversity: A Critical Comparison of Legal Tools for Augmenting Streamflows in California. *Stanford Environmental Law Journal* 15: 3.

Tomlinson, M. and Boulton, A. J. (2010). Ecology and Management of Subsurface Groundwater Dependent Ecosystems in Australia – A Review. *Marine and Freshwater Research* 61(8): 936–49.

Toze, S. (2006). Reuse of Effluent Water – Benefits and Risks. *Agricultural Water Management* 80(1–3): 147–59.

United Nations (UN) (1992). Convention on Biological Diversity (with annexes). Concluded at Rio de Janerio on 5 June 1992. *Treaty Series, No. 30619. Multilateral.* New York: United Nations.

United Nations Environment Program (UNEP) (2007). *Global Environmental Outlook GEO4 Environment for Development.* Nairobi: United Nations Environment Program.

United Nations Framework Convention on Climate Change (UNFCCC) (2007). Decision-/CP.13 Reducing Emissions from Deforestation in Developing Countries: Approaches to Stimulate Action. Bonn: UN Framework Convention on Climate Change.

van Dijk, A., Evans, R., Hairsine, P., Khan, S., Nathan, R., Paydar, Z., Viney, N. and Zhang, L. (2006). *Risks to the Shared Water Resources of the Murray-Darling Basin.* Canberra: Murray-Darling Basin Commission.

van Dijk, A. I. J. M. and Keenan, R. J. (2007). Planted Forests and Water in Perspective. *Forest Ecology and Management* 251(1–2): 1–9.

Vorosmarty, C. J., McIntyre, P. B., Gessner, M. O., Dudgeon, D., Prusevich, A., Green, P., Glidden, S., Bunn, S. E., Sullivan, C. A., Liermann, C. R. and Davies, P. M. (2010). Global Threats to Human Water Security and River Biodiversity. *Nature* 467(7315): 555–61.

Vorosmarty, C. J. and Sahagian, D. (2000). Anthropogenic Disturbance of the Terrestrial Water Cycle. *BioScience* 50(9): 753–65.

Wallis, P. J., Ward, M. B., Pittock, J., Hussey, K., Bamsey, H., Denis, A., Kenway, S. J., King, C. W., Mushtaq, S., Retamal, M. L. and Spies, B. R. (2014). The Water Impacts of Climate Change Mitigation Measures. *Climatic Change* 125: 209–20.

Witt, M. J., Sheehan, E. V., Bearhop, S., Broderick, A. C., Conley, D. C., Cotterell, S. P., Crow, E., Grecian, W. J., Halsband, C., Hodgson, D. J., Hosegood, P., Inger, R., Miller, P. I., Sims, D. W., Thompson, R. C., Vanstaen, K., Votier, S. C., Attrill M. J, and Godley, B. J. (2012). Assessing Wave Energy Effects on Biodiversity: The Wave Hub Experience. *Philosophical Transactions of the Royal Society A: Mathematical, Physical and Engineering Sciences* 370(1959): 502–29.

Wolf, J., Walkington, I. A., Holt J. and Burrows. R. (2009) Environmental Impacts of Tidal Power Schemes. *Proceedings of the ICE – Maritime Engineering* 162: 165–77.

World Commission on Dams (WCD) (2000). *Dams and Development: A New Framework for Decision-Making. The Report of the World Commission on Dams*. London: Earthscan.

WWF, GFN and ZSL. (2012). Living Planet Report 2012. Biodiversity, Biocapacity and Better Choices. Gland, WWF, Global Footprint Network and Zoological Society of London.

18

Consumers, food supply chain and the nexus

JAN LUNDQVIST, JUNGUO LIU AND JAKOB LUNDBERG

Introduction

This chapter explores the interactions between the main actors in the food supply chain, also referred to as the food system, from producers to consumers. Governance of these interactions has implications for the use of water, land, energy and the environment, as well as for the economy. The purpose of the chapter is to highlight that the inefficiencies in the system can be significantly reduced with multiple wins. Mainstream approaches focus on production and tend to overlook externalities and trade-offs. The sizeable losses, waste and conversion of the food produced are ignored. An expansion of an increasingly wealthy and urban population poses one set of challenges. Reducing a lingering poverty poses another. On the supply side, global warming and the enhanced uncertainty and variability in water availability signal an era with unprecedented predicaments. With a multitude of drivers and an increasing resource scarcity and uncertainty, a food supply chain perspective is warranted. It helps to improve knowledge about system dynamics: how much and what is produced that is beneficially used, by various socioeconomic groups, and how much goes down the drain. Water is most heavily used in connection with production, whereas significant fractions of energy are used for processing, distribution and among consumers. Greenhouse gas (GHG) emissions dominate the first segments of the supply chain. There are significant potential savings and benefits from a reduction of losses and waste of food in terms of water, energy, money and reduced GHG emissions. Effective policy for environmental health and food security needs to include incentives and demonstrations directed towards consumers apart from policies aiming at producers.

> 'Le problème avec notre époque est que
> le futur n'est plus ce qu'il était'
> Paul Valéry (1871–1945)

Purpose and contents of chapter

It is recognised that "there is no such a thing as a post-agricultural society" (Weiskel 1988; quoted in many publications, eg Postel 1999). Agriculture is a basic component of food systems, but with increasing distance between production and consumers, and with the increasing role of food processing and marketing in a rapidly urbanizing world, it is relevant to analyse the dynamics in the supply chain. The purpose is to highlight the links between production and water, land and energy systems, on the one hand, and the array of consumer needs and preferences, on the other. The inefficiencies in the system can be significantly reduced with multiple wins. With the current grave imbalances in the world in terms of undernourishment, overeating, and losses and waste of food produced, the pertinent question is not how much food is (or could be) produced but how adequate food for sustainable diets can be made available and accessed by the poor. Naturally, due regard must be taken to resource constraints and environmental implications of food system dynamics (Lundqvist 2010; Liu et al. 2013).

An estimated 842 million people are undernourished (FAO et al. 2013). A similar number of people – or about a billion, and often the same people – lack access to safe water, energy and other basic services (Collier 2007). By contrast, a billion and a half overeat, with implications for public health as well as for the environment (Beddington et al. 2012). Equally confronting, an estimated one-third of all food produced is lost or wasted (Gustafsson et al. 2011). If conversion of grains into animal feed, biofuels and overeating (which may be interpreted as a kind of waste) are included in this equation, the combined losses, waste and conversion add up to more than half of all food produced (Lundqvist 2011; Parfitt and Barthel 2010). On-going global warming, water uncertainties and other threats to food production further imbalance the inefficiencies in the food supply chain.

The staggering levels of losses, waste and conversion of food, and the unfair access to the food that is available, carry multiple and high costs. A simple fact is that, in general, all food produced has consumed water and energy, appropriated land and contributed to GHG emissions, regardless of how the produce is used or misused. At the same time, repeated pledges of an eradication of hunger are far from being realized, although the magnitude of hungry people has been reduced. Inefficiencies in the food system stand in sharp contrast to positive figures on significant improvements in increasing produce yields (ie more produce per unit area, 'more crop per drop', 'more value per drop' and higher water productivity).

This chapter first tries to capture the main features of agricultural expansion and its water, land, energy and environmental consequences. While water use in farmers' fields is associated with evaporation and transpiration of huge volumes of water, the relatively largest share of the total energy use occurs beyond the farm-gate. A large proportion of energy is used in connection with processing and distribution, and another large part is used by consumers. The largest share of GHG emissions occurs in connection with production (eg FAO 2011b).

The focus is on agriculture associated with crop and animal food products. Other parts of the food system, such as the rapidly growing aquaculture sector, also require freshwater and energy but are not included in this chapter.

Consumer attitudes and demand for goods and services are affected by higher incomes. Urban growth is an important driver that substantially changes the use of land, water and energy (Lundberg 2006). Although the agriculture sector is the largest user of water and land, the food and other commodities produced are to a large extent required by urban consumers. The human dimensions will be discussed in the latter parts of the chapter.

Dynamic and turbulent times: Past agriculture and future food systems

The latter half of the twentieth century has been described as an era of great accelerations with implications for global environmental change (Steffen et al. 2007). Human activities expanded at a historically unprecedented rate with considerable socioeconomic progress. A remarkable aspect is the high and sustained growth over time, and it seems as if trends in the demand and the production of goods and services will continue in the twenty-first century. An increasing number of countries have an annual GDP growth rate of 5 per cent or even higher. In China, the annual growth rate has been around 8 per cent or even higher for about a generation, or since the economic reforms and market principles were first introduced in 1978. A simple calculation reveals that a couple of per cent difference in annual growth rate yields significant differences in the accumulated growth over time. Two per cent annual growth rate over thirty-five years doubles the initial assets. With an annual growth rate of 5 per cent, the result after thirty-five years is five and a half times the initial value. With 8 per cent, the accumulated value is close to fifteen times bigger.

The 'easy growth era' is associated with a significant intensification in the exploitation and use of resources and environmental implications. Despite the progress and increased capacity of society to deal with various challenges, widespread poverty lingers.

On the positive side, population growth is slowing down in many countries. The fertility rate in Bangladesh, for instance, has been reduced from about 7 to 2.3 children per woman during the past twenty-five years (UNFPA 2010). But the world's population continues to increase, although at a slower rate. In the long-term, size of population may stabilise or even decrease (UN 2004; Lutz and Samir 2010). Unfortunately, most of the population growth occurs in places where resource endowment is poor, and ability to cope and adapt to global warming is constrained by poverty (see eg CGTTWFH 2014).

A continuous rapid expansion of the world's economy is expected. Jackson (2009) projects that aggregate global GDP will be eighty times higher in 2100 than it was in 1950. The validity of such long-term projections can be questioned. Even if they were to be taken seriously, they do not mean that future generations will have eighty times higher purchasing power. To what extent economic growth can be de-coupled from negative resource and environmental consequences is one major query. Similarly, socioeconomic disparities are a lingering threat to a stable and sound development. Indeed, the future is turbulent, as indicated in the quote from Paul Valéry.

Despite weaknesses of GDP as a measure of progress, these figures illustrate that economic dynamics are strong drivers in global change and that they must be seen in conjunction with demographic and social characteristics. On a finite planet, with limited and variable water resources that are becoming more unpredictable, and an uncertain energy situation, the spending of money and the use of economic strength has both positive and negative implications. As discussed below, figures on global trends mask tremendous differences between regions and social groups.

A few degrees warmer or a few latitudes greener?

In the Summary for Policymakers, IPCC (2013) states:

Each of the last three decades has been successively warmer at the Earth's surface than any preceding decade since 1850. In the Northern Hemisphere, 1983–2012 was likely the warmest 30-year period of the last 1400 years.

But the warming is not only a phenomenon at the surface. It is estimated that the uptake and accumulation of energy in the oceans accounts for more than 90 per cent of the total energy accumulated between 1971 and 2010 globally (IPCC 2013). A wide range of organisations, including those outside of the research community, continue to express grave concerns about the implications of global warming (eg World Bank 2012). Droughts, sea-level rise and storms would affect huge areas and a vast number of communities. Under these circumstances, it is hard to imagine how the world could become greener. In its current form, growth is both a cause and a victim. Pondering the links between growth and progress, Jackson (2009) elaborates on the need for an alternative development model that ensures "prosperity without growth". With reference to food security aspirations, we need to ask: How can the production potential best be used to ensure prosperity with a minimum of externalities in an era of increasing resource scarcity and uncertainty?

Here, we first discuss recommendations and projections about what are supposed to be the need and demand for additional food, water and energy. We will then scrutinise the possibilities for a decoupling of improvements in human prosperity in combination with a reduced rate of exploitation and use of scarce and vulnerable resources and the associated negative environmental effects.

Recommendations about future water, food and energy supply

A decoupling is especially important with regard to food system dynamics. Food items are comparatively resource intensive in terms of production, distribution and preparation. Up to fifteen tonnes of water is required to produce a kilogram of red meat. To grow a kilogram of grain typically requires between one and two tonnes of water. Current trends point at a shift in consumer preferences towards more animal products in the diet. About 75 per cent of total water withdrawals are allocated to irrigated agriculture, and most of it is consumptively used (ie it returns to the atmosphere as evaporation and transpiration). Energy wise, the food system consumes about 30 per cent of total, global energy use (FAO 2011b).

Estimates of the likely increase in the demand for food, water and energy during the coming decades have resulted in the following recommendations for production and supply:

- At the World Food Summit in 2009, the recommendation was that 70 per cent more food will have to be produced by 2050 as compared to the situation in 2005 and 2007. This figure has been widely circulated and used. It has recently been revised downward to

60 per cent with the same base year (Alexandratos and Bruinsma 2012). To improve food security and rural livelihoods, substantially higher increases are required in particular regions (eg sub-Saharan Africa).

- Water demand will increase for all sectors OECD (2012). For instance, estimates show that 50 per cent more water will be required by 2050. But depending on development scenarios, estimates vary. High economic growth rates are likely to give a boost to the demand for water, particularly in urban areas, while pessimistic scenarios suggest a more modest increase in the demand (CGTTWFH 2014).
- Global energy demand is expected to increase more than one-third by 2035 (IEA 2012).

Preparations for the World Food Summit in 2009 coincided with the food price crisis in 2007–08. Export restrictions had rapidly been imposed in more than thirty countries, and in other countries, subsidies and price controls were tested (Evans 2009). Food riots broke out in more than twenty countries (FAO 2009). The number of food-insecure people exploded and, according to estimates, undernourishment increased by 15 to 20 per cent or a staggering 150 million people in about a year. With these dramatic consequences, agriculture and food security issues were increasingly emphasised at international meetings such as G8, G20 and World Economic Forum.

Yet, even if, for example, rice stocks were low, the food crisis was basically one of distribution and access to food, not availability of food. It is also true that poor farmers were affected by price hikes on inputs in production (eg oil and fertilisers). In addition, adverse weather conditions hit many producers. Globally, more than enough food is produced and available at the market to feed the entire world population properly. After the introduction of the Green Revolution in the 1960s, food production increased rapidly with a parallel decrease in the number of undernourished. From 1995 to 1996, however, the number of undernourished started to increase although production continued to increase. Hence, increase of per-capita production globally is no guarantee that the number of undernourished will be reduced (Lundqvist 2010).

The argument for substantial increases in production can be related to different drivers in poor and in rich regions, respectively. The majority of undernourished and energy and water insecure people are also poor, and it is legitimate to try to boost production in areas where the poor live. Compared to investments in other sectors, investment in agriculture is regarded as the single most efficient strategy to lift the poorest out of poverty (World Bank 2008).

The drivers in rich countries and among the growing middle class in Brazil, India and China (BRIC) countries differ. Here, the demand for high-quality food items is high or growing. A common denominator in the wealthy parts of the world is an increase in the demand for animal food products. In the case of China, for instance, the shift in dietary preferences has had considerable consequences for the use of water, as shown in Figure 18.1.

The rapid increase of the water footprint in China is an illustration of the water implications of using grain for animal feed. In China as well as in other countries, an increasing share of grains produced is used to feed animals and as a feedstock in bioenergy

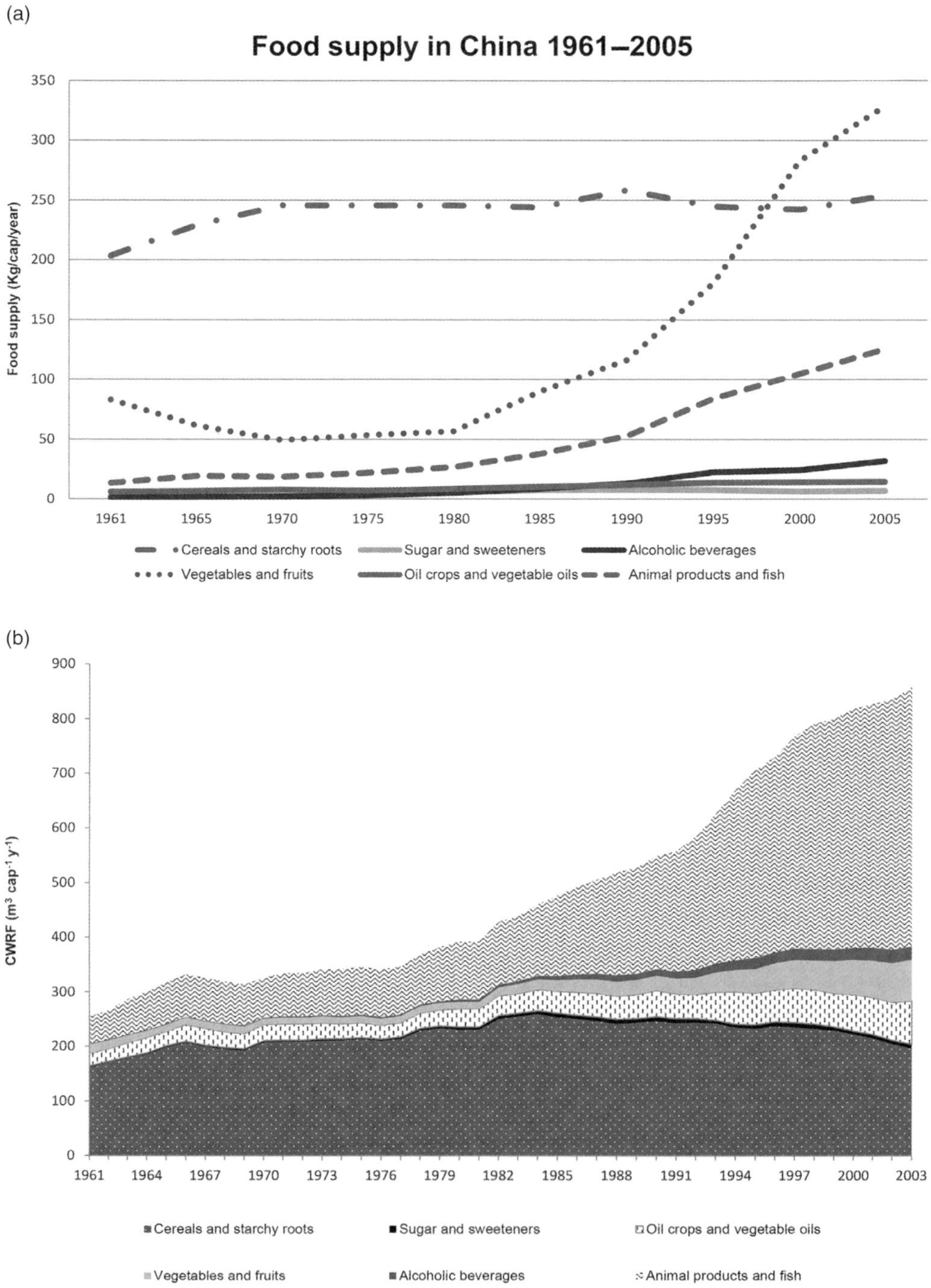

Figure 18.1 Per capita food supply and associated water footprint in China, 1961–2003 (Liu and Savenije 2008; Liu et al. 2008).

production. To meet the target for biofuel for 2020 as specified in the Medium and Long-Term Development Plan for Renewable Energy in China, 6 to 12.4 million hectares of land and 31.9 to 71.7 cubic kilometres of water would be required each year (Yang et al. 2009). Such land water requirements are equivalent to 10 per cent of China's current cultivated area and the annual average discharge of the Yellow River. With first-generation biofuel production techniques, the environmental consequences would be extremely large. As a result, there was a recent halt to biofuel expansion in China. Similar challenges are found in other countries. Almost a third of US grain harvest is, for instance, used to produce fuel for the transport sector (Brown 2012).

Globally, the amount of grains used for animal feed is much larger than the grains used for biofuels. Some 35 per cent of the grain produced is used as feed globally while about 10 per cent of grains are used for production of biofuels. In the OECD countries, the proportion of grains for feed is about 75 per cent of grain production. With a conversion ratio of about 3:1 (vegetal to animal calories), a small increase in the demand for animal food implies a much bigger increase in the demand for grain to be used for feed. Projections about future demand for animal food items show a steady increase while the demand for grain as a direct food source is levelling off or is reduced. For example, the World Bank (2008) predicted an increase in the grain demand of 50 per cent while the increase for meat is 85 per cent between 2000 and 2030.

Finally, the recommendation to increase production is motivated by the risks that are associated with a gradual reduction in the stocks of grain. According to Brown (2012), the world grain carryover stocks have dropped from an average of 107 days a decade ago to 74 days in recent years. With less food in stock in combination with water scarcity and uncertainties, and with reduction of fertile land, the possibility of withstanding shocks caused by, for instance, extreme weather events have been reduced. Reductions in stocks are one major reason for the increasing price volatility in the Middle East, as discussed by Larson (2013). When food production in major food-producing areas is affected, the effects are global. A sudden and severe drought in the Midwestern United States in 2012 was painful for the farmers in the areas hit by the calamities, but it also had repercussions on the price of food globally (Starita et al. 2012). The need for a better balance between production and demand over time and the building up of 'strategic reserves' are seen as important policies for food security and to stabilise prices at various scales, particularly in regions heavily dependent on imports such as the Middle East (Larson 2013).

There is always an uncertainty associated with estimates about future demands. For example, the projections of future water demands made in the 1970s and onwards grossly overestimated the demand in relation to the demand that actually developed (CGTTWFH 2014, p. 187; Gleick 2003). In earlier water projections, the demand from the agricultural sector played a central role. In the coming decades, the rapid expansion of urban population and activities will be relatively more important (CGTTWFH 2014). The social expectations and habits that are associated with 'urban culture' are fuelled by improvements in disposable income and give a boost to the demand for a variety of goods and services, which all require a combination of resources, notably water and energy and fertile land.

It is important to recognise the nexus character of the demand. Changes in the demand for one resource tend to have implications for the demand for other resources. For instance, bringing more water to expanding cities, industries and agriculture also requires energy. Lifting deeper-lying groundwater also requires more energy. Global warming increases the need for more water for cooling.

Three main approaches into the future

To feed a growing and generally wealthier population in an era of mounting constraints, competition for resources and significant uncertainties require a combination of approaches. Three main approaches are discussed below.

On the supply side, there are two main approaches:

1. A geographical expansion of food production at the expense of other land and water use systems;
2. An intensified food production in areas already used for this purpose. This requires a combination of improved management and additional inputs of high-quality seeds, more chemicals such as pesticides, fertilisers and energy.

Efforts related to these two main approaches can be combined with policies focusing on the demand side:

3. Making better use of what is produced (ie efforts that would reduce losses and waste of food produced). This approach could include efforts to promote sustainable diets – that is, production and intake of food items which are nutritionally sound and which have limited environmental impacts.

Constraints related to recommended increases in production

Resource constraints and competition are recognised as fundamental barriers in many regions to produce more food and to ensure water for multi-sector demands (FAO 2013). Water is often seen as the relatively scarcer and more uncertain and variable resource as compared to land. Smakhtin (2008) shows that the number of river basins that are 'closed' has increased to the extent that about 1.4 billion people are now living in basins where the remaining stream flow in the river is so limited that additional withdrawals would seriously threaten aquatic ecosystems. For transboundary rivers, the interests of downstream countries add to the constraints and challenges. The possibilities to expand irrigated agriculture are thus curtailed. Rain-fed systems will probably play a relatively more important role; but with growing variability in rainfall, the risks associated with this kind of agriculture will be more pronounced.

Land is also becoming a scarce resource. A number of reports reveal that the availability of 'unused' fertile land has shrunk. For instance, buffers of land in the United States that could be brought under cultivation are no longer available. The cropland that earlier was idle in United States has been phased out, and today, only land that is susceptible to erosion is left idle (Brown 2012, p. 4–5).

One way to cope with water and land scarcity is to rely on food imports. However, such a strategy is not an option for the poor, and it is also risky. As a result of the sudden food crisis in 2007 and 2008, export restrictions were imposed in more than thirty countries (Evans 2009). Large-scale foreign land investments or leases of land seemed to be a possibility to cope with what was perceived as a new turbulent situation. A number of contracts for land acquisitions were made, especially in sub-Saharan Africa. Referring to media reports, World Bank (2011) quotes a figure of 56 million hectares of land in less than a year. Although there was a rush for land, this figure is obviously an exaggeration (Lorenzo 2011).

One year after the contracts were signed, the actual farming had only started on an estimated 21 per cent of the announced deals (World Bank 2011, p. xiv). José Graziano da Silva, Director-General of the Food and Agriculture Organization, claims that only 10 to 15 per cent of large land deals in Africa have actually resulted in the land being developed (Tran 2012). The relatively low degree of actual cultivation is due to a combination of circumstances: lack of water, high investment requirements and, generally, poor planning due to the rush.

Buffers of land and water used to provide possibilities to expand cultivation, to balance fluctuations in supply and demand and to reduce the risk for price hikes and price volatility. With buffers gone, significantly reduced or with lands that are more expensive to develop, a continuous development of the food system will have global geopolitical implications.

Inputs for intensification of land and water use

According to Foley et al. (2011), 38 per cent of the terrestrial surface is occupied by various types of agriculture and about three-quarters of all the surface and groundwater resources that are withdrawn are allocated to agriculture, most of it for food production. In terms of the world's total energy use, the food sector accounts for about 30 per cent, and its contribution to total GHG emissions is over 20 per cent (FAO 2011b) – another 15 per cent are related to the conversion of forestry to agriculture (FAO 2011b).

A tremendous increase in total production has been achieved over many years, primarily as a result of higher yields. Compiled data for the period 1985 to 2005 shows that total global food production increased by 47 per cent, whereas areal expansion was only 2.4 per cent (Foley et al. 2011). OECD–FAO (2012) present even more striking trends. Their compilation shows that net harvested area remained almost constant from 1970 to 2010 while production about doubled. The State of Land and Water Report (FAO 2011a) shows that agricultural production increased between 250 and 300 per cent during the past fifty years, while the world's cultivated area expanded by only 12 per cent and irrigation area increased by 112 per cent.

A common view is that the seemingly negligible areal expansion of agriculture and the significant increase in total output implies that trade-offs with regard to natural ecosystems have been modest. Norman Borlaug and others pointed out that if levels of productivity had remained at the same level as they were at the onset of the green revolution, and if total production had increased to the extent recorded, the reduction of natural ecosystems would have been disastrous. However, this view does not consider all the environmental and water consequences and externalities of the production system.

As illustrated in Figure 18.2, nitrogen fertilizers increased eight times, phosphorus about two and a half times and areas under some kind of irrigation about doubled. FAO (2011a) notes that the area under irrigation doubled from 1970 to 2010.

Yield improvements and externalities

Reactive nitrogen produced by humans comes from manufacturing nitrogen for synthetic fertilizer and industrial use. Reactive nitrogen is also created as a by-product of fossil fuel combustion and by nitrogen-fixing crops and trees in agro ecosystems. As noted by The Millennium Ecosystem Assessment (2005), human activity now produces approximately as much reactive nitrogen as natural processes do on the continents. The energy required for the synthesis of nitrogenous fertilizers alone consumes approximately 5 per cent of the annual natural gas demand (around 5 exajoules per year). The energy embedded in the production of inorganic fertilizer was around 7 exajoules globally (Smil 2008; Giampietro 2002; GoS 2011). The example may be seen as an illustration of the anthropocene (Crutzen and Stoermer 2000) – that human capacity and actual effort to modify the biogeophysical systems of the earth now exceeds the natural geological forces.

The use of large quantities of fertilizers illustrated in Figure 18.2 have boosted yields but also threatened ecosystems to reach thresholds and eventually lead to tipping points and a reduction in biodiversity (Holling 1996; Weis 2007).

Consequences at national and sub-national scales

Generally, global trends hide and smoothen out changes at national and sub-national scales. Depending on water discharge and the character of the recipient, the in-stream and downstream impacts from an increasing use of nutrients and chemicals in agriculture

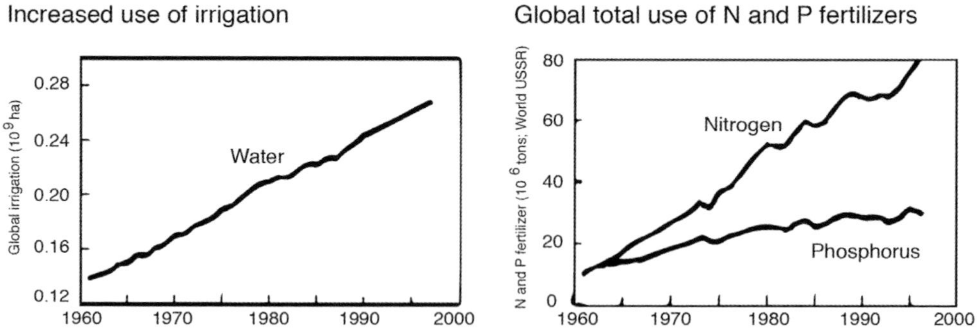

Figure 18.2 Agriculture-driven global use of irrigation, nitrogen and phosphorus fertilizers since 1961 (modified from Tilman et al. 2001).

vary. With small volumes of water discharge and limited capacity of the recipients (eg a stream and lake and with limited through flow) the concentration of polluting substances will be high with a risk for severe consequences. For a country like China, with similar water resources as the United Kingdom but with twenty times more people and a high intensity of economic activities, about 40 per cent of the streams and lakes have gradually been degraded and are now suffering from water quality problems to the extent that people are not even supposed to be in physical contact with these waters (MWRC 2009). In some cases, agriculture is a major cause, but the problem is naturally generated from a combination of feedbacks from human activities and natural processes.

To expand agriculture, water is usually diverted from rivers or lakes and conveyed to the fields. In other cases, water is diverted away from swampy areas to make it possible to cultivate areas that used to be wetlands. This global process can be illustrated by the Zhanjiang plain in China, which used to be a major marsh wetland distribution area. Due to the strong demand for more food, the marsh area sharply decreased by 77 per cent between 1954 and 2005, mainly as a result of agricultural expansion, with cropland area increased by 2.5 times in the same period (Wang et al. 2009). Locally, the increases in agricultural production can thus be associated either with intensification or with an expansion of cultivated area at a cost of losses of natural wetland resources.

Approaches (1) and (2), identified earlier in this chapter, have played a major role in past food security policies. Although the results in terms of levels of yield and total production have been commendable, the performance of the food systems as a whole is much less impressive. In the current debate, levels of losses and food waste are recognised as unacceptable (eg FAO 2013). Naturally, the producers continue to play a major role, but other actors in the food system must be duly considered. The food supply chain involves intricate interactions from farmers to consumers and vice versa. It refers to a combination of the three approaches identified in this chapter.

A supply chain perspective: From production to use and waste of the produce

The recommendations in World Food Summit and other FAO documents to increase food production by 60 per cent (2005/2007–50) would require a huge effort. Moreover, past efforts in the same direction do suggest that major social and environmental challenges may not be resolved in a satisfactory manner through a production-only approach. A pertinent question is this: Is an increase in production, *at that level of magnitude,* the best and smartest way to improve food security? What guarantees do we have that the mismatches between production, supply, access and the associated externalities in the food system – which have been documented – will not recur? Studies suggest that there is enough food to feed the entire world population (see Clendaniel 2011). The key challenge is not a lack of enough food but the uneven and unfair access to the food produced and a low efficiency in the supply chain. Formulated differently: What are the guarantees that the recommended increases in production will not result in increasing overeating and waste with implications for public health and the environment?

Losses, waste and conversions of food

About a third of food produced is lost and wasted (Gustafsson et al. 2011; Parfitt and Barthel 2011; Lundqvist et al. 2008). As illustrated in Figure 18.3, there are considerable differences between regions, and it is important to make a distinction between post-harvest losses and waste of food produced. Losses are relatively high in poor regions, and they are mainly due to a combination of outdated harvesting, storage and transport technologies. High temperature and humidity add to the hazards of storing food. Related to these technical problems, farmers in these systems have limited access to markets outside the local area. In rich countries, on the other hand, waste among consumers is relatively more common (ie throwing away food that is safe to eat).

The situation in China is naturally complex but resembles the dynamics in rich countries. The relatively biggest waste of food is in restaurants, whereas waste in households and in canteens is lower (Liu et al. 2013). With a huge population and serious water scarcity, the losses and waste of food have a high environmental consequence. About 19 per cent of grains produced are lost and wasted in supply chain in China, resulting in 135 billion cubic metres of water, equivalent to the water footprint of Canada, used to produce food not eaten, and 26 million hectares of cropland used in vain, equivalent to the total arable area of Mexico (Liu et al. 2013). A reduction in the food losses and waste will significantly release the demand pressure of natural resources, with national and even global implications.

Differences between countries in terms of the relative size of losses and waste imply that different approaches and mixes of policy are required. To curb losses, investments in physical infrastructure are necessary, while a reduction of waste refers to policies directed towards consumer attitudes and behaviour. The good news is that a reduction of waste may

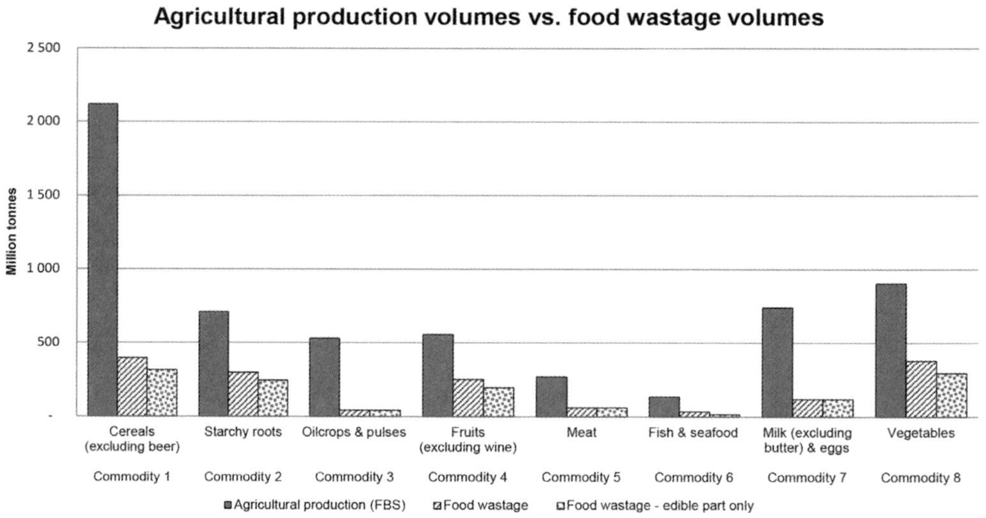

Figure 18.3 Level of food losses and waste in terms of total waste and edible parts, respectively, for selected commodities (FAO 2013).

not require much in investments, and such a change would save money for the consumers, many of whom complain that food is expensive.

There is no such a thing as a free waste of food

Needless to say, a certain level of production is necessary for food security. But if a third or more of the food produced is not reaching the people who need it, or if parts of the produce are wasted, the investments and efforts made in production are also wasted. A remark made over three decades ago is noteworthy:

> *It is distressing to note that so much time is being devoted to the culture of the plant, so much money spent on irrigation, fertilisation and crop protection measures, only to be wasted about a week after harvest.* (FAO *1981*)

In the quote above, 'waste' refers to the losses that farmers experience in the absence of proper storage, transport and market access. After the Green Revolution, it was obviously easier and relatively less costly to expand production than to reduce level of loss, especially since inputs in production were heavily subsidised. Today, the situation is different. It is often noted that the *era of cheap food* is over. But it is more correct to say that the *era of cheap food production* is over. Price on inputs will most likely continue to go up and competition for scarce resources, like water, fertile land and energy will become stiffer. The days when there was plenty of land and promises of free or subsidised and guaranteed water and energy provisions are over.

Water and energy in the food supply chain

Most of the food is produced in the open landscape. As a part of this process, huge volumes of water return to the atmosphere as transpiration and evaporation through crops and other biomass and from surfaces of the earth. As a comparison, the amounts of water used in transport and processing are quite small. Similarly, the amount for water used for cooking and food preparation by consumers is a fraction of water used for production.

As illustrated by Figure 18.4, the use of energy looks quite different. Perhaps the most surprising feature is the relatively small fraction of total energy used in connection with production of crops, livestock and fisheries. As compared to direct and indirect energy inputs of about 95 exajoules per year in the whole food supply chain, globally, the on-farm energy use is in the order of 20 per cent for cropping, livestock and fisheries production. This figure also includes energy used for pumping water, housing livestock, cultivating and harvesting crops, as well as drying and storage (OECD 2008; FAO 2011b). Another 9 exajoules per year are indirectly used to manufacture fertilizers, tractors and other farm machinery (FAO 2011b). To put these figures in context, total global primary energy use in 2008 is estimated at 491 exajoules per year, of which non-renewable energy sources such as fossil fuel contributed 427 exajoules per year (ie over 85 per cent) (IPCC 2011). Global annual primary energy has now passed 500 exajoules and end-use energy is around 300 exajoules (Ralph Sims, personal communication).

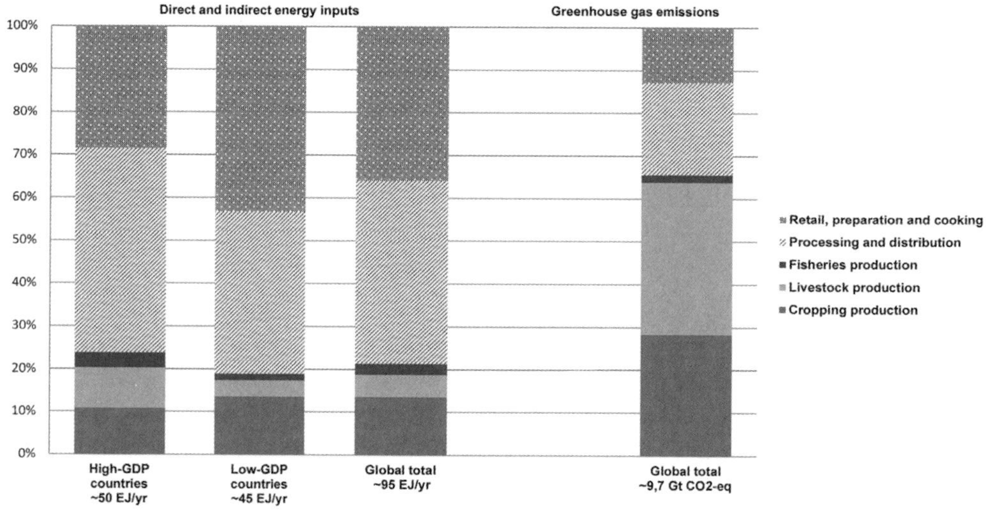

Figure 18.4 Indicative shares of final energy consumption for high- and low-GDP countries, the global total and total associated global GHG emissions for the food sector (FAO, 2011b).

Processing, distribution, retail, preparation and cooking consume about 80 per cent of the total energy used in the food chain (Figure 18.4). The energy used globally in the latter parts of the food supply chain is thus in the order of 15 per cent of total global primary energy use as compared to about 2 or 3 per cent for food production. In other words, farmers are using much less energy to feed the world as compared to the energy used in food processing, in distribution and by consumers when they prepare and refrigerate their food.

Aggregated energy use and the distribution between the various segments in the food supply chain are at about the same level for high and low GDP countries, respectively. Another interesting comparison shows that the direct and indirect energy used for the lifting of groundwater and for the manufacturing of the technical components for this operation (eg manufacturing of pump sets) is estimated to be around 0.3 exajoules per year, whereas the energy used to manufacture inorganic fertilizers is in the order of 7 exajoules per year globally (FAO 2011b). Hoff (2011) notes that some countries allocate 40 per cent of total energy use to pump groundwater, and that energy use increases exponentially with greater depth or by a factor of eighty when depth increases from 35 to 120 meters.

Climate change may increase the water demand for energy production. According to Dominguez-Faus et al. (2013), more water will be required to produce each unit of corn-based biofuel, largely due to the higher evapotranspiration in the major corn production area caused by global warming.

Water and GHG in the beginning and energy at the end of food systems

Figure 18.4 illustrates that GHG emissions are relatively higher from production, especially from livestock, whereas the last two stages in the supply chain together account for about a

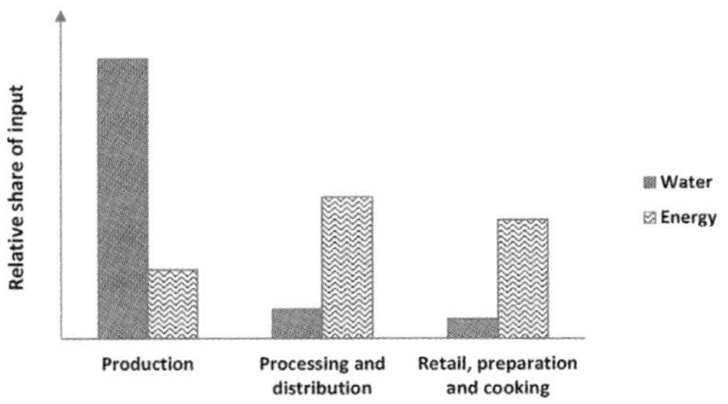

Figure 18.5 Indicative uses of energy and water in main segments of food supply chain.

third of total emissions in carbon dioxide equivalents. Apart from ruminant cattle, significant amounts of methane production are associated with paddy rice cultivation.

The general picture is that most of the total water requirements from 'field to fork' are associated with consumptive water use in the field. The amounts of water used in transport and storage are negligible in this comparison. In food processing, most of the water used can be recirculated and reused. Finally, the water required for cooking and, generally, food preparation among consumers in households and in restaurants is a mixture of consumptive and non-consumptive use of comparatively limited volumes. As discussed above, the energy use is quite different. The relatively limited use of energy in connection with production implies that policies that aim for an improved efficiency in the use of energy in connection with production imply comparatively small overall gains in total energy use in the food system. If, on the other hand, water is used more efficiently in production, the gains are comparatively greater. An illustration of the discussion is shown in Figure 18.5. An important implication from Figure 18.5 is that policies that aim for better resource use must take the whole food supply chain into account. Relatively speaking, the overall gains in energy use are dependent on improvements in the latter parts of the supply chain. However, food waste means that the accumulated water and energy inputs in the entire chain are wasted.

Reducing waste for multiple savings and benefits

In the new resources and environmental context for food security, it makes sense to point at the opportunity to feed more people *without* a *corresponding* increase in production and supply of food. The conventional slogan 'more crop per drop', which focusses on production, needs to be supplemented with another slogan that highlights the very purpose of food in society: 'feeding more from what is produced'. Estimates presented in table Table 18.1 illuminate that a reduction of losses and waste by 50 per cent means that 650 million tons of food could be used, for instance, to feed more people (ie without any additional production). At the same time, such a reduction would save water, energy and expenditure on food, and ease the pressure on land, which is also in many countries (eg China; Liu et al. 2013).

Table 18.1 Potential savings and benefits through a reduction of losses and waste of food produced by 50 per cent.

Food[1]	650 million tonnes
Irrigation water[2]	450 km3
Energy[3]	16 EJ
Monetary[4]	US$375 billion

Comments to the estimates:

[1]The total amount of food lost and wasted is estimated to be 1.3 billion tonnes (FAO 2011); a reduction by half means 650 million tons of food saved.

[2]About 2,700 km^3 of irrigation water is allocated to agriculture. The water that is used to produce the food that is lost and wasted is supposed to be one-third of this, ie 900 km^3. Halving losses and waste implies possible savings of 450 km^3 of irrigation water.

[3]The amount of energy that is used to produce, distribute, process and prepare food is in the order of 95–100 EJ (IPCC 2011; FAO 2011); with one third lost and wasted, about 32 EJ are used 'in vain'. Reducing losses and waste by 50% makes it possible to save 16 EJ. Total global use of primary energy in all sectors is about 500 EJ.

[4]The economic value of the 1.3 billion tonnes of food lost and wasted is estimated to US$750 billion (FAO 2013); half of that is US$375 billion (FAO 2013).

The figures in Table 18.1 indicate that substantial and multiple benefits are possible; farmers will be able to use a larger share of the produce for their own requirements or to sell parts of the produce on the market; for food processors and market operators, reduced losses imply that a larger share of the produce will be available for processing; and for consumers, reduced waste means less expenditure on food.

Generally, the price of the product is higher towards the end of the food chain, which reflects an accumulation of efforts and investments and, often, higher attraction. If food is lost before farmers are able to use it for their own needs, or before they are able to sell whatever amounts they like to sell, they suffer, and resources are used in vain. Waste of water is significant while the loss in energy terms would be modest. By contrast, the waste of food that is prepared and ready to eat incurs highly accumulated energy and environmental costs, including GHG emissions apart from the misuse of water.

Economically, losses and waste affect different groups in various ways. When farmers lose the crop and are unable to store or sell whatever amounts they do not need for their own direct use, they are deprived of an income and the value of the food itself. When food that is prepared and ready to eat is wasted, a range of actors have been involved in various activities in the food chain: farmers, transporters, food processing companies and retailers. Supposedly, they have all been paid for their efforts and the GDP has increased. But ironically, the benefits embedded in the product in terms of nutritional value are foregone. When food is wasted, the costs in terms of resource use and environmental implications do, however, remain. With improper disposal of food in, for example, landfills, methane gas generation is an added problem.

Policies for the supply *and* the demand sides

A main message of this chapter is that food and nutritional security requires a governance system that includes a mix of policies which address the dynamics of a food supply chain. A certain amount of food must, of course, be produced, which presumes necessary and adequate inputs of water, energy and land. For resource and environmental stewardship and to feed a burgeoning and wealthier population, the performance of the food system as a whole must improve. It is not enough to push for improved efficiency in production. New technologies and better use of existing ones (eg through better management) can help to reduce resource inefficiencies and pollution. If the current level of inefficiency prevails in downstream segments of the supply chain, progress for society and stewardship of natural resources are reduced, and negative externalities may outweigh the gains.

Generally, there's an increasing awareness about the high cost of waste and inefficiencies in the food supply chain. A number of initiatives are taken to improve the functioning of the food system. For instance, the European Parliament passed a resolution in January 2012 to halve food waste by 2015 (EU 2012).

A similar concern is also seen outside of Europe. Recently, China's new Communist Party Chief, Mr Xi Jinping, declared that "these habits of waste must be stopped immediately" after having read an article about the food waste in restaurants and especially in connection with official banquets (Hui 2013). The concern was obviously about corruption. Although this comment seems to have been incidental rather than 'planned', it has generated a substantial debate in numerous blogs in China and abroad. Colleagues in China talk about a growing worry among a representative of restaurants arranging banquets that their business is under pressure.

It is interesting to note that many actors in the private sector (eg the food industry) develop strategies to make better use of scarce natural resources, including a reduction of waste. Naturally, it is important for these industries to ensure farmers are able to produce and process enough commodities to meet consumer demand. With increasing scarcity of water and other natural resources, improved resource-use efficiency in production and processing becomes a strategic necessity for the long-term survival of a firm. Of course, businesses have a commercial interest that drives their activities. But an increasing number of them also know that businesses cannot be indifferent to resource and environmental realities. It is not smart to ignore public concerns.

The crucial role of consumers

By definition, food waste is primarily related to consumer behaviour. In industrialised countries and economies in transition, consumers are the main drivers in the demand for more food. Changing preferences in diets towards more water-intensive food items are significant, as illustrated in Figure 18.1. Consumers are part of the problem discussed above, and they must be part of the solution. Political will is important, but waste is an area where human preferences and behaviour are very strong drivers. It requires political skill to get social acceptance and compliance of sound policies, but it is crucial (Lundqvist and

Falkenmark 2010). Social and cultural habits need to be part of strategies towards a food security that tally with natural resources realities and environmental threats. Access to education, equal opportunities for men and women, and incentives and awareness about the links between food, water, energy and, generally, the environment, need to be considered in food security programmes. With more money in our pockets and in concert with players on the market, consumers can make or break the good intentions of policy makers.

Cherishing food and recognising that food in shops and on the plate require hard work, plenty of water and other scarce natural resources to produce are virtues that have been integrated in cultures and habits around the world. A revival of the timeless wisdom to respect the context for food production and its multiple values is sound and good for household budgets, health, and water and the environment.

Acknowledgements

This study was supported by the International Science and Technology Cooperation Program from the Ministry of Science and Technology of China (2012DFA91530) and the Fundamental Research Funds for the Central Universities (TD-JC-2013–2). A number of colleagues have provided valuable comments. Sincere thanks to an anonymous reviewer, Torkil Jønch Clausen, Eric Uscher, Ralph Sims and Göran Berndes. All Figures except 18.2 are drawn by Britt-Louise Andersson, SIWI.

References

Alexandratos, N. and Bruinsma, J. (2012). *World Agriculture Towards 2030/2050: The 2012 Revision*. ESA Working paper No. 12–03. Rome: FAO.

Beddington, J., Asaduzzaman, M., Clark, M., Fernandez, A., Guillou, M., Jahn, M., Erda, L., Mamo, T., Nguyen, V.B., Nobre, C.A., Scholes, R., Sharma, R. and Wakhungu, J. (2012). *Achieving Food Security in the Face of Climate Change: Final Report from the Commission on Sustainable Agriculture and Climate Change. CGIAR Research Program on Climate Change, Agriculture and Food Security (CCAFS)*. Copenhagen.

Brown, L. (2012). *Full Planet, Empty Plates. The New Geopolitics of Food Scarcity*. New York and London: Earth Policy Institute. W.W. Norton & Company Ltd.

Clendaniel, M. (2011). Infographic: People Are Starving But There Is Enough Food To Go Around. *Scientific American*, 19 October, www.scientificamerican.com/article.cfm?id=infographic-people-are-starving-but-2011–10/.

Collier, P. (2007). *The Bottom Billion: Why the Poorest Countries are Failing and What Can Be Done About It*. New York: Oxford University Press.

Calouste Gulbenkian Think Tank on Water and the Future of Humanity (CGTTWFH) (2104). *Water and the Future of Humanity*. New York: Springer.

Dominguez-Faus, R., Folberth, C., Liu J., Jaffe, A. M. and Alvarez J. J. (2013). Climate Change would Increase the Water Intensity of Irrigated Corn Ethanol. *Environmental Science & Technology* 47: 6030–7.

Evans, A. (2009). *The Feeding of the Nine Billion: Global Food Security for the 21st Century*. Chatham House Report. January. www.chathamhouse.org.uk/files/13179_r0109food.pdf.

EU Parliament (2012). *Parliament Calls for Urgent Measures to Halve Food Wastage in the EU*. Press release. 19 January, www.europarl.europa.eu/news/en/news-room/content/20120118IPR35648/html/Parliament-calls-for-urgent-measures-to-halve-food-wastage-in-the-EU/.

Food and Agriculture Organization of the United Nations (FAO) (1981). Food Loss Prevention in Perishable Crops. Agricultural Services Bulletin *43*. Rome, Italy: FAO, www.fao.org/docrep/011/ai481e04.htm.

Food and Agriculture Organization of the United Nations (FAO) (2009). The State of Food Insecurity in the World 2009. Economic crises – Impacts and Lessons Learned. www.fao.org/docrep/012/i0876e/i0876e00.htm.

Food and Agriculture Organization of the United Nations (FAO) (2011a). *The State of the World's Land and Water Resources for Food and Agriculture (SOLAW) – Managing Systems at Risk*. Food and Agriculture Organization of the United Nations, Rome and Earthscan, London.

Food and Agriculture Organization of the United Nations (FAO) (2011b). *Energy-Smart Food for People and Climate* – Issue Paper. 78 pp. Rome.

FAO, IFAD and WFP (2013). The State of Food Insecurity in the World 2013. *The multiple dimensions of food security*. Rome, FAO.

Food and Agriculture Organization of the United Nations (FAO) (2013). *Food Wastage Footprint. Impacts on Natural Resources. Summary Report*. www.fao.org/docrep/018/i3347e/i3347e.pdf.

Foley, J., Ramankutty, N., Brauman, K. A., Cassidy, E. S., Gerber, J. S., Johnston, M., Mueller, N. D., O'Connell, C., Ray, D. K., West, C. B., Bennett, E. M., Carpenter, S. R., Hill, J., Monfreda, C., Polasky, S., Rockstrom, J., Sheehan, J., Siebert, S., Tilman, D. and Zaks, P. M. (2011). Solutions for a Cultivated Planet. *Nature* 478: 337–42. doi:10.1038/nature10452.

Giampietro, M. (2002). *Energy use in Agriculture, Encyclopedia of Life Sciences*, MacMillan Publishers, Nature Publishing Group, 15 pp. www.els.net/.

Gleick, P. H. (2003). Global Freshwater Resources: Soft-Path Solutions for the 21st Century. *Science* 302: 28 November, pp. 1524–8.

Gustafsson, J., Cederberg, C., Sonesson, U., van Otterdijk, R. and Meybeck, A. (2011). *Global Food Losses and Waste – Extent, Causes and Prevention*. FAO, Rome.

Hui, N. (2013). Xi Jinping's Sudden Concern for Wasting Food. *The Atlantic*. February 27. www.theatlantic.com/international/archive/2013/02/xi-jinpings-sudden-concern.

Intergovernmental Panel on Climate Change (IPCC) (2013). *Climate Change 2013. The Physical Science Basis. Headline statements from the summary for policy makers*, www.ipcc.ch/news_and_events/docs/ar5/ar5_wg1_headlines.pdf.

International Energy Agency (IEA) (2012). *World Energy Outlook*. Paris.

Jackson, T. (2009). *Prosperity Without growth. Economics for a Final Planet*. Earthscan.

Larson, D. (2013). Introducing Water to an Analysis of Alternative Food Security Policies in the Middle East and North Africa. *At the Confluence, Aquatic Procedia* 1: 30–43. doi:10.1016/j.aqpro.2013.07.004

Liu, J., and Savenije, H. H. G. (2008). Food Consumption Patterns and Their Effect on Water Requirement in China. *Hydrology and Earth System Sciences* 12: 887.

Liu, J., Yang, H. and Savenije, H. H. G. (2008). China's Move to Higher-Meat Diet Hits Water Security. *Nature* 454: 397.

Liu J., Lundqvist J., Weinberg J. and Gustafsson J. (2013). Food Losses and Waste in China and Their Implication for Water and Land. *Environmental Science & Technology* 47(18): 10137–44. doi: 10.1021/es401426b.

Lorenzo, C. (2011). *Land Deals in Africa: What is in the Contracts?* London: IIED.
Lundberg, J. (2006). *Rethinking Urban Nature: Maintaining Capacity for Ecosystem Service Generation in a Human Dominated World.* PhD thesis Dept. of Systems Ecology, Stockholm University, Stockholm. www.diva-portal.org/su/theses/abstract.xsql.
Lundqvist, J. (2010). Producing More or Wasting Less. Bracing the Food Security Challenge of Inpredictable Rainfall. In *Re thinking Water and Food Security:* Fourth Marcelino Botín Foundation Water Workshop, eds. Martínez-Cortina, L., Garrido, G. & López-Gunn, L. London: Taylor & Francis Group.
Lundqvist, J. and Falkenmark, M. (2010). Adaptation to Rainfall Variability and Unpredictability. New Dimensions of Old Challenges and Opportunities. *International Journal of Water Resources Development* 26(4): 597–614.
Lutz, W. and Samir, K. C. (2010). Dimensions of Global Population Projections: What do We Know about Future Trends and Structures. *Phil. Trans. R Soc. B* 365: 2779–91. doi:10.1098/rstb.2010.0133.
Millennium Ecosystem Assessment (2005). *Ecosystems and Human Well-Being: General Synthesis*. Washington, DC: Island Press.
Ministry of Water Resources of the People`s Republic of China (MWRC) (2009). *China Water Resources Bulletin 2009*. Beijing, China.
Organisation for Economic Co-operation and Development (OECD) and FAO (2012). *OECD FAO Agricultural Outlook 2012–2021*. OECD Publishing and FAO. www.oecd.org/site/oecd-faoagriculturaloutlook/#publication2012.
Parfitt, J. and Barthel, M. (2010). *Global Food Waste Reduction: Priorities for a World in Transition. Science Review SR56*. London: Foresight, Government Office for Science.
Postel, S. (1999). *Pillar of Sand. Can The Irrigation Miracle Last?* A Worldwatch Book. New York: W.W. Norton.
Rockström, J., Steffen, W., Noone, K. (2009). A Safe Operating Space for Humanity. *Nature* 461: 472–5.
Smakhtin, V. (2008). Basin Closure and Environmental Flow Requirements. *International Journal of Water Resources Development*. 24(2): 227–33.
Smil, V. (2008). *Energy in Nature and Society-General Energetic of Complex Systems*. Cambridge, MA: MIT Press.
Starita, J., Dahab, M. and Lenton, R. (2012). Drought Hits the Heartland of the USA. *Stockholm Waterfront* No. 4, pp. 4 – 8. December.
Steffen, W., Crutzen, P. J. and McNeill, J. R. (2007). The Anthropocene: Are Humans Now Overwhelming the Great Forces of Nature? *Ambio* 36: 614–21.
Tilman, D., Fargione, J., Wolff, B., D'Antonio, C., Dobson, A., Howarth, R., Schindler, D., Schlesinger, W.H., Simberloff, D., Swackhamer, D. (2001) Forecasting Agriculturally Driven Global Environmental Change. *Science* 292, 281–4.
Tran, M. (2012). Land Deals in Africa Have Led to a Wild West – Bring on the Sheriff, Says FAO. Interview with Graziano. *The Guardian*. October 29. www.guardian.co.uk/global-development/2012/oct/29/land-deals-africa-wild-west/.
UK Government office for Science (GoS) (2011). *Foresight Project on Global Food and Farming Futures*, Synthesis Report C12: Meeting the Challenges of a Low-Emissions World, London. www.bis.gov.uk/assets/bispartners/foresight/docs/food-andfarming/synthesis/11-632-c12-meeting-challenges-of-low-emissions-world.pdf.
United Nations (UN) (2004). World Population to 2300. New York: UN Dept of Economic and Social Affairs. www.un.org/esa/population/publications/longrange2/World-Pop2300final.pdf.
Unver, O., ed. 1997. Water Resources Development in a Holistic Socioeconomic Context – the Turkish Experience. Special Issue: *Water Resources Development* 13(4).

United Nations Population Fund (UNFPA) (2010). *World Population Prospects, the 2010 Revision*. New York: UNFPA.

Wang, et al. (2009). Process of Land Conversion from Marsh into Cropland in the Sanjiang Plain during 1954–2005. *Wetland Science* 7(3).

Weis, A. J. (2007). *The Global Food Economy: The Battle for the Future of Farming*. Zed Books.

Weiskal, T. (1988). *The Anthropology of Environmental Decline*. Summary Statement for hearings of the Committee on Environment and Public Works. US Senate, Washington, DC, 14 September.

World Bank (2008). *World Development Report: Agriculture for Development*. http://web.worldbank.org/WBSITE/EXTERNAL/EXTDEC/EXTRESEARCH/EXTWDRS/0,,contentMDK:23062293~pagePK:478093~piPK:477627~theSitePK:477624,00.html.

World Bank (2011). *Rising Global Interest in Farmland, Can it Yield Sustainable and Equitable Benefits?* Washington, DC: The World Bank.

World Bank (2012). *Turn Down the Heat. Why a 4°C Warmer World Must be Avoided* http://climatechange.worldbank.org/sites/default/files/Turn_Down_the_heat_Why_a_4_degree_centrigrade_warmer_world_must_be_avoided.pdf.

Yang H., Zhou Y., Liu J. (2009). Land and Water Requirements of Biofuel and Implications for Food Supply and the Environment in China. *Energy Policy* 37: 1876–85.

19

Future prospects in climate, energy and water research and policy

KAREN HUSSEY, JAMIE PITTOCK AND STEPHEN DOVERS

Introduction

Combined, the chapters in this volume present a comprehensive and up-to-date analysis of the interdependencies between the climate, energy and water sectors, and the knock-on consequences of those interdependencies for other issues such as food production and biodiversity conservation. In many respects, the book makes for sobering reading, and it is worth recalling some of the most alarming facts presented by the authors, which stress the enormity and urgency of the challenges ahead.

To recall, the problem

In 2012, the International Energy Agency (IEA) predicted 35 per cent growth in annual global energy demand from 2010 to 2035, and under a forecast scenario that accounts for planned energy-related policies and announced commitments, it estimated global oil, coal and natural gas consumption to expand from 2010 to 2035 by 13 per cent, 21 per cent and 50 per cent, respectively (IEA 2012, cited in Chapter 4). Significantly, even with an anticipated increase in market share for renewable energy (eg seven-fold and twenty-five-fold projected increases in wind and solar photovoltaic generation, respectively), fossil fuels are still expected to satisfy the vast majority of future energy demand (IEA 2012, cited in Chapter 4). If the IEA projections hold true, resulting carbon dioxide emissions would rise by over 20 per cent, corresponding to a long-term average global temperature increase of 3.6 degree Celsius (IEA 2012), ushering in a suite of climate impacts.

Those impacts in turn pose major, near-term risks for energy systems, largely from episodic disruptions due to extreme weather events, especially in particularly vulnerable regions. For both energy supply and energy use, extreme weather events such as hurricanes, tornados, floods, droughts, heat waves and wildfires pose significant threats to sustainable energy services and thus to societies and economies (IPCC 2014). Both electricity demand and electricity supply will be affected by increases in both average and extreme temperatures, as heating requirements reduce and cooling requirements increase. Seasonal and chronic water supply constraints pose threats to reliable energy supplies in many regions, with many arid areas projected to become more arid, areas dependent on snowmelt for spring surface water supplies projected to see reduced flows, and many other areas likely to be subject to seasonal droughts that are more frequent, longer and more intense, in some cases coupled with seasonal heat waves that would tend to increase water demands. The knock-on consequences for energy security include chronic or seasonal reductions in hydropower potential, reduced water flows and higher ambient water temperature restricting allowable power plant withdrawals, and reduced effectiveness of thermal power plant cooling technologies. With future climate impacts likely to exacerbate current water management practices, the implications for the energy sector are, at the higher end of projections, very serious.

Geographic patterns of renewable energy supply will also be affected by changes in climate. As temperature and precipitation patterns shift, environmental conditions for renewable energy supply will also shift. In nearly every case, as environmental conditions shift, some areas will see reductions in renewable energy supply potential while others will see improvements in potential; shifting patterns of resource development will also interact with other driving forces to shape land-use decisions in localities and regions, which will add strategic challenges to increasing the renewable energy share of total energy supply. Such knock-on effects stress the need to consider energy-water-climate interactions in the context of sustainable development more broadly (we return to this theme later in this chapter). Significantly, low-carbon energy is not necessarily low-water or low-environmental impact energy, considering, for example, new hydropower reservoirs, freshwater-cooled utility-scale concentrating solar thermal and some geothermal electric generation options (Pittock et al. 2013). Further, specific carbon mitigation policies and actions may increase or decrease water consumption (Wallis et al. 2014).

Even leaving aside projected future impacts of a changing climate, the current state of the world's freshwater resources is dire, with recent estimates stating that 80 per cent of the global population now lives in catchments with threatened surface water supplies and river health, and with more than two billion people regularly facing severe water scarcity (Hoekstra et al. 2012; Vörösmarty et al. 2010, cited in Chapter 2). The situation is much the same for groundwater, which is mined at unsustainably high rates almost everywhere where livelihoods critically depend on it (eg Wada et al. 2010). The extent of unsustainable water management practices at local and regional scales now has global significance, a situation which is only going to worsen as population climbs to its anticipated 9 billion by 2030.

Larger and wealthier populations are significantly increasing demand for water in agriculture and energy generation. The business-as-usual projection is for a 70 per cent to 90 per cent increase in the amount of water consumed in agriculture by 2050 (CAoW-MiA 2007). Global demand for water withdrawals is projected to "increase by some 55% by 2050, mainly because of growing demands from manufacturing (400%), thermal electricity generation (140%) and domestic use (130%)" (WWAP 2014, p. 18). Supplying this water typically uses around 8 per cent of the energy in developed countries, and as much as a fifth in places like California (Kenway et al. 2011; WWAP 2014), contributing to greenhouse gas emissions where fossil fuels are used.

Securing water supplies with greater variability or scarcity with climate change involves two main adaptation strategies: storing more or diversifying the sources of water. Options for storing more water are often limited as readily available dam sites have already been used (with some exceptions in South America and Africa). Greater conjunctive use of water stored in aquifers offers greater security, though at the expense of the energy used in pumping. Diversification usually involves accessing water sources that require considerable energy for transmission, such as with interbasin transfers or desalination, or energy intensive treatment, as exemplified by desalination and stormwater harvesting (Pittock et al. 2013). In this context, many water adaptation strategies have perverse impacts, including increasing greenhouse gas emissions and increasing water fees for the poor, and can be considered maladaptive (Barnett and O'Neill 2010). On the other hand, there are positive synergies when water demand management reduces energy consumption (Pittock et al, 2013). Greater competition for water will further impact on freshwater biodiversity and raises questions as to the extent to which biodiversity can be conserved, and trade-offs avoided in climate adaptation options (Chapter 17; Lukasiewicz et al. 2013). The projected shifts in the socio-economic status of many people, especially those in Asia, and the impacts of climate change will put yet more strain on the earth's finite sources of land, water and energy supplies, and on the ecosystems that provide all essential functions to sustain human life.

Ironically, while there is increasing awareness of the need to manage and conserve water resources in the coming decades – and, furthermore, to develop and manage energy extraction and production processes to minimise environmental impacts – major international agencies such as the Food and Agriculture Organisation (FAO) are calling for an increase in food production to deal with declining grain stocks, the diversion of significant agricultural outputs for biofuel production, and increasing demand from a growing, changing population (FAO 2009). However, capacity to meet increased demand for agricultural outputs will be severely constrained, with evidence already indicating, with a medium level of confidence, that climate change has negatively affected wheat and maize yields for many regions and in the global aggregate (IPCC 2014). Moreover, several periods of rapid food and cereal price increases following climate extremes in key producing regions indicate a sensitivity of current markets to climate extremes, as the food riots experienced in over twenty countries in 2009 illustrated sharply. The implications of increased food production for land, water and energy resources under increasingly difficult climatic conditions are, therefore, profoundly concerning and capture perfectly the 'wickedness' of the climate-energy-water nexus. As the authors of this book have described and substantiated so fully, we find ourselves at a pivotal moment in time.

The rapid proliferation of scholarly work in the climate, energy and water domains and the significant profile afforded to that work is testament to both the enormity of the challenges ahead and the acceptance by many of the need to think differently about sustainable development. But earlier research on energy, water and climate interactions has tended to focus on one of three things: (1) a particular, singular aspect of the energy-water relationship: for example, the electricity sector's demand for water, or the energy used to extract, treat and distribute urban water supplies; (2) the implications of climate change either for the availability of water resources for consumptive uses, or for energy security; or (3) the applicability of various conceptual or modelling approaches to better account for inter-sectoral interdependencies in policy and investment decisions. Much of that excellent work has been referred to in this volume, but this book has gone further, putting many of the pieces of the puzzle together to offer a range of solutions that both the public and private sectors can readily pursue to manage the trade-offs between climate, energy and water, and to seize the opportunities that they present. In this way, it is a book about solutions, not problems.

A suite of solutions

The identification of solutions must begin with a clear articulation of the underlying problems. The chapters in this volume have articulated many problems through the lens of the climate, energy and water nexus that go far beyond a simplistic analysis of the interdependencies between and trade-offs within the three sectors. Indeed, distilling the analysis in the preceding chapters reveals four fundamental problems that must be solved, and we explore each here, including strategies that can be employed to tackle them.

Governance

The first, overarching problem is disjointed, inefficient and often opaque governance structures and processes that fail to account for the relationships between sectors in decision making. The result is confused government objectives that drive inefficient or perverse outcomes in the design and/or implementation of market and regulatory arrangements in the climate, energy and water sectors, or which overlook synergistic opportunities that could be exploited. The lack of co-ordination in public policy and administration is by no means unique to the climate, energy and water sectors, and there has been significant scholarship in the field of policy integration in recent years (Ross and Dovers 2008; US National Research Council 2013). Nevertheless, in this volume, several authors critically assess the reasons for a lack of cross-sectoral co-ordination in the climate, energy and water nexus and point to the institutional 'logic' that is unique to each sector; comprising history, management culture, policy settings, statutory requirements, markets and operational practices (Chapter 11). Add to this the increasing complexity and disaggregation of governance processes, the increasing emphasis on markets and market-based instruments to achieve public goods, the divestment of key water and energy assets to private entities, and the blurring of operational imperatives in utilities to include both economic and environmental objectives, and the result is an institutional landscape that is a veritable maze of actors and competing objectives. Perhaps surprisingly, the solution to this disjointed, inefficient and often opaque governance landscape is more targeted and authoritative steering from the state (Chapter 11). As we have seen in the case studies from California and Texas in the United States, and in the State of Victoria in Australia, where better co-ordination across the sectors has been achieved it has been because of purposeful intervention from state agencies or actors.

For example, the state government in California has demonstrated that it is possible to successfully reform governance to better manage links and trade-offs inherit in the nexus. As a co-ordinating governance instrument, the Californian law AB 32 made significant gains and established:

- An executive-level, inter-agency committee in the form of the 'Climate Action Team' (CAT) tasked with responsibility for integrating across the climate, energy and water sectors, reporting directly to the most senior political appointee in the jurisdiction;
- Sector-specific working groups reporting to the interagency team;
- A cross-sector working group (the Water–Energy Team [WET] of the CAT [the 'WET-CAT']), which became the primary vehicle for the investigations into the climate-energy-water nexus, particularly water-related energy consumption; and
- The preparation of a Scoping Plan enunciating how to achieve reductions in greenhouse gas (GHG) emissions.

This co-ordination was achieved through direction by key actors in the Californian State Government and its agencies. Intentional and purposeful actions were required to co-ordinate across more than thirty government agencies, energy and water utilities and quasi-government organisations (Chapter 11). Using the authority of the office of the Governor and the Californian state legislature, specific co-ordination instruments were deployed within the energy and water sectors and State Government authority was used to establish statutory targets, implement mandatory cross-agency co-ordination processes and adapt regulatory market rules. It is worth noting that the Californian case represents the 'best case scenario' for undertaking cross-sectoral integration: water and energy issues are acute and the state enjoys significant informational and institutional capacity. The fact that California is the only state in the United States to embark on such reforms suggests the lessons are important but are also likely to be difficult to transfer except under similar conditions.

Notwithstanding the significance of the Californian experience, the process of developing increased levels of cross-sectoral co-ordination between climate-energy and water was highly incremental. Specifically, the experience in California represents a case of how water and energy management embodied new objectives and adapted existing governance mechanisms to new demands and objectives (water and energy security, and GHG emission reductions). So, while these initiatives represented new, strong, steering institutions at a higher level, cross-sectoral co-ordination has primarily been achieved through the layering of amendments, revisions or additions to existing energy and water institutional processes. It is to these existing mechanisms that we now turn our attention.

Pricing of resources

The second significant problem underpinning the 'energy-water' nexus is that administered prices for energy and water do not accurately reflect their true scarcity, their value to society and/or the environmental and economic costs of their production and consumption (Chapters 4 and 13). A further problem exists where water, energy and carbon markets are designed independently and not harmonised, leaving gaps that encourage inefficient resource exploitation. Examples in Australia include the failure of many state governments to include within the cap-and-trade water market some major users, such as miners, energy generators and carbon sequestration plantations (NWC 2011; Pittock et al. 2013). The consequence is reduced incentive for users to conserve resources and use them efficiently, or for producers and suppliers to innovate or expand supply to meet industrial and consumer demands. There are practical solutions, including the following:

- Establishing mechanisms to allocate sustainably water that is not otherwise allocated;
- Facilitating the transfer of water from non-energy sectors (eg through water markets or the temporary purchase of water, noting that the underlying regulatory arrangements for water purchasing and long-term contracts may need to be reformed);
- Incorporating environmental externalities into pricing wherever possible, for instance, by ensuring that water, energy and carbon markets are harmonised;
- Setting up a robust system that continuously generates price signals, in real time, to producers and consumers about the true value to society of the energy and water that they produce or use;
- Monitoring the outcomes of new pricing structures to ensure that there are no unintended or inequitable outcomes in society; and
- Establishing a regulatory and supervisory capacity that intervenes only when there is a clear justification for doing so and conviction that the intervention will actually improve on the outcomes delivered by the market (because regulatory failure is even more common than market failures).

The socioeconomic benefits of these measures means that economic reform can be used as a policy window to enhance environmental and natural resources management, as demonstrated by Australia's water reforms from the 1990s (Grafton et al. 2012). Indeed, the Australian and Californian examples point to the importance of persistence in the trajectory of policy reform, the need to build on incrementalism and to seize policy windows.

Regulation

The third problem is that states' environmental regulation and associated risk management frameworks are not sufficiently robust and flexible as to account for cross-sectoral impacts (where those impacts are known). This is exacerbated by the rapid pace and scale of technological development, which sees new technologies developed to enhance energy production, water supply and/or to mitigate or adapt to climate change, but in the absence of regulatory 'triggers' to account for trade-offs or exploit opportunities. Well-established sustainability mechanisms, such as project-level environmental impact assessment (EIA) and strategic environmental assessment (SEA) for policies, plans and programs, offer significant potential for integrating decision making across the climate, energy and water sectors by providing a decision-making framework within which climate, energy and water impacts and tradeoffs can be evaluated (Hussey and Pittock 2012). However, stronger integration of regulation at the climate-energy-water nexus will likely require the use of deep forms of SEA, characterised by a focus on the actions, policies and strategic planning processes of government, rather than private actors, and it considers not just direct environmental impacts but also the economic and social structures that underpin unsustainable patterns of development (Chapter 12). Implementation of this kind of SEA for interlinked climate-energy-water policies would require attention to issues of institutional capacity, such as whether suitable organizations exist or can be retasked for the purpose, and whether there are suitable people and knowledge available to undertake integrated assessments of this kind. Similarly, and as mentioned in the governance section earlier, the effectiveness of procedural tools such as SEA require strong forms of institutional steering to ensure appropriate policy settings are implemented and enforced.

In addition to the more rigorous implementation of procedural mechanisms such as EIA and SEA, other solutions involve implementation of better standards, licenses and targets, such as the following:

- Establishing minimum efficiency standards for water and energy use across the economy (combined with new pricing and market incentives). Specifically in relation to the energy and water sectors, this could include mandating the use of power plant cooling technologies that withdraw less water for cooling (eg combined-cycle gas turbines), or setting energy efficiency targets or rebate schemes for water utilities to decrease energy use.
- Changing permitting and licensing requirements to better consider water and energy use, for example, with respect to the sourcing of water and energy supplies under seasonable variability. This could include placing a moratorium on thermoelectric power generation in water-stressed regions or the requirement to generate energy resources co-located and on-site at water-related sites to reduce grid dependency of the water facilities. The incorporation of projected future impacts from climate change into the assessments process for new licenses could also be mandated.
- Establishing and enforcing demand management strategies to reduce peak demand in the water and energy sectors, and by so doing, reduce peak demand in the embedded energy and water. Even in jurisdictions where water scarcity is less of a problem, water conservation and efficiency strategies could be used as an effective means of saving energy.
- Establish a greenhouse gas emission target, which may induce energy and water efficiency outcomes. As Peel describes in Chapter 12, new climate change legislation in Victoria has provided an opportunity to improve the integration between climate-related and decision-making in other sectors through interjecting climate change considerations into existing regulations for those sectors.
- Periodic relicensing of energy and water infrastructure (as exemplified by the relicensing of non-Federal hydropower dams in the United States) provides a key reform window to reassess these systems and upgrade them in light of new knowledge and higher environmental and social standards (Pittock and Hartmann 2011).

Inadequate knowledge

The fourth problem is insufficient, incomplete or badly communicated knowledge and information on the cross-sectoral impacts of actions taken or technologies deployed, which results in decisions being made in a position of ignorance or uncertainty. Responsibility for generating and disseminating knowledge and information on the linkages between key sectors will lie largely with governments, inter-governmental agencies and researchers, though key industry actors have already committed significant resources in support of those research efforts, and there is certainly not an absence of activity in this space. Indeed, in many respects a deficiency in knowledge is not the problem; rather, it is the dearth of 'fit for purpose' data and decision-support tools that can be readily used in policy and investment decision making. Certainly, there are tools and models available, and a valuable contribution of the nexus approach is in the analytical framing of multiple resource metrics, with some of those framings described in this book. However, there is no consistent, common approach to the collection and publication of data, nor any standard for compatible formats to inform the viability and feasibility of various solutions. To that end, significant and concerted attention could be dedicated to:

- The development of a publicly available, user-friendly integrated assessment tool, which can account for the trade-offs and opportunities between sectors and be maintained on an annual basis (similar tools are used by the IEA, World Bank and IMF, and Shell Pty. Ltd. to develop forecasts);
- The development of industry-supported standard water and energy protocols and codes to inform standalone and accessible datasets but also to inform integrated assessments undertaken by relevant stakeholders from the public and private sectors;
- Research and development to bring new, beneficial technologies to market;
- Education, training and dissemination of knowledge and information on the connections between energy and water, in all relevant professional educational offerings; and
- Data on climate, energy and water resources and use should be collected in compatible formats and publicly available to enable cross-sectoral decision making (Hussey and Pittock 2012).

As importantly, there is an urgent need to better communicate and educate business on their sector's specific exposure to nexus risks and to elucidate the ways in which they can improve their competitive advantage by addressing many of these risks proactively (even in the absence of the better targets, standards and licensing requirements identified above). For example, competitive advantage can be improved by cutting operational costs through increased water and energy efficiency; reducing exposure to a carbon price by mitigating greenhouse gas emissions; reducing exposure to risks of extreme weather events through cost-effective climate change adaptation measures; and innovating in the form of new 'greener', more eco-efficient products (see particularly Chapter 14). The incorporation of these positive synergies into business planning is, arguably, the 'low hanging fruit' in the suite of solutions identified above.

Prospects from here

The two final chapters in this volume highlight the inevitable knock-on consequences of the nexus for other sectors, and while these chapters focus on just two sectors (food and biodiversity), they raise the possibility of using a nexus 'framing' to draw us back to the overall goal of practical sustainability with a whole-systems focus (Chapters 15 and 16). Indeed, there is a very real need to develop an economy-wide appreciation of the nexus beyond the more obvious domains of energy, water and climate, so as to extend to a second generation analysis of the nexus including all sectors of the economy. This must include analysis of the economic gains that will be derived from appreciation of the nexus – in other words, the business case. While not exhaustive, Table 19.1 illustrates the depth and breadth of analysis required, as well as early (albeit somewhat sporadic) signs that 'nexus framing' has been incorporated into policy and investment decision making.

Table 19.1. Analysis of the energy–water nexus, climate change impacts and the business case for nexus management in three major economic sectors.

Sector in the economy	**Manufacturing**	**Defence**	**Construction**
Case studies	The Coca Cola Company agreed from 2007 to more sustainably manage its energy and water use globally, and source more sustainably produced agricultural inputs after controversy over its water use in India (WWF and TCCC 2013). A water sustainability program by major Australian poultry producer, Inghams reduced water consumption (Water Stewardship Australia 2010) after supplies were jeopardised during a 2002–2010 drought.	US Department of Defence in its most recent review labels climate change as a 'threat multiplier' and promises to respond through "further actions to increase energy and water security, including investments in energy efficiency, new technologies, and renewable energy sources" (Department of Defence 2014, p. vi). Case studies in Africa have just been initiated by the US Department of Defence to understand this new form of 'non-traditional' security threat. A 2013 review by the Australian Security Policy Institute identifies how climate change will exacerbate security threats as well as the risks posed to Australian Defence Force operating capacity (Press et al 2013). One recommendation is for a 'green team' to see where further energy efficiencies can be achieved.	Following a controversial quarry proposal in Scotland, in 2000 the global extractive industry supplier LaFarge started a partnership with WWF to reduce its footprint on climate change, persistent pollutants, water, biodiversity, and promoted more sustainable construction (LaFarge and WWF 2009).

Table 19.1. (cont.)

Sector in the economy	Manufacturing	Defence	Construction
Examples of better practices for energy and water use	Identify where water and energy use is high in manufacturing processes and where efficiencies could be optimised. Set targets for reducing or replenishing or offsetting energy and water use. Locate new facilities where sustainable water and energy supplies are more secure.	Identify where energy and water use is high in military operations and where efficiencies could be optimised. Reduce reliance on fossil fuels from non-aligned nations or regions in conflict. Identify where energy and water transport creates vulnerabilities in defence that would be reduced through development of more local or autonomous supplies. Identify where vulnerabilities to climate impacts could be reduced	Identify where water and energy use is high in construction materials and practices, and where efficiencies could be optimised. Substitute construction materials made using less energy and water. Use more resilient materials that can better withstand climate impacts. Construction of flexible facilities that may be readily renovated or strengthened to deal with a changing climate.
Examples of vulnerability to climate impacts	Scarcity reduces physical or legal access to water needed for production. Water quality is diminished increasing industry costs, for instance, through eutrophication or higher salinity levels. Water scarcity or heatwaves force thermal power stations to close.	Extreme climatic events disrupt supplies of water and energy to military forces. Climatic extremes exceed the design tolerances of military equipment. Hotter climates limit the effectiveness of military personal, increasing demand for health services, energy for cooling and water. Climate change aggravates conflict and results in displacement of people increasing reliance on military forces to maintain security and aid recovery from natural disasters.	Extreme climatic events disrupt supplies of water and energy to material suppliers and construction sites. Extreme weather disrupts materials production (eg flooding of extractive industries) and construction (eg wet weather delays). Climatic extremes exceed the tolerances of materials used in construction. Previously low-risk lands become vulnerable to sea-level rise, flooding and other hazards.

Table 19.1. (cont.)

Sector in the economy	Manufacturing	Defence	Construction
Significant links to other sectors	Manufacturers rely on suppliers of agricultural and other raw materials suppliers who themselves are dependent on and consume a lot of energy and water. Often water use by suppliers dwarfs the volume consumed in processing by manufacturers. Increasingly customers of manufactured products – secondary processors and major retailers – require suppliers to demonstrate good practices in their energy and water use. Governments respond to public pressure and increasingly seek to regulate or tax energy and water use where industry is not applying good practices.	Military technological innovation usually results in growth of new industries in the civilian economy. More autonomous energy generation and water supply technologies are particularly likely to benefit the lives of people and industries in regional and remote areas.	Construction draws on a wide range of material suppliers. The capital cost and time taken for construction usually means that infrastructure cannot readily be changed creating path dependency in many respects, including energy and water consumption. Other societal sectors rely on infrastructure and will be disrupted if construction is delayed or infrastructure fails.
Business case	Secure supplies of energy and water. Reduce energy, carbon and water costs through demand management. Increase resilience to price shocks. Secure contracts with major customers and retailers who seek Enhance social licence to operate.	Reduce operating costs to governments. Defence force transport lines are less exposed to climatic disruption or attack, and units can operate more autonomously. Loss of fewer personnel. Flow on benefits to the civilian economy from military technological developments.	Reduce energy and water costs in construction and operations. Increase resilience of infrastructure to climate impacts.

Conclusion

From this analysis and others' in this volume, five summary considerations on effective cross-sectoral integration can be distilled. First, it is clear that it is the enhanced, consolidated or incrementally expanded use of existing mechanisms rather than the development of completely new mechanisms that is required to facilitate more effective cross-sectoral co-ordination and alignment, and this in turn demands persistence in policy reform. Second, with the complexity of existing governance arrangements, there is seldom a single solution; rather, a combination and diversity of mechanisms operating at different scales, in different institutional domains, and within both the public and private sectors, is required. However, such diversity must also be co-ordinated and connected to ensure consistence and avoid perverse outcomes. Third, a combination of government-led and market-led mechanisms and transition processes is likely to be needed, and understanding the links between those mechanisms is crucial. Fourth, the mere presence of co-ordinating mechanisms is insufficient: successful cross-sectoral co-ordination requires sufficient capacity, authority, community legitimacy, resourcing, transparency and accountability to be effective. Finally, even then, achieving the desired cross-sectoral co-ordination outcomes is highly relational to the appropriate strength of the measures employed and of the government agencies that will implement them. Indeed, it is interesting to note that, to date, much of the effort to understand and account for the links between climate, energy and water has been undertaken by intergovernmental agencies or researchers – neither of which have the requisite institutional steering capacity to pursue the solutions outlined above. What is needed from here are strong policy leaders, and visionary leaders of industry, to systematically implement the strategies described in this book.

References

Barnett, J. and O'Neill, S. (2010). Maladaptation. *Global Environmental Change* 20: 211–13.

Comprehensive Assessment of Water Management in Agriculture (CAoWMiA) (2007). *Water for Food, Water for Life. A Comprehensive Assessment of Water Management in Agriculture.* London & Colombo: Earthscan & International Water Management Institute.

Department of Defense (2014). *Quadrennial Defense Review 2014.* Washington, DC: US Department of Defense.

Grafton, R. Q., Pittock, J., Davis, R., Williams, J., Fu, G., Warburton, M., Udall, B., McKenzie, R., Yu, X., Che, N., Connell, D., Jiang, Q., Kompas, T., Lynch, A., Norris, R., Possingham, H. and Quiggin, J. (2012) Global Insights into Water Resources, Climate Change and Governance. *Nature Clim. Change* 3: 315–21.

Hoekstra, A. Y., Mekonnen, M. M., Chapagain, A. K., Mathews, R. E. and Richter, B. D. (2012). Global Monthly Water Scarcity: Blue Water Footprints versus Blue Water Availability. PLoS ONE 7: e32688.

Hussey, K. and Pittock, J. (2012). The Energy-Water Nexus: Managing the Links between Energy and Water for a Sustainable Future. *Ecology and Society* 17: 31 [online].

International Energy Agency (IEA) (2012). *World Energy Outlook 2012.* Paris: OECD/IEA.

Intergovernmental Panel on Climate Change (IPCC) (2014). *Climate Change 2014: Impacts, Adaptation and Vulnerability, Working Group II Contribution to the 5th Assessment Report of the Intergovernmental Panel on Climate Change*. Cambridge University Press.

Kenway, S. J., Lant, P., Priestley, A. and Daniels, P. (2011). The Connection between Water and Energy in Cities – A Review. *Water Science and Technology* 63: 1983–90.

LaFarge and WWF (2009). *10 Years. Lafarge WWF. The Partnership Explained in 10 Answers*. Paris and Gland: LaFarge and WWF.

Lukasiewicz, A., Finlayson, C. M. and Pittock, J. (2013). Identifying Low Risk Climate Change Adaptation in Catchment Management While Avoiding Unintended Consequences. Gold Coast: National Climate Change Adaptation Research Facility.

National Research Council of the National Academies (NRC) (2013). Sustainability for the Nation: Resource Connections and Governance Linkages. Washington, DC: The National Academies Press.

National Water Initiative (NWC) (2011). The National Water Initiative—securing Australia's water future: 2011 assessment. Canberra: National Water Commission.

Pittock, J., Hussey, K. and McGlennon, S. (2013). Australian Climate, Energy and Water Policies: Conflicts and Synergies. *Australian Geographer* 44: 3–22.

Pittock, J. and Hartmann, J. (2011). Taking a Second Look: Climate Change, Periodic Re-licensing and Better Management of Old Dams. *Marine and Freshwater Research* 62: 312–20.

Press, A., Bergin, A. and Garnsey, E. (2013). Heavy Weather: Climate and the Australian Defence Force. Canberra: Australian Security Policy Institute.

Ross, A. and Dovers, S. (2008) Making the Harder Yards: Environmental Policy Integration in Australia. *Australian Journal of Public Administration* 67: 245–60.

Wallis, P. J., Ward, M., Pittock, J., Hussey, K., Bamsey, H., Denis, A., Kenway, S., King, C., Mushtaq, S., Retamal, M. and Spies, B. (2014). Integrating the Governance of Water and Carbon Mitigation, Climatic Change 125, 209–220

United Nations World Water Assessment Programme (WWAP) (2014). The United Nations World Water Development Report 2014: Water and Energy. Paris: UNESCO.

World Wildlife Fund and The Coca-Cola Company (WWF and TCCC) (2013). A Transformative Partnership to Conserve Water. Gland and Atlanta: 2012 Annual Review, WWF and The Coca-Cola Company.

Vorosmarty, C., Mcintyre, R., Gessner, M., Dudgeon, D., Prusevich, A., Green, P., Glidden, S., Bunn, S., Sullivan, C. and Liermann, C. (2010). Global Threats to Human Water Security and River Biodiversity. Nature 467: 555–61.

Index